Materials Forming, Machining and Tribology

Series Editor

J. Paulo Davim, Department of Mechanical Engineering, University of Aveiro, Aveiro, Portugal

This series fosters information exchange and discussion on all aspects of materials forming, machining and tribology. This series focuses on materials forming and machining processes, namely, metal casting, rolling, forging, extrusion, drawing, sheet metal forming, microforming, hydroforming, thermoforming, incremental forming, joining, powder metallurgy and ceramics processing, shaping processes for plastics/composites, traditional machining (turning, drilling, miling, broaching, etc.), non-traditional machining (EDM, ECM, USM, LAM, etc.), grinding and others abrasive processes, hard part machining, high speed machining, high efficiency machining, micro and nanomachining, among others. The formability and machinability of all materials will be considered, including metals, polymers, ceramics, composites, biomaterials, nanomaterials, special materials, etc. The series covers the full range of tribological aspects such as surface integrity, friction and wear, lubrication and multiscale tribology including biomedical systems and manufacturing processes. It also covers modelling and optimization techniques applied in materials forming, machining and tribology. Contributions to this book series are welcome on all subjects of "green" materials forming, machining and tribology. To submit a proposal or request further information, please contact Dr. Mayra Castro, Publishing Editor Applied Sciences, via mayra.castro@springer.com or Professor J. Paulo Davim, Book Series Editor, via pdavim@ua.pt

More information about this series at http://www.springer.com/series/11181

Preface

Abrasive based machining processes are advanced machining and finishing procedure and it is widely used as finishing solutions in high-end industries such as aerospace, automobile, tools and dies, and prostheses. Abrasive flow finishing processes are mainly used to polish, remove molten layers, and finish complex shapes on the outer surfaces of the workpiece. Abrasive flow finishing processes have a wide range of applications and could be used in every shop floor as a finishing solution. The proposed subject of the book is advances in abrasiv based machining and finishing in broad sense. However, specifically, the proposed book will cover the novel machining and finishing strategies implemented in various advanced machining processes for improving machining accuracy and overall quality of the product. The main aim and scope of the book is to present the capability of advanced machining processes using abrasive grain and manufacture ability of product by implementing various innovative machining procedures for enhancing the production rate as well as quality. It will also fulfill the gap between the production of any complicated components and successful machining with abrasive particles. In this book the technologies for successful improvement in advanced machining and finishing processes with abrasive particles have been discussed. It can be a valuable reference to students and R&D researchers involved in advanced machining domain. This book can be used as a reference material for MS and Ph.D. level students. In addition, this book can serve as an important reference book for students, researchers, engineers, educators, and industry professionals carrying out research on abrasive based machining processes. Altogether, the book contains 11 chapters considering all the major aspects of abrasive based machining processes to make the book more valuable source of knowledge to all.

This book offers a broad reference for abrasive based machining processes, its varieties, technologies, and applications. The book aims to contain extensive state-of-the-art research on abrasive based machining processes in a complete collection, which are currently available in journal articles, conference proceedings, and in the form of book chapters. The goal is to provide a complete reference to the undergraduate and graduate students, as well as researchers, educators, and industry personnel working on the abrasive based machining and finishing processes area.

Chapter 1 deals with discussion of fundamental aspects of abrasive based machining and finishing processes. This chapter mainly divided into mainly four sections. These are conventional abrasive processes, advanced abrasive machining processes, hybrid advanced abrasive machining processes, and advanced finishing processes. Each process is presented with material removal mechanism, overview of machine setup, process parameters, and applications. Chapter 2 deals with the USM process principle, mechanism of material removal, varieties of USM set up, tool development of USM process, improvement and production of 3D profile by USM process. It also highlights the effects of different parameters of USM process on performance and development of USM process. In Chap. 3 the development, principle, mechanism, and setup details of rotary ultrasonic machining have been discussed. It also highlights the effects of diverse input parameters on performance of RUM. Various advancements in RUM are also discussed. The application area of rotary ultrasonic machining (RUM) in micro domain has also been highlighted. Chapter 4 highlights the Abrasive water jet machining (AWJM) process and modern technology developed the AWJM process into a complete machining process that can produce precise and consistent. A wide variety of materials ranging from stone, glass, titanium, steel, ceramics, composites, etc. can be machined by the AWJM process. Chapter 5 discusses the research trends and developments in the field of abrasive assisted EDM and micro-EDM. The parameters that control the performance of the abrasives in the hybrid system of abrasive mixed EDM have been explored. This chapter will provide a comprehensive overview of this highly efficient and productive manufacturing technique from the mechanism and application perspective. Chapter 6 talks about the Electrolytic In-Process Grinding (ELID) process that carries out in situ dressing of the grinding wheel using an electrochemical reaction, i.e., anodic dissolution. This chapter will shed light on the history, fundamental, and the current status of the ELID grinding related researches. Chapter 7 deals with the fundamentals of abrasive flow finishing, which is applied for nanofinishing of simple as well as complicated components. The chapter also presents the research advancement that has been carried out globally in the direction of nanofinishing utilizing AFF process. Chapter 8 discusses the Magnetic abrasive finishing (MAF) for finishing, deburring, and precision chamfering of components. During MAF process very low magnitude of finishing forces is applied on the magnetic abrasive particles to obtain nanometer surface finish. In order to understand the physics of the process considerable amount of work has been carried out in the form of experimental study with few attempts in modeling of the MAF process. Chapter 9 talk about the magnetic field-assisted finishing processes. The most controlled and effective way to produce surface roughness up to nanometer by using magneto-rheological fluid (MRF). With the help of MRF high-grade surface finish can achieve with convenient tolerances and without ruining the surface contour. Working principle and its advantages of MR fluid and rotational magneto-rheological abrasive flow finishing are discussed in this chapter. Chapter 10 explains that the CMP process has been deployed as planarization technique for the fabrication of IC in the semiconductor industries. This chapter presents an overview of CMP technology, working principle, its recent advancement status in terms of green slurry development, abrasives development, new polishing method, research trend, and challenges. Chapter 11 represents the feasibility and applications of the rotary abrasive float polishing (RAFP) process which is a novel polishing technique.

The Editors acknowledge Springer Nature for this opportunity and for their professional support. The Editors would also like to sincerely thank all the authors of book chapters for their valuable contributions to this book.

January 2020

S. Das
G. Kibria
B. Doloi
B. Bhattacharyya

Acknowledgement

This book has become a reality due to the constant inspirations and encouragement received from the Senior Professors such as Late Professor S. K. Sorkhel of Jadavpur University, Prof. S. K. Basu of COE, Pune, Prof. V. K. Jain of IIT Kanpur and also from Colleagues such as Dr. D. Banerjee, Dr. S. Chakraborty, Dr. S. Mitra, Dr. S. Sarkar, Dr. A. S. Kuar, Dr. B. R. Sarkar of Production Engineering Department, Jadavpur University, Kolkata, India. The editors would like to convey warm regards to Dr. A. Manna, Dr. S. Dhobe, Dr. V. U. Rathod, Dr. Sandip S. Anasane, Dr. Mukandar Sekh, Dr. Shamim Haidar, for constant support and active participation in preparing the manuscripts of this book.

The financial support from Jadavpur University, Aliah University and Swami Vivekananda Institute of Science and Technology for utilizing research outcomes on the Advances in Abrasive Based Machining to enrich this book is also acknowledged.

The Editors acknowledge Springer Nature for this opportunity and for their enthusiastic and professional support. The team members of Springer Nature, Dr. Guido Zosimo Landolfo, Editorial Director and Dr. Hubertus von Riedesel, Vice President, Physical Sciences & Engineering, Springer Nature, Dr. Petra Jantzen, Editorial Assistant, Applied Sciences of Springer have put their constant effort in transforming this book into its final shape. Finally, the Editors would like to thank all the contributors of chapters for their availability for this work.

S. Das
G. Kibria
B. Doloi
B. Bhattacharyya

Contents

Editors and Contributors

About the Editors

Dr. S. Das is Assistant Professor in Mechanical Engineering Department in Swami Vivekananda Institute of Science and Technology, Kolkata, India. He completed his M.E. in Production Engineering in 2011 from Production Engineering Department, Jadavpur University, Kolkata. Then he joined Haldia Institute of Technology as Assistant Professor in Production Engineering Department in July 2011. After that, he joined as Research Fellow under "UGC-BSR Research Fellowship in Science for Meritorious Students" 2011–2012 in April 2012. He has done his research work in the area of ultrasonic machining during entire duration. Then he joined Swami Vivekananda Institute of Science and Technology as Assistant Professor in Mechanical Engineering Department in August 2013 and working in the same department till date. Author has published one international book chapter, 7 research papers in national and international referred journals as well as presented more than 13 research papers in reputed national and international conferences related to advance machining processes. His research areas include metal cutting, non-traditional machining, finishing, advanced manufacturing systems, etc.

Dr. G. Kibria is Assistant Professor in Department of Mechanical Engineering in Aliah University, Kolkata, India. He completed his M.Tech. in Production Engineering from Jadavpur University, Kolkata in 2008 and Ph.D. from Jadavpur University, Kolkata in 2014. He has worked as Senior Research Fellow (SRF) in Council of Scientific & Industrial Research (CSIR) sponsored project from 2008 to 2011. His research interests include non-conventional machining processes, micromachining and advanced manufacturing and forming technology. He is a life member of The Institution of Engineers (IEI), India. He is the author of several book chapters on internationally recognized book publishers such as Elsevier, Springer, and Nova Publishers. He has also published 26 International and national research papers in various reputed journals and 37 research papers in reputed national and international conferences proceedings. He is editorial board member as well as reviewer of a number of reputed international journals, namely, Optics and Laser Technology, International Journal of Advanced Manufacturing Technology, Manufacturing Review, International Journal of Physical Sciences, etc. In his credit, he has received I.S.T.E. National Award for Best M.Tech. Thesis in Mechanical Engineering in 2008. He also received the Institution Prize (Gold Medal) of 2008–2009 from The Institution of Engineers (India) for the best paper.

Dr. B. Doloi is Professor and ex-Head of Production Engineering Department of Jadavpur University, Kolkata. He is actively engaged in teaching and research work in the area of advanced manufacturing technology. He has guided several M.E. and Ph.D. (Engg.) theses on production engineering. He has published about 160 research papers in national and international conferences and journals and presented his research work in reputed conferences. He has completed a research project under the DST-SERC Fast Track Project for Young Scientists and also CSIR Extramural research project on laser micromachining. He is a member of various professional bodies such as the Institution of Engineers (India) and Indian Society for Technical Education, New Delhi. Several Ph.D. theses have been completed under his guidance. He has completed several research projects. Recently, he published a book "Modern Machining Technology", published from Academic Press. His research areas include non-traditional machining, micromachining, advanced manufacturing systems, etc.

Dr. B. Bhattacharyya is fellow of Indian National Academy of Engineering (INAE) and Professor and Ex-Head of the Production Engineering Department and Coordinator of Center of Advance Study (CAS) Programme under University Grants Commission and Quality Improvement Programme under AICTE of Jadavpur University. His research areas include non-traditional machining, micro-machining, advanced manufacturing systems, etc. He has published 140 research papers in National and International Journals and 271 research papers in National and International Conferences. He has published several book chapters and conference proceedings. Recently, he has published one book titled "Electrochemical Micromachining for Nanofabrication, MEMS and Nanotechnology", William Andrew Applied Science Publishers, Micro & Nano Technologies Series, Elsevier Inc, USA. Recently, he published a book "Modern Machining Technology," published from Academic Press. Several Ph.D. theses have been completed under his guidance. He has completed several research projects. He is recipient of various awards, e.g., Gold Medal and Certificate of Achievements for research papers and thesis as well as the Career Award of the UGC, New Delhi.

Contributors

Rubina Bahar Department of Mechanical and Materials Engineering, Lee Kong Chian Faculty of Engineering and Science, University Tunku Abdul Rahman (UTAR), Sungai Long, Malaysia

B. Bhattacharyya Production Engineering Department, Jadavpur University, Kolkata, India

Dipankar Bose Mechanical Engineering Department, NITTTR, Kolkata, India

S. Das Mechanical Engineering Department, Swami Vivekananda Institute of Science and Technology, Kolkata, India

B. Doloi Production Engineering Department, Jadavpur University, Kolkata, India

Vishal Gupta Department of Mechanical Engineering, Thapar Institute of Engineering and Technology, Patiala, India

Muhammad P. Jahan Departmental of Mechanical and Manufacturing Engineering, Miami University, Oxford, OH, USA

Neelesh Kumar Jain Discipline of Mechanical Engineering, Indian Institute of Technology Indore, Simrol, MP, India

G. Kibria Department of Mechanical Engineering, New Town Campus, Aliah University, Kolkata, India

S. Kumar Production Engineering Department, Jadavpur University, Kolkata, India

Alakesh Manna Departmental of Mechanical Engineering, Punjab Engineering College (Deemed to be University), Chandigarh, India

Arpan Kumar Mondal Mechanical Engineering Department, NITTTR, Kolkata, India

Muslim Mukhtarkhanov Department of Mechanical and Aerospace Engineering, School of Engineering and Digital Sciences, Nazarbayev University, Nur-Sultan, Kazakhstan

Samad Nadimi Bavil Oliaei Department of Mechanical Engineering, Çankaya University, Ankara, Turkey

Asma Perveen Department of Mechanical and Aerospace Engineering, School of Engineering and Digital Sciences, Nazarbayev University, Nur-Sultan, Kazakhstan

Anand C. Petare Discipline of Mechanical Engineering, Indian Institute of Technology Indore, Simrol, MP, India

Deopkant Prasad Mechanical Engineering Department, NITTTR, Kolkata, India

Atul Singh Rajput Mechanical Engineering Department, NITTTR, Kolkata, India

Vivek Rana Discipline of Mechanical Engineering, Indian Institute of Technology Indore, Simrol, MP, India

Asif Rashid Departmental of Mechanical and Manufacturing Engineering, Miami University, Oxford, OH, USA

Tanveer Saleh Department of Mechatronics Engineering, International Islamic University Malaysia, Kuala Lumpur, Malaysia

M. R. Sankar Department of Mechanical Engineering, Indian Institute of Technology Tirupati, Tirupati, India

I. Shivakoti Department of Mechanical Engineering, Sikkim Manipal Institute of Technology (SMIT), Sikkim Manipal University (SMU), Majhitar, East Sikkim, India

Sachin Singh Department of Mechanical Engineering, Thapar Institute of Engineering and Technology, Patiala, India

Chapter 1
Introduction to Abrasive Based Machining and Finishing

G. Kibria, I. Shivakoti, S. Das, B. Doloi, and B. Bhattacharyya

Abstract To manufacture intricate shaped engineering components with high degree of surface finish and tolerances, it is always needed reliable machining and finishing processes. Since last two decades, various machining and finishing processes have been developed for manufacturing or processing wide variety of end products for their successful engineering as well as biomedical applications. The present chapter deals with discussion of fundamental aspects of abrasive based machining and finishing processes. The chapter mainly divided into mainly four sections. These are conventional abrasive processes, advanced abrasive machining processes, hybrid advanced abrasive machining processes and advanced finishing processes. Each process is presented with material removal mechanism, overview of machine setup, process parameters and applications.

Keywords Abrasive particles · Grinding · Surface finish · Machining rate · Advanced machining · Finishing processes · Magnetic field

---

G. Kibria (✉)
Mechanical Engineering Department, Aliah University, Kolkata 700160, India
e-mail: prince_me16@rediffmail.com

I. Shivakoti
Department of Mechanical Engineering, Sikkim Manipal Institute of Technology (SMIT), Sikkim Manipal University (SMU), Majhitar, East Sikkim, India
e-mail: ishwar.siwa@gmail.com

S. Das
Mechanical Engineering Department, Swami Vivekananda Institute of Science & Technology, Dakshin Gobindapur, Sonarpur, Kolkata 700145, India
e-mail: somnath96@gmail.com

B. Doloi · B. Bhattacharyya
Production Engineering Department, Jadavpur University, Kolkata 700032, India
e-mail: bdoloionline@rediffmail.com

B. Bhattacharyya
e-mail: bb13@rediffmail.com

S. Das et al. (eds.), *Advances in Abrasive Based Machining and Finishing Processes*, Materials Forming, Machining and Tribology,
https://doi.org/10.1007/978-3-030-43312-3_1

1.1 Introduction

Since last two decades, abrasive based machining and finishing is found to be key area and important category of manufacturing in broad sense and most particularly machining and finishing technologies for the effective and efficient method of product manufacturing [1]. The basic of abrasive machining is to remove material from workpiece from desired portion to obtain the desired product shape as well as for achieving the desired geometry and surface integrity for the purpose of desired functionality of that particular product in real life applications. Basically, in abrasive machining processes, material is removed from desired workpiece by the multitude of hard abrasive particles by the applications of external mechanical energies [2]. These abrasive particles can be or may not be in the bonded form. Not only is that, the loosely available abrasive particles mixed with magnetic particles also being used for nanofinishing applications [3]. Thus, abrasive particles also play important role in different nanofinishing processes. Thus, machining and finishing with abrasives covers with a vast areas ranging from conventional abrasive processes (such as grinding, honing, lapping, buffing, etc.) to advanced abrasive machining processes (such as abrasive water jet machining, abrasive jet machining, ultrasonic machining, electro discharge grinding, etc.) to advanced micro and nano finishing processes (such as magnetic abrasive finishing, abrasive flow machining, magneto rheological finishing, etc.). These processes are being used commercially in different industrial sectors for product manufacturing. Certainly, there are several important advantages of each of these processes over another. However, most of the processes are applied for machining and finishing of different work material and also it depends on the level of surface integrity and accuracy to obtain [4]. Figure 1.1 shows the list of machining and finishing processes that use abrasive particles for the manufacturing of usable components. In the following sections, the brief overview of these processes with basic schematic representation is described.

1.2 Conventional Abrasive Processes

1.2.1 Grinding

It is one of the oldest surface (flat and curve) generation and material removal processes for obtaining the accurate shape and surface finish from raw materials. It is noted that in comparison of plain milling or simple turning, the level of surface finish and precision is obtained as high as ten times [5]. Basically, in this process, a solid abrasive wheel is brought in contact with workpiece to be machined at controlled speed and feed. Each of the abrasives at the periphery of grinding wheel acts as single point cutting tool and little amount of material is removed from work sample creating tiny chips in micrometer to millimeter scale. The material removal mechanism in grinding is shearing, ploughing, rubbing, fracturing and crushing, etc. [6]. Using

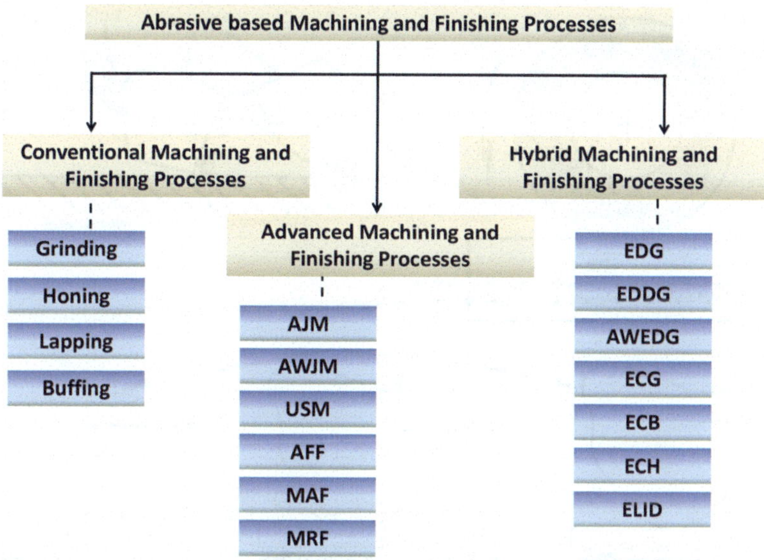

Fig. 1.1 List of machining and finishing processes utilizing abrasive particles

various type of binder, the abrasive particles are bonded together to form a grinding wheel of various shapes and sizes. With long machining environment, the sharp cutting edge of abrasives is broken and becomes dull. Therefore, the cutting/machining ability of the grinding wheel reduces. However, the grinding wheel is re-sharpened automatically by the rubbing action of peripheral surface of the wheel and job by broken away of dull abrasive particles. Figure 1.2 shows the schematic representation of grinding process with material removal mechanisms.

There are several abrasive grains available and used for manufacturing of grinding wheel. Aluminum oxide (Al_2O_3) is principally used for the grinding of ferrous metals due to several outstanding properties such as quite hard, chemically stable, available in different grades, etc. [7]. Silicon carbide (SiC) is also used for the production of grinding wheel and it is used mainly for machining of non-ferrous and density materials. Cubic boron nitride (cBN) and diamond are also used for very high precision requirements as these abrasives are costly. Some other outstanding properties of these abrasives are hard, tough and sharp, high heat and wear resistance, etc. [8]. Diamond is chemically stable with Fe. Thus, grinding wheel made of diamond abrasive particles is not recommended for machining ferrous materials. The grinding performance of the wheel also largely depends on the average size of abrasive particles. Larger particles remove more material from the surface; however, the quality of machined surface becomes very rough. In comparison, using fine abrasives, surface finish as well as geometrical tolerance obtained is outstanding. The type of binder also play important role for the durability of grinding wheel.

The pore structure or density of abrasives in grinding wheel also has major role in the performance of grinding. The pores in the grinding wheel structure mainly

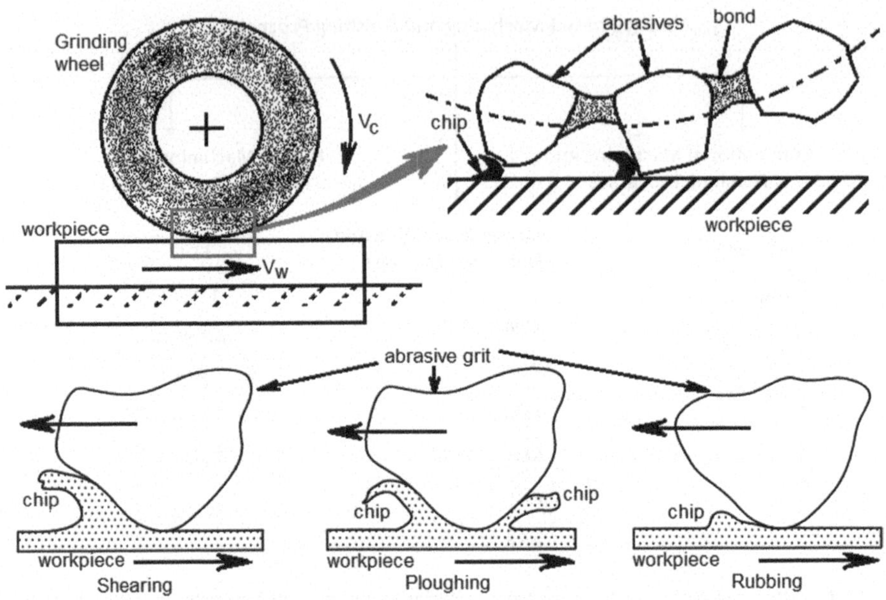

Fig. 1.2 Schematic representation of grinding process with material removal mechanisms

act as void space for the coolant retention as well as space to accommodate the tiny chips formed by grinding [9]. However, for machining of hard materials, it is always recommended to use dense wheel. On the other hand, open dense structure of wheel is recommended for machining of soft materials. Due to huge amount of grinding force during machining, several unwanted aspects comes in picture such as higher energy consumption, dimensional inaccuracies, grinding temperature, vibrational effects, etc. [10]. Thus, to reduce the grinding force, several controllable measures are adopted such as use of coolant, proper selection (size and shape) of wheel, proper conditioning, proper selection of grinding wheel rotational speed, etc.

1.2.2 Honing

Honing is one of the key processes for surface finishing as well as for improving the bore or hole alignment. After drilling or boring operations, generally lot of surface irregularities and scratches are left out on the machined surface. At this instant, honing operation play important role for making the surface even and useful. The process is also used for finer surface finish. In addition, in many cases, the form characteristics like sphericity, cylindricity, etc. are improved [11]. In this process, abrasive (like aluminium oxide, silicon carbide, diamond etc.) protruded sticks attached on a rotating wheel periphery as shown in Fig. 1.3 is rotated over the workpiece surface with a controlled feed and speed. This specially designed rotating and reciprocating

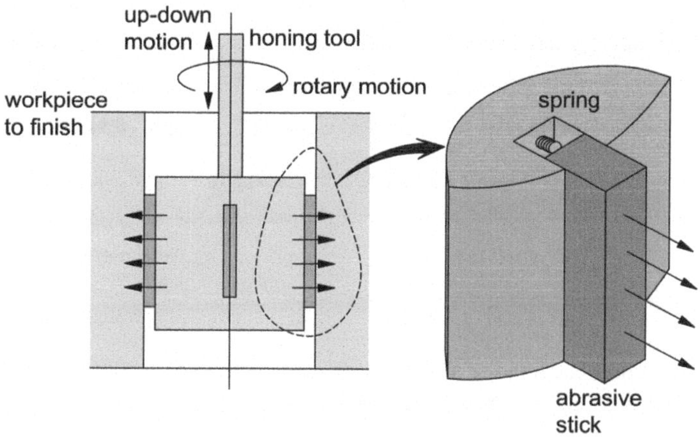

Fig. 1.3 Schematic of honing process

tool is known as hone. As depicted in figure, this hone tool needs to provide rotary and reciprocating motion against the stationary work sample for removal of tiny layer and finish the desired surface. For effectively use of the process, in recent times, diamond and carbon boron nitrite abrasives are drawing much attention as use of these abrasives makes the honing process much faster (may be in one stroke of hone tool). There is several process parameters involved in honing process. With increase in rotational speed of honing tool, generally it is found that the removal rate of material increases. However, at the same time, precision surface finish is obtained. Honing time is another important factor in this process. With longer honing time, it is found that surface roughness first decreases to some extent; then after, the roughness of machined surface again increases [12]. Honing stick pressure and stroke length are another two important parameters which directly control the level of surface finish as well as required tolerances of the bore or hole to be finished.

1.2.3 Lapping

Another important conventional surface finishing process is known as lapping. It works in the principle of abrasive wear by three bodies. In this process, very high tolerance with microscopic level surface finish is achieved with greater accuracy and precision [13]. In Fig. 1.4, the schematic view of lapping process is depicted. Loose abrasive particles carried by carrier fluid are passed between the workpiece surface to be finished and a lap (soft counter surface). The workpiece holding against the lap has random movement with a little pressure and depending upon the finishing time, abrasive concentration and rotational speed of the lap, various level of surface finish is achieved. Compared to honing, lapping removes less material from finished

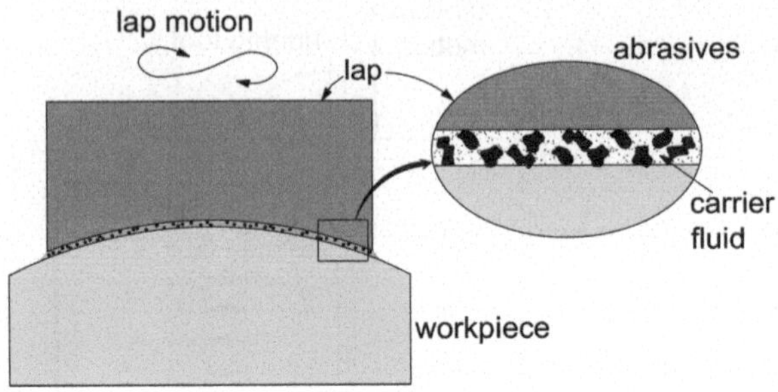

Fig. 1.4 Schematic view of lapping process of curved workpart

surface and the process has the ability to produce concave or convex surfaces (with specially designed lap) and three dimensional shapes [14].

1.2.4 Buffing

Buffing is another frequently used surface finishing method using abrasive particles. In this process, abrasive compound is impregnated on a cloth wheel and when this cloth wheel rotates against the workpiece surface, mirror like finish is obtained [15]. The abrasive compound is made with greaseless compound-based matrix and abrasives. This specially made compound is sprayed or pressurized into the rotating wheel and the rotating wheel is acting as carrier medium of that compound. The process is mainly divided into two types (i) cut buff and (ii) color buff. Cut buff is complicated process and this process is applied to generate smoother lined finish. Cut buff requires high efforts and pressure [16]. Thus, the fatigue of the operator is needed to consider. On the other hand, color buff (also known as finish buff) is applied in case where fine lines are left by cut buff. This buff process is quicker and simple.

1.3 Advanced Abrasive Machining Processes

1.3.1 Abrasive Jet Machining (AJM)

Abrasive jet machining (AJM) is advanced mechanical type material removal process in which mechanism of material removal is impact of abrasive particles. A mixture of high pressurized carrier gas and abrasive particles (with required concentration) is prepared in several stages and the mixture is made to impinge on the surface of material which to be machined [17]. A discharge nozzle (kept at some distance from working surface) with inner diameter in the range of 0.3–0.5 mm is employed to flow out the abrasive-carrier gas mixture to erosion of material. The ejection of

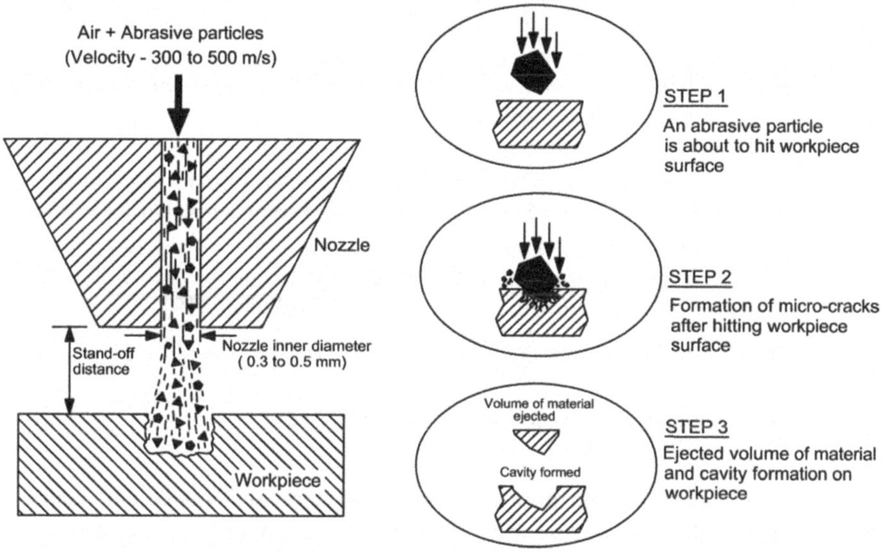

Fig. 1.5 Scheme of material removal in AJM process

tiny debris with sharp edges is occurred due to formation of several brittle fractures (micro-cracks). Since these ejected debris are mixed with abrasive particles and it is very costly to separate out the debris and abrasives, thus, reuse of abrasives are not recommended. In Fig. 1.5, the schematic representation of material removal mechanism in AJM process is depicted.

For effectively utilized the process, a complete abrasive jet machining setup is required. The setup consists of several sub-components such as air compressor, pressure control valve, air drier, filter, mixing chamber and machining chamber. Such components ensure that fresh and dried air is mixed with dry abrasives for the elimination of contamination of abrasives and also for increasing the maximum efficiency in the process. In AJM, air, nitrogen or carbon dioxide is used as propelling gas. For economy of machining, the gas should be non-toxic, cheap and easily available. For homogeneously mixing of abrasives with dry air or gas, electro-magnetic shaker is employed. The vibrational amplitude controls the amount of abrasives to mix with the gas or air. Using CNC controlled X–Y stage, the movement of workpiece in specific path is carried out depending upon the cut profile required. For the prevention of wear out of nozzle material, generally sapphire and tungsten carbide nozzle are used. The life span of sapphire nozzle is 200 h approximately. Several type of abrasive particles (such as Silicon carbide (SiC), Aluminium oxide (Al_2O_3), glass beads, sodium bicarbonate, crushed glass) can be used and the abrasive particles are chosen based on the material hardness, cut depth to obtain, accuracy and precision of cut, etc. [18]. In addition, size of abrasive also controls the material removal rate and surface finish of cut. Other process parameters which also control the process characteristics in AJM are stand-off distance, impingement angle, pressure and velocity of carrier gas, etc.

Among several striking advantages of AJM process, the most important feature is that the process is employed to machine heat sensitive, brittle and fragile materials. In addition, the process is applied in case of machining of inaccessible surface or position of work sample. Among the drawbacks, the most important issue is conglomerate among the abrasive particles. Flaring of discharging jet is also deteriorating the accuracy and precision cut. Health hazard and pollution of surrounding air may occur. The process is used to cut nylon, Teflon components, to clean metallic mould cavities and also to drill or engrave of thin sectioned workpiece [19].

1.3.2 Abrasive Water Jet Machining (AWJM)

In AWJM, high velocity and high pressurized mixture of abrasive and water is directed towards the workpiece surface and allowed to impinge on the workpiece. Due to conversion of the kinetic energy of water and abrasive into pressure energy, material starts to erode from impinging zone of work material [20]. Therefore, the kinetic energy of abrasive particles play important role in this process to erode material. Here, high pressurized water is considered to be the carrier medium of abrasive particles. The mechanism of material erosion in AWJM process is depicted schematically in Fig. 1.6. The machine setup consists of several sub-components namely booster pump, intensifier pump, abrasive tank and metering valve, mixture nozzle, XY motion controller, working chamber, etc. In AWJM process, high pressure water as high as 2000–4000 bars is required and the intensifier along with LP booster pump has sufficient capability to produce this huge pressure for cutting variety of materials in AWJM. Intensifier is to amplify the pressure using hydraulic cylinders of different cross-sections. The most important component of the AWJM setup is the mixing chamber where required quantity of dry abrasives is mixed with high velocity water. Due to high velocity of water jet, the abrasives receive momentum to proceed towards the exit of the cutting head along with the flowing water. The mixing process takes place by series of interactions by abrasives with water jet and inside wall of mixing chamber and focusing tube and ultimately, abrasive mixed water is discharged from tip of cutting head. Therefore, the velocities of water and mixed abrasives are equivalent. Mixture of water and abrasives are discharge using ON-OFF valve controlled electronically. The workpiece to be cut is moved or the jet cutter is moved in XY plane at different cut profiles using CNC controller connected. After cutting of material at required depth and profile, the high velocity water-abrasive mixed jet is broken by using catcher. There are mainly three type of catcher configurations used in AWJM. In Fig. 1.7, the catcher configurations are depicted schematically. For effective and accurate AWJM cutting, abrasive feed nozzle is applied. This abrasive feed nozzle is of two types, namely single jet side feed nozzle and multiple jet central feed nozzle. For the first type, mixing efficiency is less and wear of nozzle is high. The multiple jet central feed type design, the mixing process of abrasive with water is efficient and

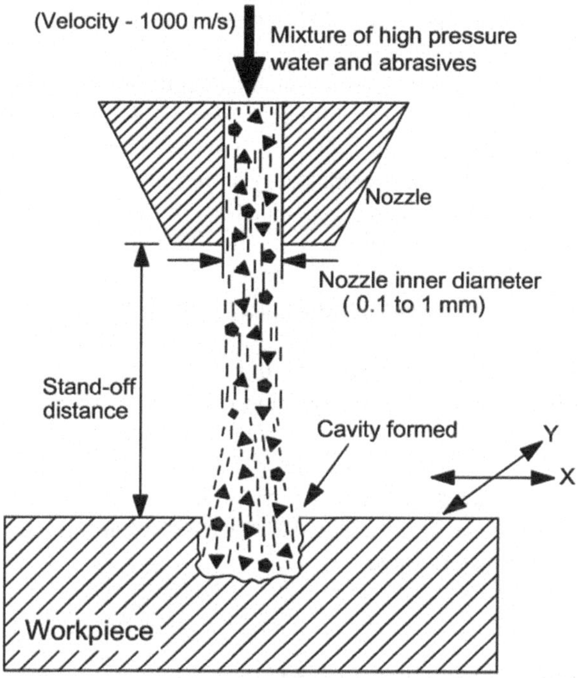

Fig. 1.6 Schematic view of mechanism of material removal in AWJM

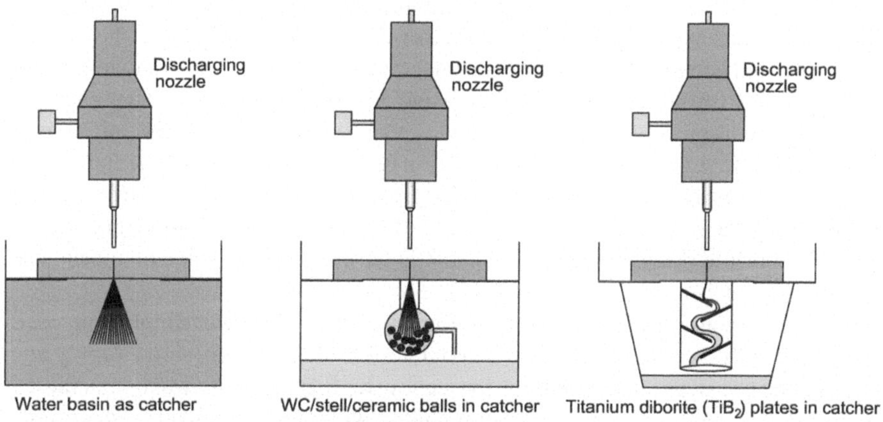

Fig. 1.7 Representation of catcher types in AWJM

life span of mixing section is longer [21]. However, it is very costly due to design complicacy.

In AWJM, the process characteristics such as depth of cut, surface finish of cut surface, accuracy and tolerances are influenced by number of process parameters. These process parameters are orifice internal diameter, length and type of material, focusing tub diameter and length, abrasive type and mess size, abrasive flow rate, stand-off distance, feed rate of workpiece, etc. The cutting ability of the jet is largely depends on mixing ratio (ratio of mass flow rate of abrasive particles to mass flow rate of water). Generally, depth of cut is proportionally increased with water pressure and abrasive flow rate. However, the increasing trends are different for different abrasive materials. For cutting of thick work sample, it is always requires to cut the material in several passes. The accuracy of cut profile is also depends on feed rate of workpiece. Since, the cut profile is designed in software in computer, thus, complex geometrical cut is always easy process. AWJM can be applied to cut wide range of materials ranging from metals, ceramics and composites and even stones/granite [22]. In addition, since multi-jet cutting facility is available in AWJM, thus, thick materials up to 200 mm can be cut easily. However, stray cutting effect always exists due to which accuracy deteriorated.

1.3.3 Ultrasonic Machining (USM)

In ultrasonic machining (USM), a ductile and tough tool is vibrating in ultrasonic vibration (vibrational frequency in the range of 18–40 kHz, amplitude in the range of 10–50 μm) and between the tool and workpiece surface (kept at very small machining gap), the slurry (mixture of water and abrasives) is fed [23]. While pressing the tool towards the workpiece via slurry, it provides a static feed force. During pressing of abrasive particles by vibrating tool, the erosion action takes place by Indentation process by abrasives and materials are removed from tool and workpiece surface. Thus, in USM, material removal is in sequence of indentation, crack initiation, crack propagation and brittle fracture [24]. The schematic view of material removal mechanism in USM process is represented in Fig. 1.8.

USM setup consists of some sub-units namely ultrasonic power supply unit, oscillating unit, transducer, acoustic horn, coupler, tool feeding system, slurry supply unit, etc. To generate ultrasonic vibration frequency, high power sine wave generator is employed. The power supply unit also included with automatic load compensation and frequency controller unit. It is very much important to monitor the overload for protecting the system from conditions that could normally result in failure. The ultrasonic vibration with a frequency above 20 kHz and amplitude range of 10–50 μm is generated through a power generator and ultrasonic transducer. The electrostrictive converter (called as transducer) is highly efficient and the degree of energy conversion is 96% [25]. The acoustic head (compilation of transducer and primary acoustic horn) is an important part of USM machine. There are mainly three functions of acoustic horn such as (i) amplify the amplitudes to practical limits (ii) transmit the

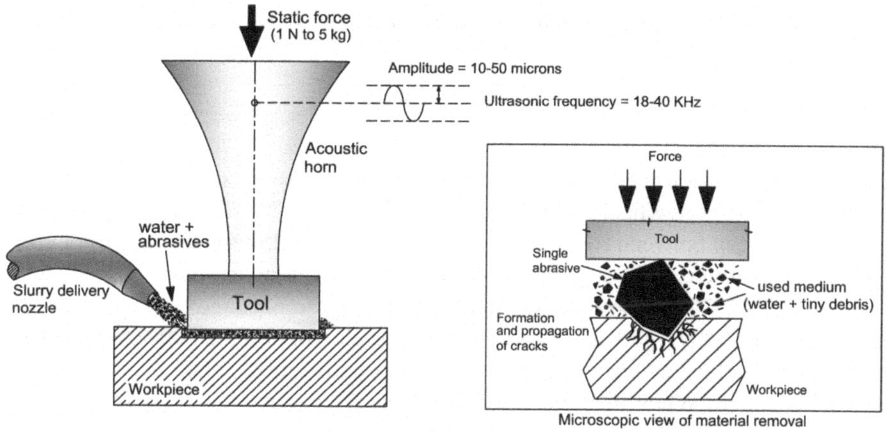

Fig. 1.8 Mechanism of material removal in ultrasonic machining (USM)

mechanical energy to the tool, and (iii) concentrate the power on a small machining area. The coupler is fastened between the horn and transducer. Depending upon the efficiency of tool feeding mechanism and the gap between tool and workpiece, the performance of machining varies. In USM, the slurry circulating unit is an important sub-component of the setup as this unit control the amount of slurry to deliver in the machining zone. For re-circulating the slurry, a pumping system with required tubing is employed. The pump takes slurry from this reservoir and supply this slurry as slurry jet in the machine bed where machining is taken place. The used slurry is taken back to the reservoir again by gravity feed. It is very much important to properly select the type, size and concentration of abrasives before actual machining of materials. In USM, the abrasive particles mostly used in order of hardness are boron carbide, silicon carbide and aluminum oxide. The average size of particles is kept in the range of 3–15 μm. Coarse type abrasives are used in USM when higher value of material removal rate is required. However, at the same time, surface finish is degraded. On the contrary, fine type of abrasives are used when surface finish is preferred than material removal rate.

The performance of USM process is controlled by several process parameters. The most important parameters are amplitude and frequency of vibration, abrasive type, size and concentration, slurry flow rate, tool material type, shape and roughness, static load and tool feed, etc. Depending upon the design of concentrator such as circular, rectangular or hollow cylindrical cross section, the efficiency of frequency transmission is varied [26]. The penetration rate in USM is directly increased with amplitude and frequency of vibration of tool material. However, there is a particular critical value of grit size with which maximum penetration rate is obtained. Moreover, machining rate in USM is obtained as maximum at critical value of concentration of abrasive particles mixed in carrier medium. The USM process is applied to machine hard, brittle, fragile materials. For fabrication of dies for wire drawing, punching and blanking, USM process is employed.

1.4 Hybrid Advanced Abrasive Machining Processes

1.4.1 Electro Discharge Grinding (EDG)

EDG uses a rotating wheel (usually made of graphite due to inexpensive and machinability and maximum wheel surface speed is 180 m/min) [27]. The conductive workpiece and the gap between rotating wheel surface and workpiece surface (inter electrode gap) is immersed in dielectric fluid inside a machining chamber. A pulse generator, which can supply voltage range of 30–80 V, current range of 0.5–200 A and pulse frequency of 50–250 kHz is used to generate electrical spark between the wheel surface and workpiece surface in IEG. A servo control mechanism is used to keep the IEG constant. When the wheel and workpiece is connection to pulse generator and DC pulse power is provided, electrical sparks are generated at the IEG in the location where minimum gap exists. Due to generation of sparks, huge temperature in the range of 12,000–20,000 °C is produced and materials are removed from both rotating wheel and the workpiece. During pulse-off-time of discharging cycle, there is sufficient duration to fresh dielectric fluid to capture the space of IEG and debris generated in previous sparks are removed [28]. The schematic representation of EDG process arrangement is shown in Fig. 1.9.

1.4.2 Electro Discharge Diamond Grinding (EDDG)

EDDG process uses a metal bonded diamond wheel instead of metallic wheel as employed in EDG. In the wheel, the bonding materials which are used are brass or bronze. In Fig. 1.10, the schematic view of EDDG process setup with enlarged

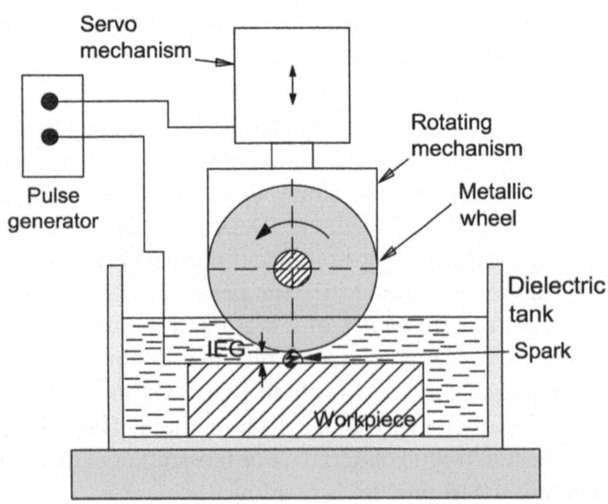

Fig. 1.9 Mechanism of material removal and configuration of EDG process setup

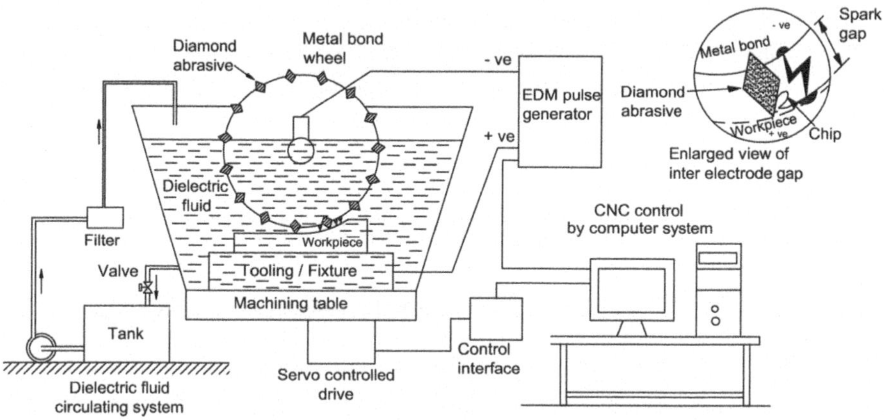

Fig. 1.10 Schematic of EDDG process with enlarged view of material removal mechanism

view of IEG is shown. In EDDG process, when DC pulse power is supplied between the wheel and workpiece, the spark is developed at the minimum gap between the wheel and the workpiece. The properties of workpiece surface where discharges were created become softer and the diamond abrasives easily remove the soft portion of materials from workpiece surface. Thus, material removal occurs at low value of grinding force. Therefore, in EDDG, material removal is caused by simultaneous occurrences of electrical discharge as well as abrasion by diamond abrasives [29]. The protrusion height of the diamond abrasives is kept higher than the value of inter electrode gap (gap between metallic surface of wheel and workpiece surface). During machining by this diamond protruded abrasive wheel, the workpiece surface is always in contact with protruded diamonds on outer periphery of the wheel. Thus, the chances of unnecessary short-circuiting between the wheel and workpiece are avoided. In addition, the property of high thermal conductivity of protruded diamond particles helps to reclaim the ambient temperature when the sparking is ended. By utilizing this hybrid process of EDDG, the limitations of diamond grinding and electrical discharge grinding. During sparking, the material is softening due to formation of electrical discharge and due to this, simultaneous grinding force which is required is less. Along with this advantage, the abrasion process by the protruded diamond abrasives on the wheel are able to remove recast layer and surface damages (surface and sub-surface micro-cracks) developed on the workpiece surface [30]. EDDG process has several process parameters which can be controlled for achieving required surface quality and machining rate. Because of uninterrupted in-process dressing and de-clogging of diamond protruded wheel, enhanced grinding efficiency is obtained. Thus, the wheel never becomes dull and grinding ability of the wheel is maintained.

The process has several advantages in comparison with conventional EDDG. Force required for material removal per grit is reduced. Furthermore, problems of wheel loading and glazing encountered in EDG process is eliminated in EDDG. One of the striking advantages is the process does not left out any re-solidified layer and cracks formed by electro discharge phase. However, due to thermal erosion of bonding material of wheel and abrasion action of diamond abrasives, loss of diamond abrasives occurs continuously. Also, abnormal short-circuiting occurs which reduced the efficient discharge between the bond material and workpiece.

1.4.3 Wire Electro Discharge Grinding (WEDG)

By introducing this wire electrode, the wire EDG process is capable of generating micro features with high accurate features and geometrical dimensions [31]. The wire EDG process is carried out by supplying DC pulse power between the wire electrode and rotating cylindrical job. The electrical discharges occur between these anode and cathode immersed in dielectric fluid. This process is applied to fabricate cylindrical macro and micro shaft with accurate geometrical dimensions and surface finish [32]. Figure 1.11 shows the schematic view of wire-EDG process. The workpiece is held and rotated with work holding mandrel held vertically and while rotating, the mandrel along with workpiece has a downward feed motion at Z axis. The wire is travelling using a wire guide support and the position of this wire guide is controlled in x and y directions using positional control drives. Due to repetitive discharges, material is eroded from rotating electrode workpiece as well as from travelling wire. The wire is continuously fed from fresh reel of wire. After discharge, wire used is taken by take-up reel.

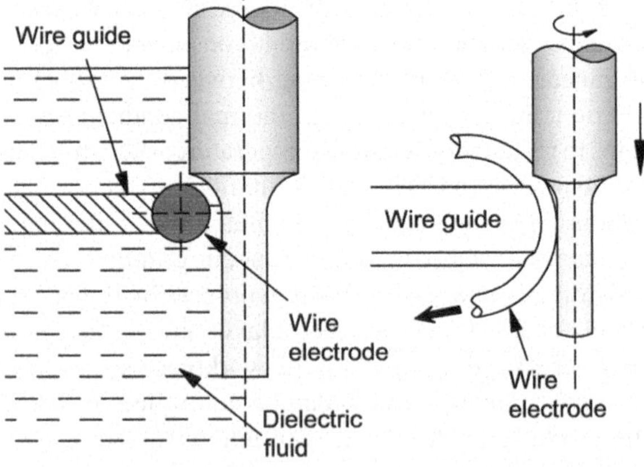

Fig. 1.11 Schematic representation of wire EDG process

1.4.4 Powder Mixed Electrical Discharge Machining (PMEDM)

Electrical discharge machining (EDM) is advanced type thermo-electrical material removal process which is widely used for manufacturing wide range of macro and micro components. In this process, repetitive electrical discharges are created between two electrodes (tool is connected to one terminal and workpiece is connected to other terminal). These two electrodes are kept in such as way that very small gap exists in between. The discharge gap is known as inter electrode gap (IEG). The discharge zone is immersed in a dielectric fluid. Due to generation of huge amount of energy at the discharge zone followed by several steps of plasma generation and due to melting and evaporation of material from workpiece, material start to remove from work sample [33].

Due to various drawbacks (such as degradation of kerosene at high discharge temperature, generation of toxic gases, adhesion of carbon particles on tool electrode, etc.) of use of hydrocarbon dielectrics in EDM and to promote the safe and pollution free machining environment, one of the best choice of dielectric fluid is de-ionized water. In addition, for effective utilization of discharge energy and to reduce the recast layer formation at the machining zone in EDM, various types of abrasive particles (conductive or semi-conductive) is mixed in dielectric fluid at various concentrations [34]. With the application of voltage between the tool electrode and workpiece, an electric field is generated. With abrasive particles mixed with dielectric fluid in the IEG, the abrasives get energized and arrange themselves in zigzag format. Thus, in the discharge zone, the powder particles lie like chain and when spark discharge occurs, and then the whole discharge energy is distributed over the chain structure due to early breakdown of dielectric strength under the tool. The powder particles changes the structure of plasma channel created between the electrodes and the size of plasma channel is enlarged. Thus, with powder particles added in dielectric fluid, uniform discharge occurs and surface finish of machined components is improved [35]. The schematic representation of completed EDM system with enlarged view of chain structure of powder additives at the IEG is viewed in Fig. 1.12.

Although, it is obvious that the machining rate and quality of surface machined in PMEDM is improved, there are several important factors on which the stability of discharges between the electrodes depends on. These factors are debris and powder concentration and distribution at the IEG under the tool electrode, formation of bubbles, deionization of dielectric fluid and initial surface condition of the job surface. Generally, in EDM, there are lots of influencing factors which directly affect on various performances like material removal rate, tool wear, surface quality as well as overcut. In addition to these process parameters, there are also some important parameters related to characteristics of abrasive powder. These parameters are abrasive material, size, concentration of powder abrasive in dielectric, particle density,

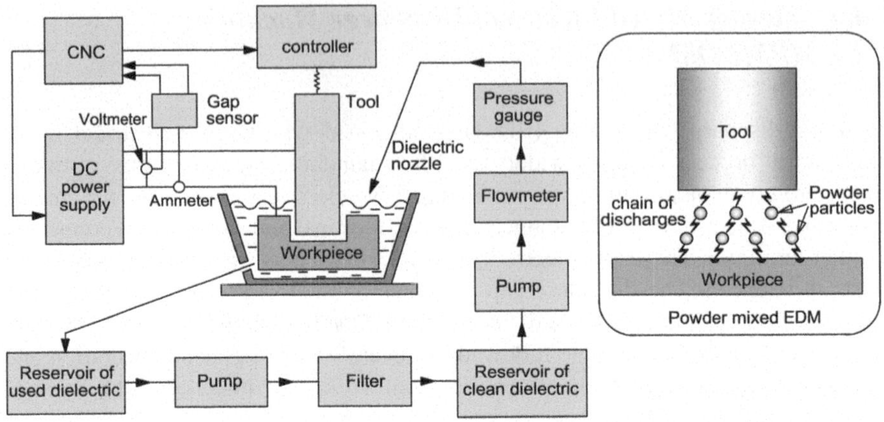

Fig. 1.12 Schematic of EDM setup and enlarged view of machining zone in PMEDM

melting point, specific and latent heat, etc. In the literature, it has been found that several researchers have used number of abrasive powders and explored the usefulness of abrasive particles in dielectric fluid during machining of wide range of materials [36].

1.4.5 Electrochemical Grinding (ECG)

ECG has the capability to machine any tough-to-cut materials and the process does not rely on the material's hardness or strength. This process employs a rotating grinding wheel with a conductive bonded material and with the use of electrolytic solution between the grinding wheel and workpiece surface, the anodic dissolution of the metallic workpiece occurs [37]. In most cases, aluminium oxide and diamond are used as abrasives on grinding wheel in ECG. For the aid of electrochemical reaction, the electrically conductive bond materials (copper, brass, and nickel) are used. The required gap distance in ECG is controlled and maintained by the height of abrasive grits exist at machining zone. The electrolyte exists between consecutive grains carry out electrolysis process. Material removed from the workpiece is carried by the electrolyte. In ECG, more than 95% of the material is removed by anodic dissolution and remaining 5% or less removed by the abrasive action of the grinding wheel [38]. While rotation of grinding wheel, the material removal occurs simultaneously by electrochemical dissolution and mechanical removal by abrasive grinding. The schematic illustration of ECG process with the view of material removal zone (I, II and III) is depicted in Fig. 1.13.

The material removal is purely by electrochemical dissolution at zone I. Because of rotation of grinding wheel, fresh electrolyte enters to the IEG. The by-products of EC reactions increase the electrolyte conductivity. However, gas bubbles formed

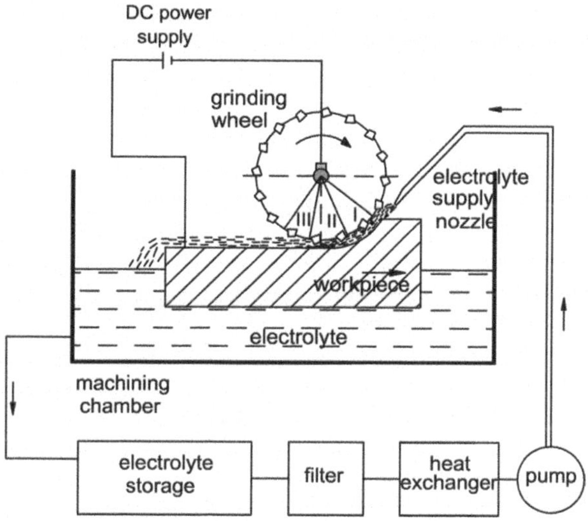

Fig. 1.13 Schematic illustration of ECG process

due to EC reactions decreases the electrolyte conductivity. Ultimately, decrease in electrolyte conductivity occurs and reduction of IEG occurs. Thus, abrasive particles protruded on grinding wheel come in contact with the workpiece surface. Thus, material removal occurs due to abrasion action. In zone II, simultaneous removal of material by abrasive particles as well as non-reactive oxide layer is removed from the zone. The material removal is done completely by electrochemical dissolution in zone III. For effective material removal and to achieve quality surface, continuous flow of electrolyte at machining zone is provided. Thus, a separate electrolyte circulation system (with pressure regulating valve, pump and filter) is provided with machining setup. The influencing process parameters in ECG are voltage, current, wheel rotating speed, type and concentration of electrolyte, feed rate of workpiece, etc. Since, most material removal takes place by EC dissolution, thus the wheel life is higher than other advanced grinding process like EDG.

1.4.6 Electrochemical Honing (ECH)

ECH is a kind of material removal process in which electrochemical dissolution as well as abrasive action takes part into. The process minimizes the limitations of mechanical honing process. These limitations include reduced productivity as a result of periodic breakdown of honing sticks, smaller tool life, difficulty in finishing hardened workpiece and chances of mechanical deterioration to the workpiece surfaces. ECH involves cyclic sequence of machining by electrochemical dissolution, followed by mechanical honing. Most of the removal process (about 90%) is done by anodic dissolution [39]. A small DC electric potential is applied across the workpiece and honing tool to acts as an anode and cathode respectively. Small inter-electrode

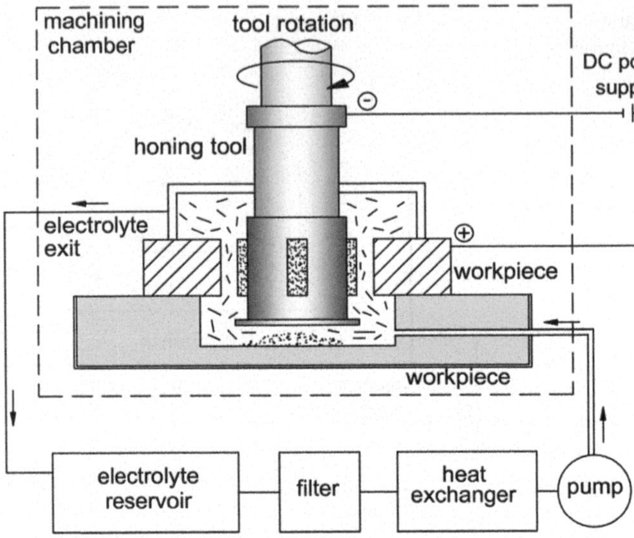

Fig. 1.14 Representation of electrochemical honing process

gap between tool and workpiece is maintained, and is filled with electrolyte. When DC power is applied to the cathode and anode, the electrochemical reactions start between tool and workpiece and after some moment of material removal, passivating oxide layer is developed on workpiece surface and EC reactions are stopped. The honing action by the honing sticks attached to the cathode tool then removed the passivating oxide layer and the workpiece surface is ready for further EC reactions. In this cycle, material removal takes place and accurate and quality surface is obtained in EC honing. Figure 1.14 shows the representation of EC honing process with tool-workpiece arrangement.

The EC honing setup consists of some important sub-units such as power supply system (range of 6–30 V and current up to 110 A across the electrodes), electrolyte flow and circulating unit (includes electrolyte reservoir, positive displacement pump, pressure gauges, flow meter, heat exchanger, magnetic filters, chamber drains and settling tank), machining chamber, tool-workpiece arrangement, tool rotational unit, etc. The characteristics of electrolyte namely type, concentration, flow rate, temperature, etc. play major role in rate of material removal from workpiece surface. In addition, precision and surface finish of workpiece machined largely depend on the size and hardness of abrasive grits in honing sticks and rotary and/or reciprocal motion of the honing tool [40].

1.4.7 Electrochemical Buffing (ECB)

Electrochemical buffing (ECB) is one of the methods of super finishing process which uses loose abrasive particles to finish workpiece surface. In electrochemical machining (ECM), one of the important drawbacks is generation of passive layer on the work surface. This layer restricts further anodic dissolution of the workpiece. This problem can be eliminated mostly in ECB process. ECB combines electrochemical anodic dissolution and conventional buffing using free abrasive particles [41]. During the cycle time, when anodic dissolution of work material occurs, passive film of electrolyte is formed, which is completely removed by the abrasion action of loose abrasive particles provided between the workpiece surface and buffing pad. Thus, the work surface is ready for further electrochemical reactions and further material removal. In Fig. 1.15, the schematic view of EC buffing process is depicted. The EC buffing system consists of DC power supply, buffing pad, rotating and feeding system of buffing tool, electrolyte flow and circulating system, etc. The surface finish and accuracy is mostly depends on influencing process parameters such as applied voltage, current density, abrasive materials, size and concentration, flow rate and concentration of electrolyte, rotating speed and pressure of buffing pad, etc. ECB can be successfully applied to finish almost all types of metals and their alloys such as steel and its alloys, stainless steel, aluminum, copper, titanium, molybdenum etc. used for manufacturing industrial products, vacuum chambers, space chambers, gas cylinders, plant plumbing etc. [42].

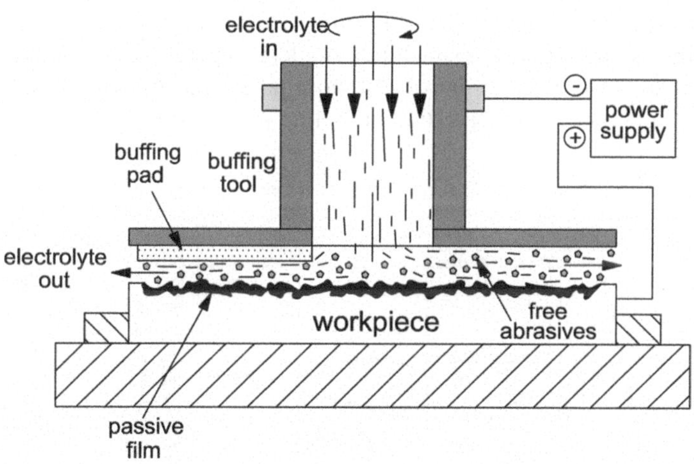

Fig. 1.15 Schematic representation of electrochemical buffing

1.5 Advanced Finishing Processes

1.5.1 Abrasive Flow Finishing (AFF)

Several limitations in conventional finishing methods have enforced to develop the advanced finishing processes. The AFF process is applied to finish components made of wide range of materials such as superalloys, non-ferrous alloys, carbides, ceramics, refractory materials, semiconductors, composites, quartz, etc. [43]. In Abrasive Flow Finishing (AFF), the medium applied is a mixture of viscoelastic polishing medium (which is made of polymer, plasticisers and additives) and small sized abrasive particles. Depending upon the shape and size of workpiece, configuration of tooling is made for holding the workpiece. The polishing medium is held at both sides of tooling and using two vertical cylinders. The polishing medium is made to move backward and forward over the workpiece. The hydraulic unit provides required pressure for finishing the workpiece (different complex geometries) in nanometer level. There are several process parameters on which the assessment of polishing the workpiece is made. These process parameters are extrusion pressure, flow rate of working media, viscosity of working media, number of cycle per minute, size and shape of abrasives mixed, concentration of abrasive particles and rheology properties of polishing medium [44]. Depending upon the initial surface finish of the workpiece and finishing requirement of final products, the values of these process parameters are selected. The schematic diagram of abrasive flow finishing process with machine setup required to carry out the process is shown in Fig. 1.16.

As mentioned in the figure, the AFF setup requires some basic sub-components such as vertical opposed working medium cylinders, upper and lower hydraulic cylinders, hydraulic power packs, control unit, hose pipes and structure of the machine. In addition, workpiece holding fixture or tooling is needed for properly holding the

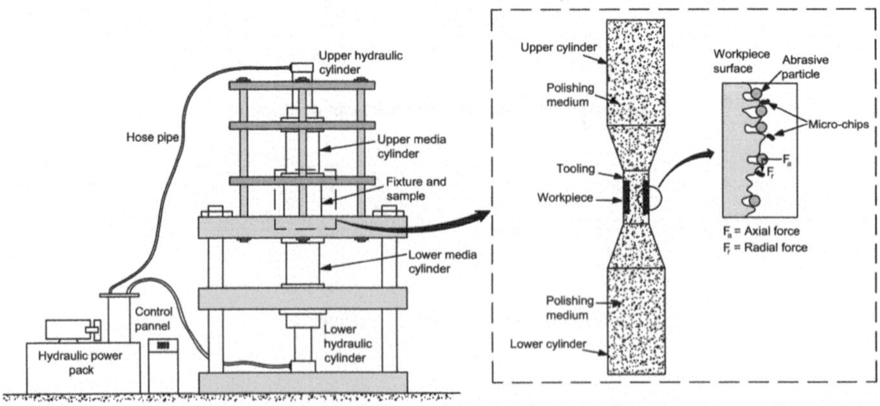

Fig. 1.16 Schematic view of AFF setup with sub-components and enlarged view of mechanism of material removal

workpiece to be finished. The level of surface finish obtained is also largely depending on the polishing medium, which is prepared by using polymer, plasticisers, fine abrasive particles and additives [45]. Depending on the value of working media flow rate, amount of abrasion, the size of edge radius and homogeny of stock removal rely on. Boron carbide, silicon carbide, aluminium oxide and diamond are some frequently used abrasive particles in AFF process [46]. Other process parameters which influence the quality of surface finish are extrusion pressure, number of cycle, size and concentration of abrasive particles, initial condition of the surface of workpiece, etc. Based on the geometrical complexity of the workpiece to be finished, the AFF process is categorized into four types such as one-way AFF, two-way AFF, orbital AFF and micro-AFF. In addition, several hybrid finishing processes have also been developed utilizing AFF process. These are namely magnetic assisted AFF, electrochemical assisted AFF, ultrasonic assisted AFF, centrifugal force assisted AFF, spiral polishing, etc.

1.5.2 Magnetic Abrasive Finishing (MAF)

MAF process is a unique and effective advanced finishing process for finishing both internal and external surfaces of workpiece. The schematic view of MAF process with enlarged view of a single abrasive (with forces exerted) is depicted in Fig. 1.17. A homogeneous mixture of abrasive particles, magnetic iron powders and some carrier fluid (like mineral oil) is produced and employed for finishing. This homogeneous mixture is kept at the gap of bottom face of the magnet's north-pole and workpiece

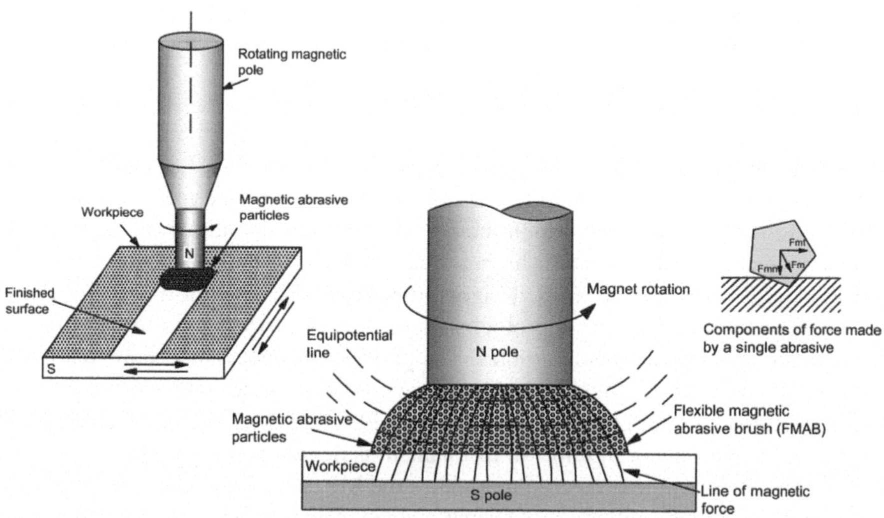

Fig. 1.17 Representation of MAF process with enlarged view of material removal mechanism

surface to be finished. South-pole of the magnet is kept in touch with bottom of workpiece surface. Upon activation of electromagnet, the magnetic iron particles in the mixture are aligned along the line of magnetic force and a chain-like structure is fashioned. This structure is known as flexible magnetic abrasive brush. In the mixture, the non-magnetic abrasive particles get entrapped within and in between the chains of aligned iron particles. Due to the soft and flexible bonding of magnetic medium, it is easy to adapt any type of non-uniformity in shape of workpiece surface which to be finished. A single abrasive grit exerts force in μN range. Instead of using the mixture of magnetic iron particles, non-magnetic abrasives and mineral oil, sintered ferromagnetic abrasives can be used. As seen in figure, a single abrasive exerts force Fm, which has two components, i.e. Fmt and Fmn. The normal force, Fmn exerts at vertical direction of the surface to be machined and is responsible for indenting the workpiece surface and the axial force, Fmt exerts at horizontal direction to the workpiece surface and this force is responsible for removing the micro-chips from the rough surface of the workpiece [47]. Depending on the performance of magnetic abrasive particles (MAPs), the level of finishing of given workpiece is decided.

Utilizing the MAF process, several researchers have developed a number of hybrid finishing processes namely magnetic abrasive deburring (MADe), ultrasonic assisted MAF, electrolyte MAF, etc. There are several important process parameters in MAF which directly affect the level of surface finish and precision. These process parameters are magnetic field, rotational speed of magnet, abrasive particle's shape, size and concentration, viscosity of homogeneous mixture, feed rate of workpiece, etc. The process is self-adaptability and controllability. However, depending on the shape and profile of work surface, different design of electromagnet can be made. The process is applied for precision finishing of various biomedical devices and automotive components (shafts and crank shafts), etc.

1.5.3 Magnetic Rheological Finishing (MRF)

In conventional finishing technologies such as grinding, lapping, honing, etc., fixed or bonded abrasive protruded on wheel or shaft cannot produce nano finished surface due to several problems which include wearing out the abrasives, large value of indentation and wide scratch, higher value of friction between fixed abrasives and workpiece surface, etc. [48]. Magnetorheological (MR) fluid-based finishing techniques can overcome these problems as this novel process uses loose abrasive particles along with magnetic particles. The process uses Magnetorheological fluid (called MR fluid also) which is consists of magnetic particles and loose abrasives mixed in carrier medium (water or additives) [49]. The schematic representation of MRF process with material removal mechanism is depicted in Fig. 1.18.

As seen in figure, the delivery pump takes the MR fluid from MR fluid reservoir and supplies the fluid to the rotating carrier wheel. The electromagnets are kept just below the rotating carrier wheel. When electrical connection is established to the electromagnets, the MR fluid becomes rigid and the carrier wheel is so designed that

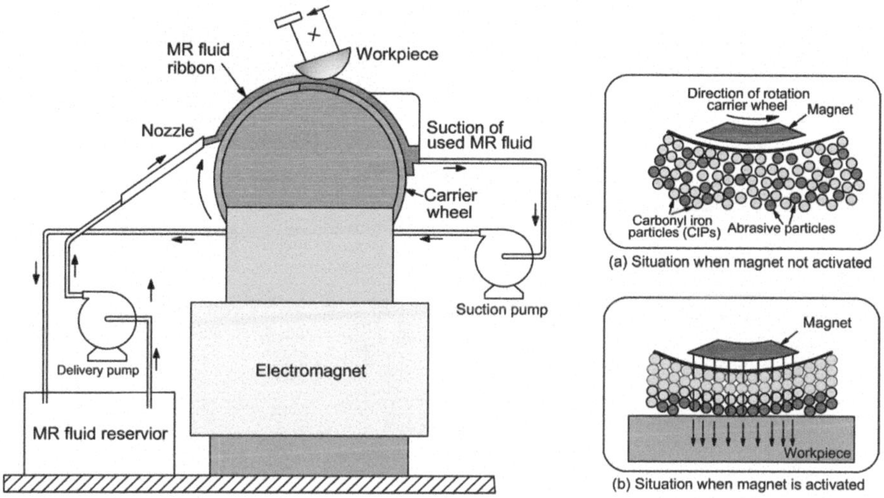

Fig. 1.18 Schematic representation of MRF process

the MR fluid spread on the top surface of carrier wheel. The workpiece to be finished (of different shapes and size) is kept over the carrier fluid just keeping in touch with MR fluid ribbon. Upon activation of electromagnets, when the carrier wheel is rotated and MR fluid is supplied by the nozzle, the workpiece which is in contact with MR fluid ribbon is finished due to the relative motion of workpiece surface and MR fluid. Swinging and rotating motion is provided to the workpiece to finish the workpiece surface completely. The MRF process has several process parameters which greatly influence the performance characteristics such as surface finish, material removal rate, surface flatness, etc. The important process parameters in MRF are viscosity, pH value of MR fluid, type, size, hardness and concentration of abrasive particles, rotational speed of wheel, gap between wheel surface and workpiece surface, type, size, hardness and concentration of magnetic particles, fluidity and viscosity of carrier fluid, etc. The process can finish the optical glasses and surface finish as low as tens of angstroms can be obtained. In addition, MRF process can finish freeform surfaces (internal and external) efficiently [49].

1.5.4 Magnetorheological Abrasive Flow Finishing (MRAFF) and Rotational-MRAFF

MRAFF is an advanced type nano finishing process which combines AFF and MRF processes. Thus, in MRAFF, one can take the advantages of both processes and also to carryout nano finishing of complex geometrical parts and used for a wide range of industrial applications [50]. The material removal phenomena in MRAFF is surface

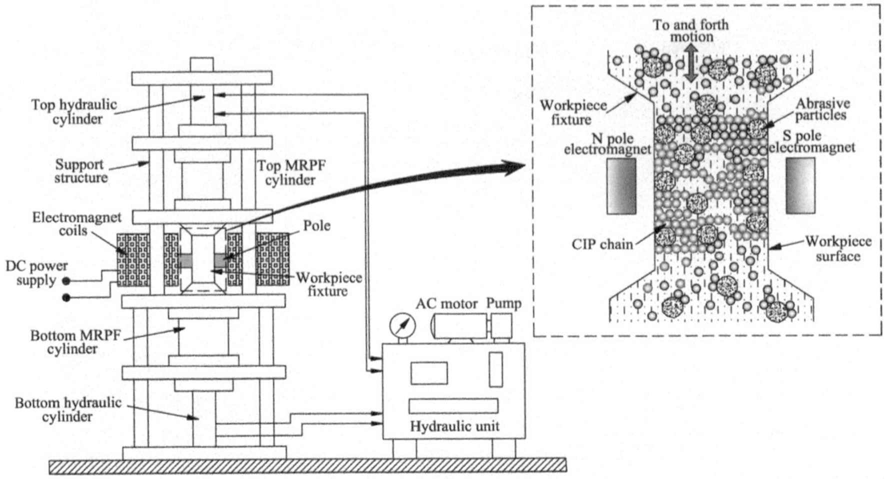

Fig. 1.19 Representation of magnetorheological abrasive flow finishing (MRAFF) setup with finishing mechanism of worksurface

abrasion process by back and forth movement of MR polishing fluid in an electromagnetically dense region. The magnetorheological polishing (MRP) medium is the combination of carbonyl iron particles, abrasive particles and additive medium (paraffin or grease). These ingredients are homogeneously mixed in appropriate proportion or concentration for effective utilization of the ability of the mixture to finish complicated profiles of workpiece [51]. The representation of MRAFF machine with enlarged view of finishing mechanism is depicted in Fig. 1.19.

As shown in figure, the cylindrical workpiece is held between two magnetic poles (N and S poles) and through the hollow workpiece, the MR polishing medium is directed to flow. When magnetic field is applied, depending upon the magnetic field strength, the CIPs obtain dipole moments. At a moment of time, the dipolar interface among the CIPs goes above their thermal energy. In this moment, the CIPs are aggregated in a fashion of chain like structure along the field direction. Thus, in presence of magnetic field, the rheological behavior of MR polishing fluid transformed from weak viscoplastic to strong Bingham plastic. The abrasive particles are embedded between adjacent chains of CIPs. The carbonyl iron particles along with non-magnetic abrasive particles provide required finishing force to the workpiece surface. In MRAFF, the quality of surface finished is influenced by number of process parameters namely extrusion pressure, magnetic flux density, number of finishing cycle, type, size and concentration of abrasive and magnetic particles, etc.

In rotational MRAFF process, finishing of workpiece surface is carried out by additionally rotating the magnets provided. In Fig. 1.20, the schematic view of R-MRAFF setup with enlarged view of finishing mechanism is depicted. Due to rotation of magnets and extrusion pressure of top and down hydraulic cylinders, the MR polishing fluid is rotated in spiral motion as indicated in figure. Thus, the MR fluid has two motions simultaneously such as axial motion and circular spiral motion.

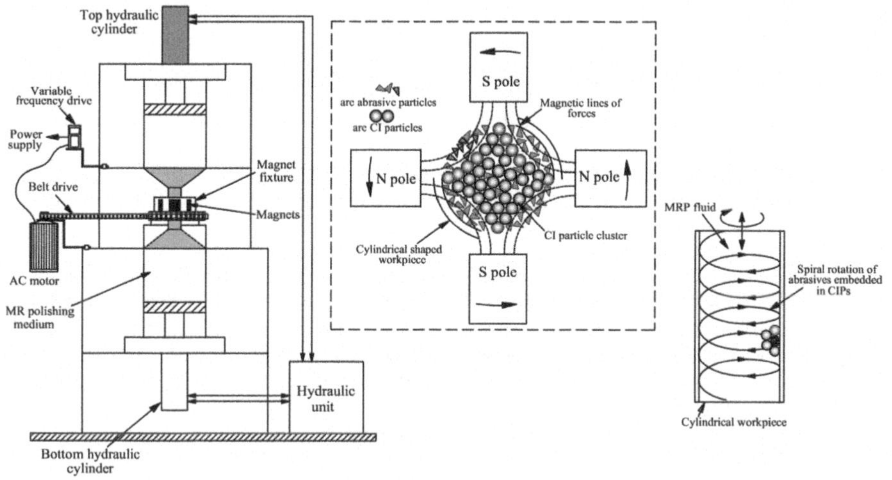

Fig. 1.20 Rotational MRAFF setup with enlarged view of finishing mechanism

During movement of polishing medium in helical form, the rough peaks on the workpiece surface are sheared off by the motion of abrasive particles embedded by carbonyl iron particles. If the finishing mechanism is compared for MRAFF and R-MRAFF, then it is revealed that the movement of MR polishing fluid is due to the extrusion pressure (reciprocating type) delivered by the movement of top and bottom MRPF cylinders and pistons [52]. However, in R-MRAFF, the abrasive particles along with CI particles move in helical path inside the workpiece wall due to the combined forces of extrusion (because of back and forth movement of MR fluid) and rotation of magnets. Therefore, the interaction area between the abrasive particles and workpiece surface is high. In addition to the process parameters in MRAFF process, R-MRAFF has added parameter i.e. rotational speed of magnets. Compared to MRAFF, R-MRAFF process can generate nano level surface finish since this novel process has more controlled finishing mechanism [53].

1.6 Summary

Abrasive based machining and finishing processes have important role in manufacturing as well as finishing macro and micro products with required surface finish, form accuracy and geometrical tolerances. In this chapter, each of the developed processes that utilize abrasives (applied in bonded or loosely) has been discussed with fundamental mechanism, process characteristics and important applications in various sectors ranging from automotive, aerospace to engineering and biomedical sciences.

References

1. Guo L, Zhang X, Chen S, Hui J (2019) An experimental study on the precision abrasive machining process of hard and brittle materials with ultraviolet-Resin bond diamond abrasive tools. Materials 12(1):125–135
2. Shaw MC (1995) Precision finishing. CIRP Ann Manuf Technol 44(1):343–348
3. Taniguchi N (1983) Current status in, and future trends of ultraprecision machining and ultrafine material processing. Ann CIRP 32(2):573–582
4. Shaw MC (1996) Principles of abrasive processing. Clarendon Press, Oxford
5. Malkin S, Guo C (2008) Grinding technology: theory and applications of machining with abrasives. Industrial Press, New York
6. Chen X, Brian Rowe W (1996) Analysis and simulation of the grinding process. Part II: mechanics of grinding. Int J Mach Tools Manuf 36(8):883–896
7. Bhowmik S, Naik R (2018) Selection of abrasive materials for manufacturing grinding wheels. Mater Today Proc 5(1):2860–2864
8. Pande SJ, Halder SN, Lal GK (1980) Evaluation of grinding wheel performance. Wear 58(2):237–248
9. Sun X, Stephenson DJ, Ohnishi O, Baldwin A (2006) An investigation into parallel and cross grinding of BK7 glass. Precis Eng 30(2):145–153
10. Koshy P, Zhou Y, Guo C, Chand R, Malkin S (2005) Novel kinematics for cylindrical grinding of brittle materials. CIRP Ann Manuf Technol 54(1):289–292
11. Fischer H (1986) Handbook of modern grinding technology. Chapter 13, King RI, Hahn RS (eds). Chapman and Hall, New York
12. Lee J, Malkin S (1993) Experimental investigation of the bore honing process. J Eng Ind 115(4):406–414
13. Evans CJ, Paul E, Dornfeld D, Lucca DA, Byrne G, Tricard M, Klocke F, Dambon O, Mullany BA (2003) Material removal mechanisms in lapping and polishing. CIRP Ann 52(2):611–633
14. Chandrasekar S, Shaw MC, Bhushan B (1987) Comparison of grinding and lapping of ferrites and metals. J Eng Ind 109(2):76–82
15. Sax DJ (2007) Buffing wheels and equipment. Met Finish 105(10):30–43
16. Dickman A Jr (1995) Polishing and buffing: theory and practice. Met Finish 93(12):34–36
17. Jadav SM, Lekurwale RR (2018) Effect of process parameters while machining using abrasive jet machine (AJM). In: Proceedings of international conference on intelligent manufacturing and automation, pp 575–581
18. Prasad SR, Ravindranath K, Devakumar MLS (2016) A research review on advance approaches in abrasive jet machining. In: International conference on recent innovations in civil & mechanical engineering, pp 57–62
19. Rani RM, Seshan S (1998) Studies on abrasive jet machining through statistical design of experiments. Exp Tech 22(2):28–30
20. Nair A, Kumanan S (2017) Multi-performance optimization of abrasive water jet machining of Inconel 617 using WPC. Mater Manuf Process 32(6):693–699
21. Supriya SB, Srinivas S (2018) Machinability studies on stainless steel by abrasive water jet-review. Mater Today Proc 5(1):2871–2876
22. Zhu H, Huang C, Wang J, Zhao G, Li Q (2009) Modeling material removal in fracture erosion for brittle materials by abrasive waterjet. Adv Mater Res 76–78:357–362
23. Thoe TB, Aspinwall DK, Wise MLH (1998) Review on ultrasonic machining. Int J Mach Tools Manuf 38(4):239–255
24. Kumar J (2013) Ultrasonic machining—a comprehensive review. Mach Sci Technol 17(3):325–379
25. Komaraiah M, Reddy PN (1991) Rotary ultrasonic machining—a new cutting process and its performance. Int J Prod Res 29(11):2177–2187
26. Benedict GF (1987) Non-traditional manufacturing processes. Marcel Dekker Inc, New York
27. Shih HR, Shu KM (2008) A study of electrical discharge grinding using a rotary disk electrode. Int J Adv Manuf Technol 38(1–2):59–67

28. Abothula BC, Yadava V, Singh GK (2010) Development and experimental study of electrodischarge face grinding. Mater Manuf Process 25(6):482–487
29. Koshy P, Jain VK, Lal GK (1996) Mechanism of material removal in electrical discharge diamond grinding. Int J Mach Tools Manuf 36(10):1173–1185
30. Yadav SKS, Yadava V, Narayana VL (2008) Experimental study and parameter design of electro-discharge diamond grinding. Int J Adv Manuf Technol 36(1–2):34–42
31. Masuzawa T, Fujino M, Kobayashi K, Suzuki T, Kinoshita N (1985) Wire electro-discharge grinding for micro-machining. CIRP Ann 34(1):431–434
32. Periyanan PR, Natarajan U, Elango A (2011) Optimization of parameters on material removal rate in micro-WEDG process. Int J Eng Sci Technol 3(9):66–76
33. Kibria G, Bhattacharyya B (2011) Investigation into micro-hole geometrical accuracy during micro-EDM of Ti-6Al-4 V employing different dielectrics. Int J Mach Mach Mater 10(4):310–325
34. Kibria G, Shivakoti I, Bhattacharyya B (2014) Experimentation and analysis into micro-hole machining in EDM on Ti-6A1-4 V alloy using boron carbide powder mixed de-ionized water. Int J Manuf Mater Mech Eng 4(1):22–41
35. Kibria G, Sarkar BR, Pradhan BB, Bhattachayya B (2010) Comparative study of different dielectrics for micro-EDM performance during microhole machining of Ti-6Al-4 V alloy. Int J Adv Manuf Technol 48(5–8):557–570
36. Talla G, Gangopadhayay S, Biswas CK (2016) State of the art in powder-mixed electric discharge machining: a review. Proc Inst Mech Eng Part B J Eng Manuf 231(14):2511–2526
37. Zaborski S, Łupak M, Poroś D (2004) Wear of cathode in abrasive electrochemical grinding of hardly machined materials. J Mater Process Technol 149(1–3):414–418
38. Łupak M, Zaborski S (2009) Simulation of energy consumption in electrochemical grinding of hard-to-machine materials. J Appl Electrochem 39(1):101–106
39. Jain NK, Naik LR, Dubey AK, Shan HS (2009) State-of-art-review of electrochemical honing of internal cylinders and gears. Proc Inst Mech Eng Part B J Eng Manuf 223(6):665–681
40. Dubey AK (2008) Experimental investigations on electrochemical honing. Proc Inst Mech Eng Part B J Eng Manuf 222(3):413–426
41. Tailor PB, Agrawal A, Joshi SS (2015) Parametric understanding of electrochemical buffing (ECB) using current-voltage characterization. Mach Sci Technol 19(3):440–459
42. Tsui HP, Yan BH, Wu KL, Wu WC (2007) Optimizing electrochemical buffing control parameters for surface finishing of ultrahigh purity components. Adv Mater Res 24–25:109–116
43. Kavithaa TS, Balashanmugam N (2016) Nanometric surface finishing of typical industrial components by abrasive flow finishing. Int J Adv Manuf Technol 85(9–12):2189–2196
44. Petare AC, Jain NK (2018) A critical review of past research and advances in abrasive flow finishing process. Int J Adv Manuf Technol 97(1–4):741–782
45. Jain VK, Adsul SG (2000) Experimental investigations into abrasive flow machining. Int J Mach Tools Manuf 40(7):1003–1021
46. Sambharia J, Mali HS (2017) Recent developments in abrasive flow finishing process: a review of current research and future prospects. Proc Inst Mech Eng Part B J Eng Manuf 233(2):388–399
47. Saraeian P, Mehr HS, Moradi B, Tavakoli H, Alrahmani OK (2016) Study of magnetic abrasive finishing for AISI321 stainless steel. Mater Manuf Process 31(15):2023–2029
48. Jain VK (2008) Abrasive-based nano-finishing techniques: an overview. Mach Sci Technol 12(3):257–294
49. Sidpara A, Das M, Jain VK (2009) Rheological characterization of magnetorheological finishing fluid. Mater Manuf Process 24(12):1467–1478
50. Das M, Jain VK, Ghoshdastidar PS (2008) Analysis of magnetorheological abrasive flow finishing (MRAFF) process. Int J Adv Manuf Technol 38(5–6):613–621
51. Jha S, Jain VK, Komanduri R (2007) Effect of extrusion pressure and number of finishing cycles on surface roughness in magnetorheological abrasive flow finishing (MRAFF) process. Int J Adv Manuf Technol 33(7–8):725–729

52. Nagdeve L, Jain VK, Ramkumar J (2018) Development of inverse replica fixture for nano-finishing of knee joint using R-MRAFF process. J Micromanuf 2(1):35–41
53. Das M, Jain VK, Ghoshdastidar PS (2012) Nanofinishing of flat workpieces using rotational–magnetorheological abrasive flow finishing (R-MRAFF) process. Int J Adv Manuf Technol 62(1–4):405–420

Chapter 2
Advancement in Ultrasonic Machining for 3D Profile Cutting

S. Das, S. Kumar, G. Kibria, B. Doloi, and B. Bhattacharyya

Abstract Hard and fragile materials for example ceramics, glass and quartz crystals are getting extra consideration in modern years owing to their higher characteristics for example high strength, high hardness, chemical durability and low density. Ultrasonic machining is an abrasive based advanced machining with non-chemical, non-electrical and non-thermal process that is particularly suitable for those brittle and hard materials. The USM process principle, mechanism of material removal, varieties of USM set up, tool development of USM process, improvement and production of 3d profile by USM process and various research issues are studied and summarized in this chapter. It also highlights the effects of different parameters of USM process on performance and development of USM process.

Keywords Ultrasonic machining · Profile accuracy · Ceramics · Surface roughness · Material removal rate

---

S. Das (✉)
Mechanical Engineering Department, Swami Vivekananda Institute of Science and Technology, Kolkata 700145, India
e-mail: somnath96@gmail.com

S. Kumar · B. Doloi · B. Bhattacharyya
Production Engineering Department, Jadavpur University, Kolkata 700032, India
e-mail: santosh14fiem@gmail.com

B. Doloi
e-mail: bdoloionline@rediffmail.com

B. Bhattacharyya
e-mail: bb13@rediffmail.com

G. Kibria
Department of Mechanical Engineering, New Town Campus, Aliah University, Kolkata 700156, India
e-mail: prince_me16@rediffmail.com

S. Das et al. (eds.), *Advances in Abrasive Based Machining and Finishing Processes*, Materials Forming, Machining and Tribology,
https://doi.org/10.1007/978-3-030-43312-3_2

2.1 Introduction

From the beginning of the human race, researchers have developed state-of-the-art products with the help of various energy sources and tools to meet the demands of making life easy and enjoyable. Since the last decade, manufacturing industries will face some new challenges. The main driver of industrial innovation is better quality of life and higher work efficiency. Modern improvements in the properties of materials have resulted in high durability, high rigidity and extremely hard and brittle materials to meet the requirements of advanced industry. As a result, engineers and technologists are facing more and more problem to produce complex profiles on these materials. They are tremendously complicated to machine by manual or traditional machining processes. This also requires the improvement of better tooling materials to maintain productivity. Such materials can be machined economically by non-traditional processes based on direct use of various energy sources, such as ultrasonic machining (USM), electro-chemical machining (ECM), electron-beam machining (EBM), electro-discharge machining (EDM) etc. [1, 2]. Ultrasonic machining has great applications for machining hard and brittle materials, specifically ceramics, tungsten, carbides and glass etc. those are eclectically poor conductors. USM can easily machine workpiece harder than 40 HRC [3, 4]. During the ultrasonic processing process tool of preferred shape oscillates at an ultrasonic frequency from 19 to 25 kHz at a working amplitude of 15–50 μm [5, 6]. The desired shape tool is vibrated on the machining zone and the transmitted the kinetic energy to the abrasive particles which act as an indenter. In USM material is removed due to initiation of crack, propagation of this crack and brittle fracture of the material [7–9]. Production of non-circular holes and cavities on ceramics, metal and other components through superior geometric and dimensional accurateness is a difficult task [10–15]. It is shown that the material removal rate increases with increasing tool tip amplitude diameter of abrasive particles [16]. The purpose of this chapter is to improve the 3D profile and product presentation through the USM process principle, mechanism of material removal, tool development of USM process. The effects of various parameters of USM process on performance of ultrasonic machining and process development of USM process also describe into this section.

2.2 Basics of Ultrasonic Machining Process

The fundamental of USM process involves tool, normally prepared of a ductile and strong material is vibrating by small amplitude and extremely high frequency and nonstop flow of slurry mixed with abrasive between the tiny gap of tool and the workpiece. USM process is non-traditional mechanical process employed for machining conductive as well as non-conductive materials; by preference those by means of small ductility and hardness higher than 40 HRC for example inorganic glasses, ceramics, bio-ceramics and quartz etc. Water mixed with abrasive grains flush into the gap between the tool and workpiece during machining. The desired shape tool is vibrated at ultrasonic frequency of 20 kHz and amplitude of 8–30 μm. The tool is pushed to workpiece with definite static power. The energized abrasive grains which are transmitted by the vibrated tool are directly hammering on the workpiece. In the machining area, wash out of slurry refreshes the abrasive particles. It also takes away the debris and damaged abrasive elements left from the gap.

2.3 Different Types of USM Setup

In 1927, the cutting and drilling action through the assist of ultrasonic vibration was explained by A. L Loomis and R. W. Wood. After a long period, in 1942 L. Balamuth was first proposed ultrasonic machining, while his investigation was stated that the dispersal of solid into liquid by magneto-strictively vibrated nickel tube. The primary announcement of whole apparatus and techniques in ultrasonic machining was published in 1953–54 [5]. All through suitable research work and progress on USM equipment, various types of USM processes have been introduced.

2.3.1 Stationary USM Process

In this process the abrasive slurry is injected into the space between a vibrating tool and stationary workpiece. Material is abraded away until a mirror image of the tool is cut into the workpiece. A hard metalic tool is pressed with high frequency and low amplitude of oscillation perpendicular to workpiece, which convey high velocity to fine particles of abrasive presented between the workpiece and tool. These particles strike the workpiece, chipping away small particles, and the tool is gradually fed into the workpiece. The "chips" are carried away from the workpiece by a constant flow of cooled slurry. The workpiece is abraded into a mirror image of the tool. Figure 2.1 shows the schematic of stationary USM system.

2.3.2 Rotary USM Process

Rotary type USM is an invention of two machining process that merges diamond grinding elements with stationary USM process, increasing higher material removal rate (MRR) than diamond grinding or stationary USM. In rotary USM process a revolving drill bonded with diamond abrasive particles are vibrated ultrasonically on the axial whereas spindle is supplied at a constant pressure in the direction of the workpiece. The coolant is pumped with the drill core, preventing jamming of the drill and keeping it cool. Using directly bonded abrasives tool in RUM process and simultaneously combining rotation and vibration, it provides very fast and high

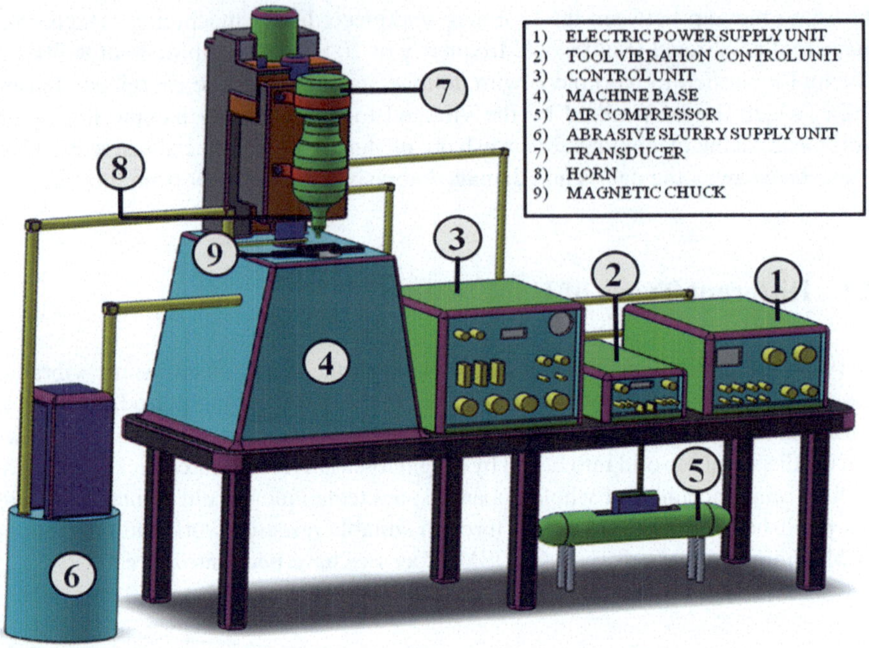

Fig. 2.1 Diagram of USM set up

quality machining technique for various glass and ceramic applications. It is easier to drill deep holes with RUM than USM and improve hole accuracy with higher surface finish. Spindle speeds in rpm are programmable using a CNC controller for controlling the speeds of up to 8000 rpm. Different types of tool shapes are used for RUM process on ceramic and glass machining applications are commonly used in diamond-synthesized tool. Figure 2.2 illustrates the schematic of Rotary Ultrasonic Machining process.

2.4 Detail Description of Mechanism of Material Removal in USM Process

In this process the material of workpiece is removed from the machining zone by the grazing action of liquid based abrasive slurry, which is circulating between the vibrated tool and workpiece. The tool does not contact the workpiece directly, so that the machining pressure is rarely used. For that reason this operation is perfect for machining tremendously brittle and hard materials, for example ceramics and glass with producing no heat.

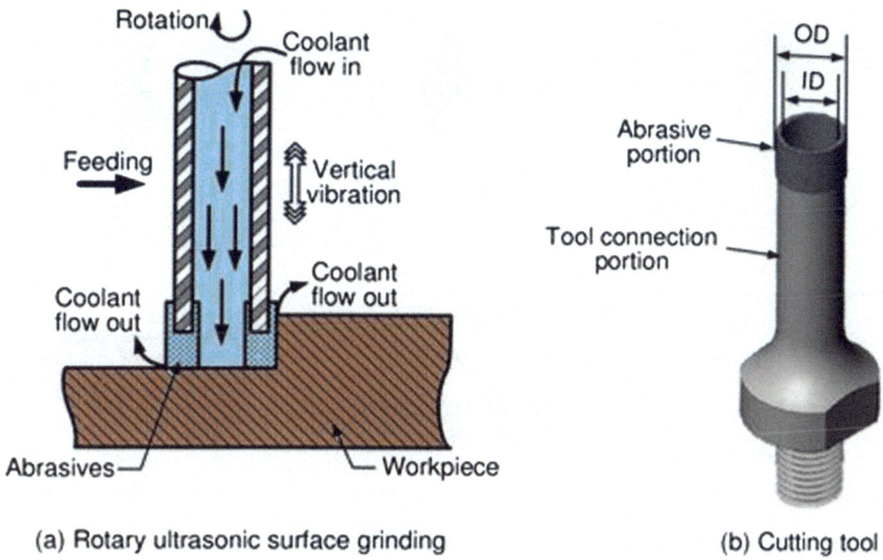

Fig. 2.2 Schematic of Rotary Ultrasonic Machining (RUM) and cutting tool

2.4.1 Material Removal Mechanism

Owing to micro-cracks generation through the fracturing in brittle nature of the workpiece, the material is detached from the machining region of the workpiece during USM process. Mechanism of crack production has been investigated in USM in order to develop machining effectiveness and accuracy. The collision of a particular abrasive grain for the duration of USM process is able to very essential for proper crack production of the workpiece. The most important reasons for taking away the material from the workpiece have been documented as follows:

(i) The main principle is hammering effect. This phenomenon is happened by abrasive particle when the workpiece and tool get in touch with each other.
(ii) The another reason is impact action by abrasive particle with free-moving in the machining zone.
(iii) The cavitations erosion plays a role for removal of materials by abrasive slurry concentration [5, 6].

In general material removal mechanism of USM, cavitations erosion is not significant all time. The schematic diagram of step by step mechanism of material subtraction in USM is shown in Fig. 2.3. The mostly responsible reason is first two mechanisms for material elimination owing to micro cracks. These are produced owing to the quick hammering of all coarse particles. The striking stroke is largely established as a main reason in crack production [7].

The material is taking away from the workpiece in the USM process depends largely on radial cracks start and proliferation of lateral cracks. The beginning of

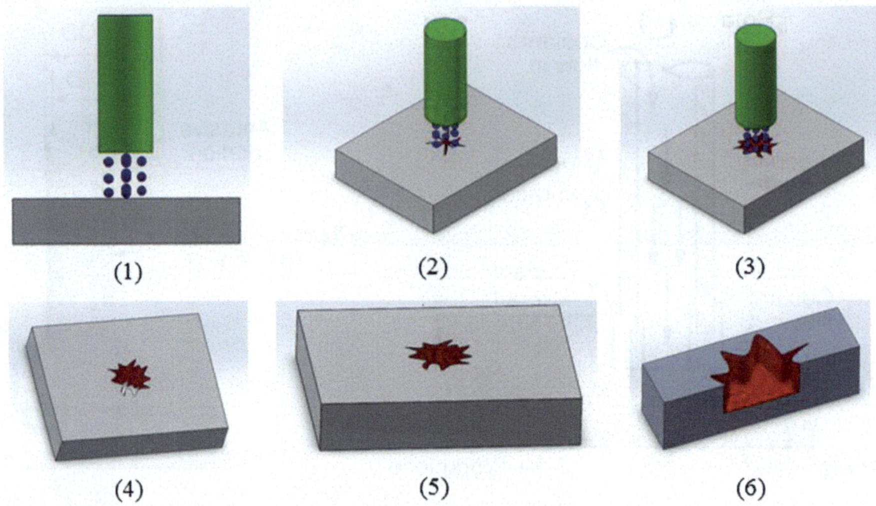

Fig. 2.3 Schematic diagram of step by step mechanism of material subtraction in USM

localized deformation and cracks formation of any hard and brittle workpiece material is shown in Fig. 2.4. Abrasive particles are produced these cracks and perform as an indenter [8, 9]. The fracture begins on the upper surface of the workpiece by striking the abrasive grains in the machining region. Various researchers have been studied in these regions.

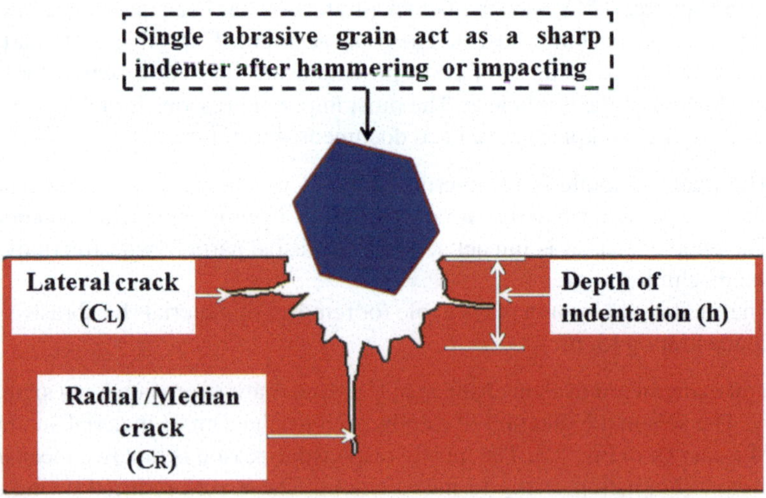

Fig. 2.4 Fractural deformation of brittle material

2.4.2 Review Work of MRR Model

Various investigations have been done in the region of material removal mechanism. Following section includes the details of research work carried out by various researchers in the mechanism of material removal in USM.

M. C. Shaw explored the material removal mechanism in USM process. Material was removed primarily the nonstop hammering by abrasive grains and secondarily striking by free exciting particles. Every abrasive particle was assumed as rigid, identical and round in shape. Every single one impacts were considered as identical. Depth of penetration was also considered as inversely proportional to work material flow stress. For the tool face given area, the numbers of active grains were inversely proportional to the square of the average grain diameter [16]. G. E. Miller proposed a model of the material removal mechanism in ultrasonic machining. Primary physical principles and equation of machining rate were derived from this model. In this case circular solid tools were used with puddle slurry. A relationship of the calculations of the theory was made with the experimental data that the theory gave a right correlation between the machining speed and the variables involved in the machining process [17]. L. D. Rozenberg et al. proposed that the material was removed from the workpiece by brittle fracture. All abrasive grains were incompressible and unequal contour but could be considered as spherical shape [18]. Based on the investigational confirmation, the statistical calculation of abrasive grain diameter size d was given by:

$$\varnothing(d) = 1.095\frac{N}{d_m}\left[1 - \left(\frac{d}{d_m} - 1\right)^2\right]^3 \tag{2.1}$$

where, N is number of active abrasive grains and d_m is the mean dia. of grains.

N. H. Cook proposed the material removal mechanism in USM. As per discussion the workpiece material was removed due to indentation fracture by the hemispherical shape abrasive. All abrasive grains were assumed as spheres of uniform radius. Viscosity effects of slurry concentration were considered insignificant. According the MRR model, material removal rate was considered as proportional to square root of grain radius [19]. G.S. Kainth analyzed the mechanism of material removal in ultrasonic machining. According to proposal the direct impact on the workpiece by abrasive grains was main reason for removing material from the workpiece. The investigation had been applied to compute material removal during machining on glass materials using abrasive of 400 mesh size. In that experimentation the mild steel tool with different static force and amplitude of tool oscillation was used [20]. E. V. Nair et al. performed one hypothetical study into the mechanics of USM. The periodic hammering by the cutting tool on the workpiece material through rigid spherical abrasive particles is responsible for material removal. The dislodging of material from the workpiece during USM process was owing to brittle fracture. The cavitations erosion was negligible for removal of the material. Finally, the differences of machining rates were depended by mean diameter of abrasive grains, amplitude and frequency of vibration [21]. K. P. Rajurkar et al. suggested that USM process was an efficient advance machining process for ceramic materials. This research presented the experimental model of the MRR mechanism during machining of alumina ceramic [22]. T.C. Lee et al. explained the basic mechanism of the USM of ceramic composites. The effects of the MRR and the surface roughness were measured and discussed with changing the static load applied, the amplitude of the tool tip and the size of the abrasive. It was concluded that if the static load applied, amplitude of the tool tip and the grit size of the abrasive was increased, the material removal rate also increased and it was roughening of the workpiece machined surface [23]. M. Wiercigroch et al. suggested that the improvement of material removal rate (MRR) in USM was very much associated with high amplitudes forces produced by develop tool on the workpiece and it also developed micro-cracking in the cutting zone. Mechanism of material removal was observed as micro-cracking produced on the workpiece due to impact of grains [24]. C. Nath et al. suggested that due to quick mechanical indentations the micro-chipping via micro-cracks is generated by abrasive grits. The basic material removal mechanism during USM of hard–brittle materials like glass and ceramics was observed [25].

2.5 Tool Development of USM Process

The final form and dimension of the developed products depends on the tool during ultrasonic machining (USM). The production of micro tool is in actuality immense challenge. Holding the micro tool appropriately with good accuracy is not an easy task of the job owing the purpose of vibrations at the tool ending. New technique was launched to conquer this difficulty. This tool is prepared on the machining time in other technique. In the beginning the macro tool was set to head of the tool. Wire electro discharge grinding (WEDG) technique was utilized to manufacture micro tool. Then this tool was utilized for machining [26]. Using this method micro tool of less than 20 μm can be attained [3]. Figure 2.5 shows micro tool used in micro USM process.

The main problem is to design and build up the tool for ultrasonic machine for producing any shape. A cylindrical rod of stainless steel of grade 304 has been selected for fabricating the tool. After that the stainless steel tool has been fabricated with the help of lathe as per design of CAD model. Figure 2.6 shows the CAD model of stepped tool steps having both in circular cross section. Figure 2.7 shows the CAD model of stepped tool having one step of square cross section and another step of circular cross section. Figure 2.8 shows the CAD model of hemispherical tool. Next the tool holding hexagonal bolt and the bottom face of the developed tool has been properly cleaned. Then this tool holding bolt and tool have been joined by silver brazing. Silver brazing is a joining process whereby a non-ferrous filler metal, alloy is heated to melting temperature above 800 °C. Silver brazing can employ by flame heat sources. The copper filler metal is used in silver brazing. Flux is necessary for brazing to remove and prevent reformulation of surface oxides on the base metals.

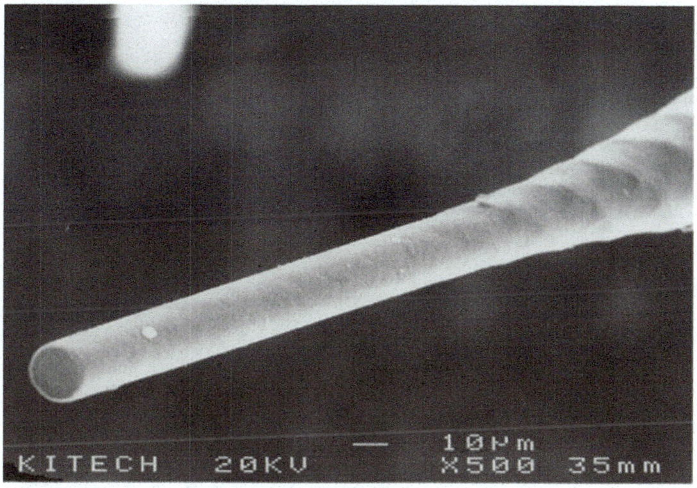

Fig. 2.5 Micro tool prepared for Micro USM [27]

Fig. 2.6 CAD model of stepped tool of circular cross-section

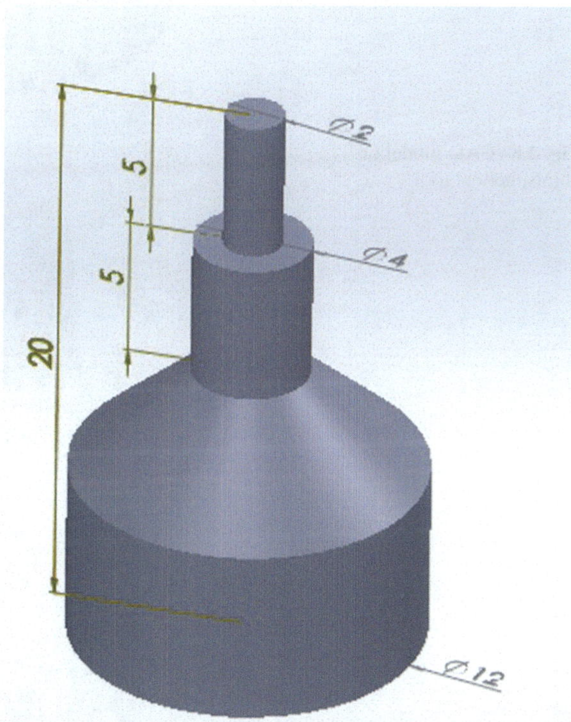

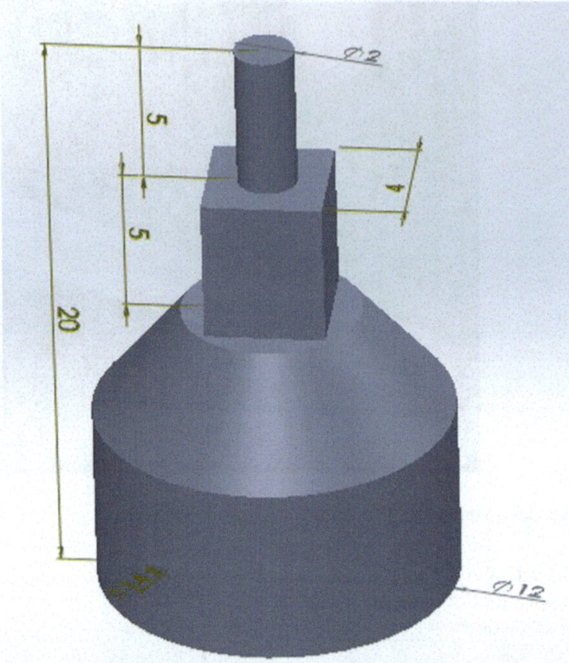

Fig. 2.7 CAD model of stepped tool having both square and circular cross-section

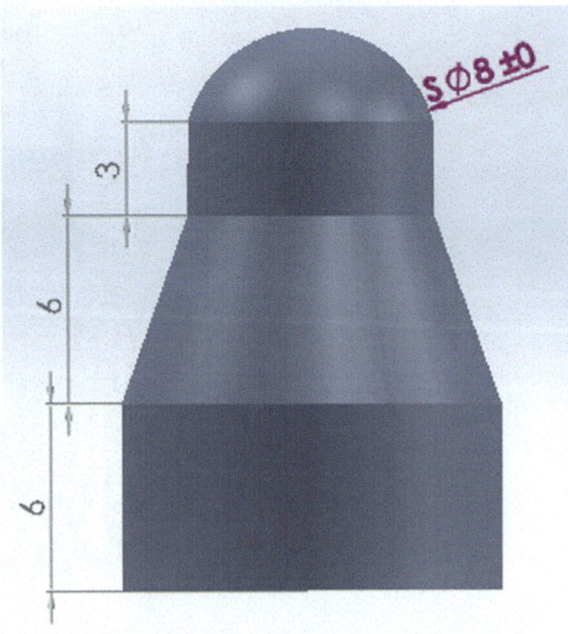

Fig. 2.8 CAD model of hemispherical tool

Fig. 2.9 Photographic view of stepped tool of circular cross-section for USM

Proper brazing is very much essential for good joint design. The photographic views of the developed tools are shown in Figs. 2.9, 2.10 and 2.11.

2.6 Influences of USM Process Parameters on Responses During Machining

The performance of USM process depends upon process parameters, which is shown in Fig. 2.12. The amplitude of vibration, frequency of vibration, power rating and static load are the process parameters related to energy input during USM. One of the most effective process parameter is abrasive particle. The types of material, hardness, strength of the workpiece, abrasive grain size, abrasive slurry concentration and abrasive slurry flow rate are the process parameters related to effectiveness of abrasive slurry during USM. The tool material, tool geometry and property are tool related process parameters in USM. The controllable process parameters are selected during this present research work i.e. power rating, abrasive grain size, slurry concentration and tool feed rate.

Fig. 2.10 Photographic view of stepped tool having both square and circular cross-section for USM

So many abrasive materials are available in different grain diameters for USM process. The selecting criteria for a particular machining condition depend on hardness, useable life, cost and particle size of the abrasive. Boron carbide, silicon carbide and aluminum oxide are the commonly used abrasive for ultrasonic machining process. Boron carbide abrasive is very costly and it has high hardness and useable life. Abrasive grain sizes are in the range from 240 to 800. In this present research work boron carbide abrasive with different grain sizes are used.

Abrasive slurry concentration is one of the process parameters of USM. The abrasive grains are mixed with water to form the slurry. The concentrations of abrasive slurry are varied from 30 to 60%.

Power rating is another important controllable process parameter of USM. High power is required for drilling very hard material and more penetration.

Tool feed rate is one another process parameter of USM. Tool feed rate is the velocity at which the tool is fed, that is, advanced against the workpiece. It is expressed in units of distance per minute.

The influences of ultrasonic machining process parameters on various responses have been investigated and explained using various graphs.

Fig. 2.11 Photographic view of hemispherical tool for USM

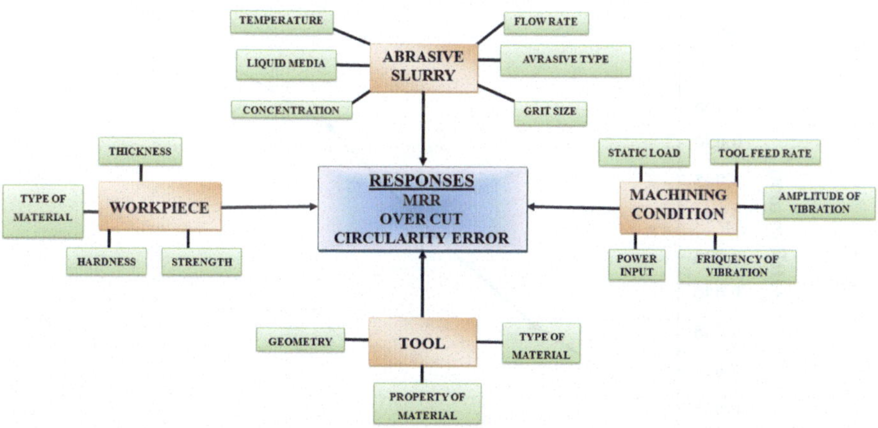

Fig. 2.12 Process parameters of USM

2.6.1 *Influences of USM Process Parameters on MRR*

The significant parameters in the USM process are abrasive grain diameter. The experiments have been carried out with varying abrasive grain diameter. Figure 2.13 illustrates the outcome of abrasive grain diameter on MRR. From this graph, it is found that small abrasive grain i.e. 14 μm grain diameter provides the less MRR however coarser abrasive i.e. 64 μm grain diameter offers high MRR. Coarser abrasive have spiky corner in every side of grain particle, therefore more MRR is found. This active abrasive grains are taken an influencing part in material removal process whose diameters are large.

Figure 2.14 presents the outcome of concentration of abrasive slurry on MRR of production of square stepped type hole on zirconia bio-ceramics. Hence, the material removal rate depends only how much effective particles of abrasive grains are present in the machining region. When the slurry concentration is more, the density of the slurry is high. So the extra effectual abrasive grains move toward to the effective gap. As a result the more MRR is achieved using high percentage of abrasive slurry concentration.

Figure 2.15 exhibits the outcome of power rating on MRR. In the graph, it is evidently exposed that with enhance in power MRR increases. At that time extra power is used, subsequently the abrasive particles hit on the workpiece surface through high value of force. The active grains are striking through more momentum and rapid crack of the workpiece surface is found owing to quicker transmission of cracks. Therefore the workpiece material takes away from the workpiece at quicker rate as a result improving MRR.

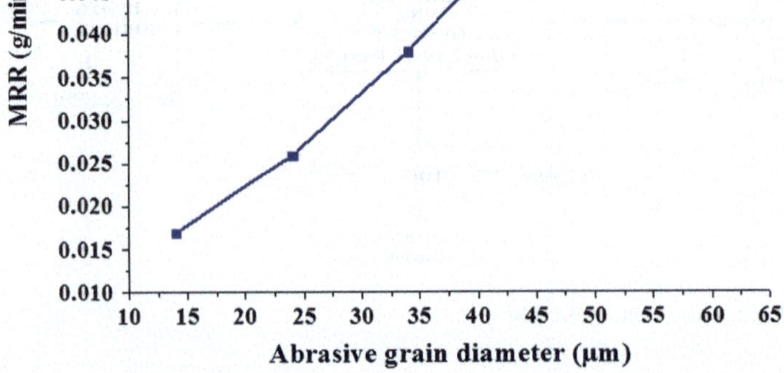

Fig. 2.13 Result of abrasive grain diameter on MRR for square stepped hole generation

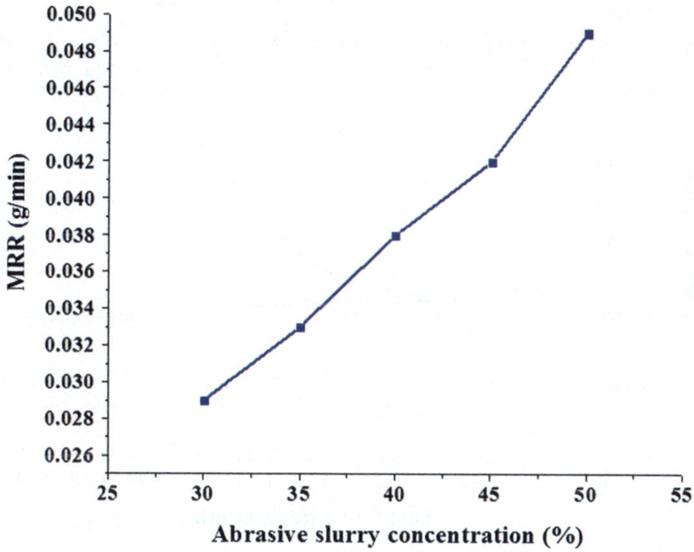

Fig. 2.14 Result of abrasive slurry concentration on MRR for square stepped hole generation

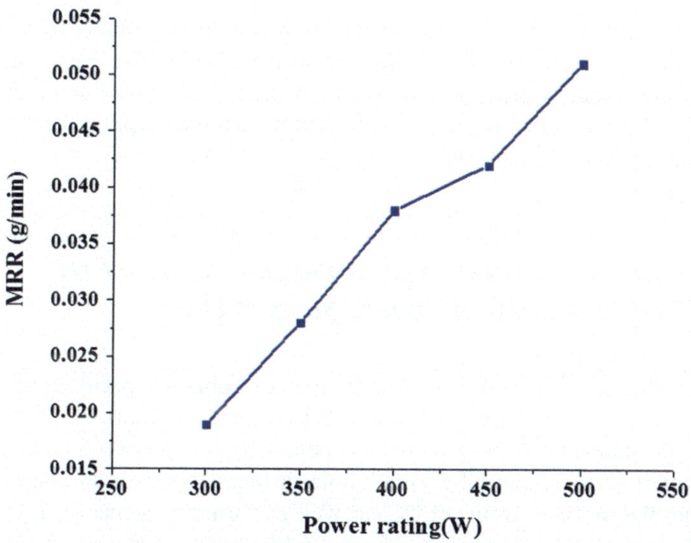

Fig. 2.15 Result of power rating on MRR for square stepped hole generation

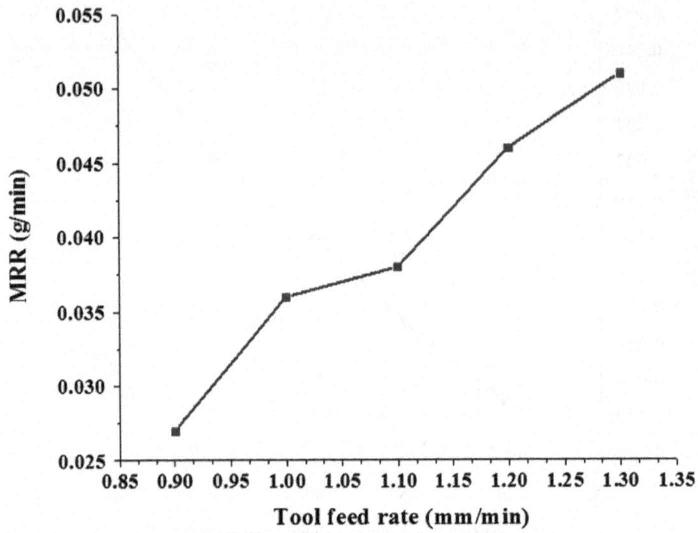

Fig. 2.16 Result of tool feed rate on MRR for square stepped hole generation

It is observed from Fig. 2.16 that material removal rate is influenced by tool feed rate during the square stepped type hole generation by USM. The crack transmission rate on the workpiece is efficiently more when the tool is stirred down at superior tool feed rate. Very quickly material is detached from the workpiece. Therefore high tool feed rate gives highest MRR.

2.6.2 Influences of USM Process Parameters on FFOC, CCOC and DOC of Square Stepped Hole

Figure 2.17 illustrates the result of grain diameter of abrasive particles on corner to corner overcut (CCOC), flat to flat overcut (FFOC) and diametrical overcut (DOC) of square type stepped hole on zirconia bio-ceramics. It is observed from the graph, that very small abrasive particles create hole through less overcut and also found least amount flat to flat overcut (FFOC) as well as corner to corner overcut (CCOC) of square hole during USM process. Normally bigger abrasive grain diameter is 64 micron and regular lesser abrasive grain diameter is 16 micron. So that the cutting region of average bigger grain diameter is high compare to the average lesser grain diameter. The smaller abrasive grain diameter has less contact surface area in the machining zone. So that small overcut i.e. DOC, corner to corner overcut (CCOC) and flat to flat overcut (FFOC) of this square hole are achieved by low value of abrasive grain diameter.

Figure 2.18 shows the result of abrasive slurry concentration on corner to corner

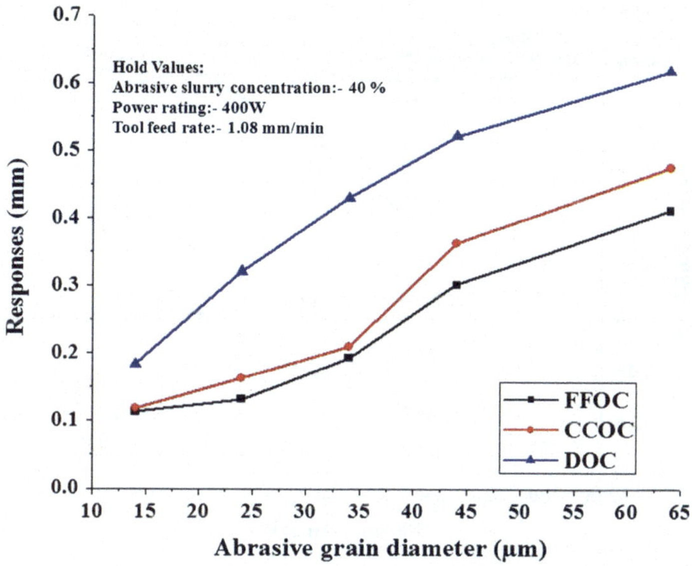

Fig. 2.17 Result of abrasive grain diameter on FFOC, CCOC and DOC for square stepped hole generation

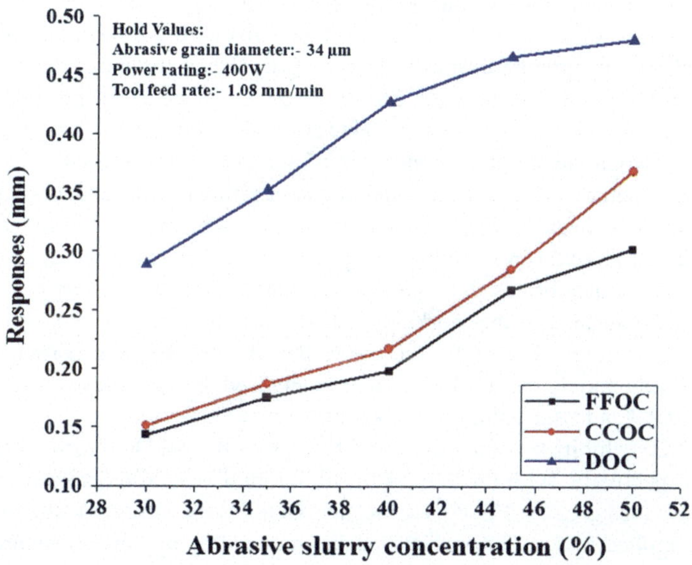

Fig. 2.18 Result of abrasive slurry concentration on FFOC, CCOC and DOC for square stepped hole generation

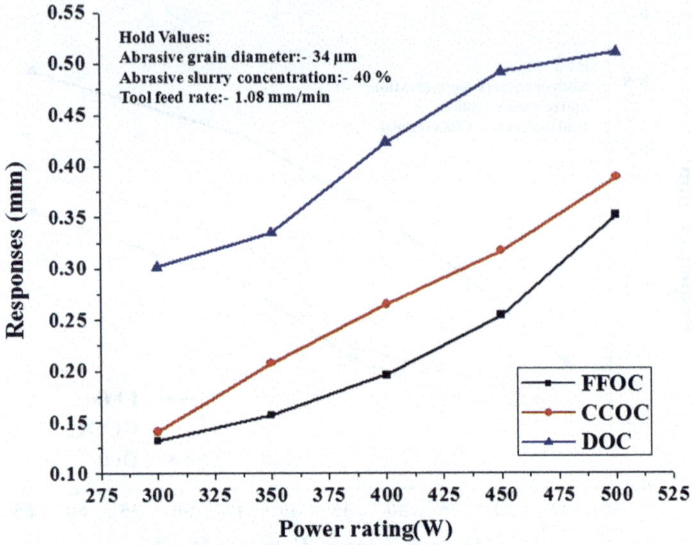

Fig. 2.19 Result of power rating on FFOC, CCOC and DOC for square stepped hole generation

overcut (CCOC), flat to flat overcut (FFOC) and diametrical overcut (DOC) of square form stepped hole on zirconia bio-ceramics respectively. If a lower concentration is used, a high quality precision profile is achieved. When the concentration of the abrasive grain is extremely low, the hitting force is less because the total mass of abrasive in the operating area is small. Therefore, the values of FFOC, CCOC and DOC are less when the small concentration of abrasive slurry is used.

Figure 2.19 demonstrates the outcome of power rating on corner to corner overcut (CCOC), flat to flat overcut (FFOC) and diametrical overcut (DOC) of square shaped stepped hole on zirconia bio-ceramics respectively. In fact, the high applied power means that the tool vibrates with great force and supply high energy to abrasive particles on the surface of the workpiece. Therefore, the material is removed from the workpiece quickly. Therefore, the flat to flat overcut (FFOC), corner-to-corner (CCOC) overcut and diametrical overcut are increased. So to achieve a small FFOC, CCOC and DOC a lesser value of the power is favored.

Figure 2.20 exhibits that the effect on FFOC, CCOC and DOC with varying tool feed rate respectively. With the help of this diagram it is acquired so as to high tool feed rate provides greatest FFOC, CCOC and DOC. With a higher value of the feed rate of the applied tool, additional material is reduced by the abrasive particles just below the tip of the tool. Therefore, the all overcut raise by enhancing the feed rate of the applied tool. Less tool feed rate is chosen to attain lower value of FFOC, CCOC and DOC of square stepped hole.

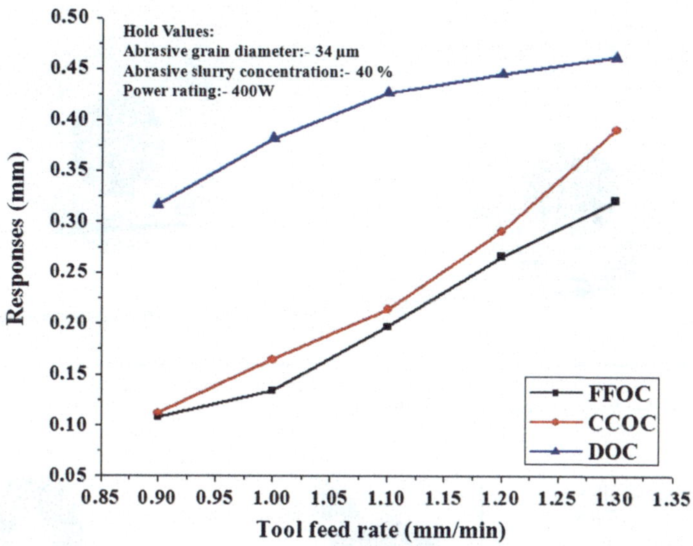

Fig. 2.20 Result of tool feed rate on FFOC, CCOC and DOC for square stepped hole generation

2.7 Improvement and Production of 3D Profile by USM Process

An accurate circular stepped type hole has been created on alumina and zirconia bio-ceramics material during USM process as showing Figs. 2.21 and 2.22. The actual photograph of square stepped hole on the workpiece after machining on shown in Fig. 2.23. The real snap of hemispherical cavity on hydroxyapatite bio-ceramics after machining is shown in Fig. 2.24.

One new technique was anticipated for ultrasonic micro machining set up. It was producing micro tools through wire electro discharge grinding (WEDG) system. With the help of this technique, micro holes of 20 μm diameter on a silicon plate and quartz of 50 μm in depth could be obtained [3, 4].

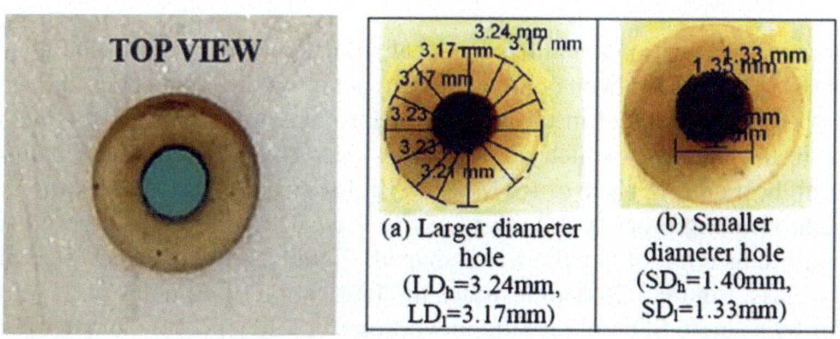

Fig. 2.21 Alumina workpiece after machining

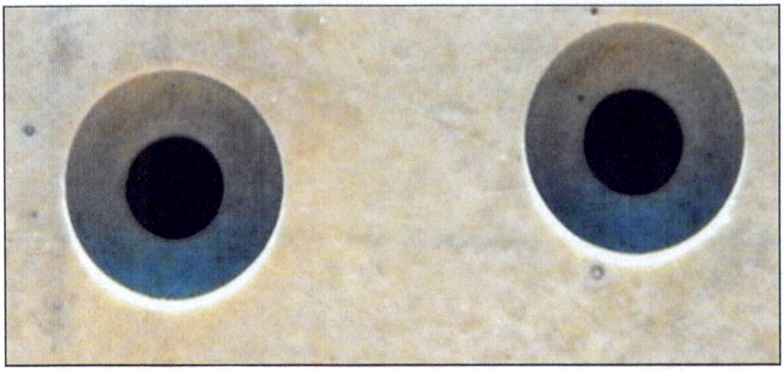

Fig. 2.22 Zirconia workpiece after machining

Fig. 2.23 Square stepped hole on zirconia workpiece after machining

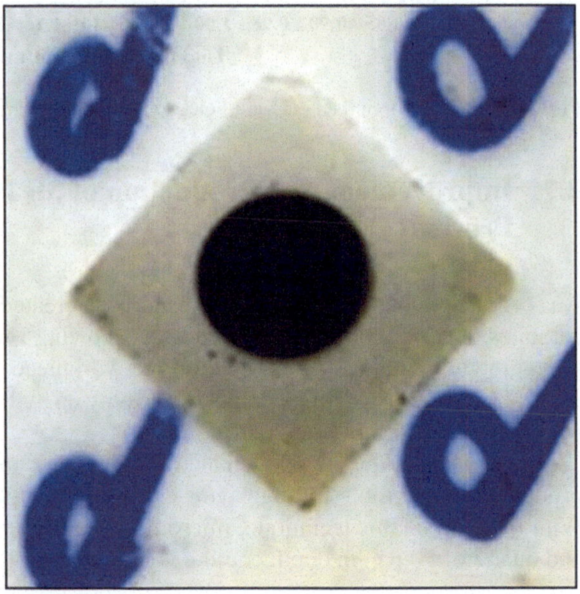

Different investigational and hypothetical studies on rotary ultrasonic micromachining (RUSMM) have been conducted for machining different ceramic materials. Sapphire is usually related in the field of electronics, mainly in the production of various circuits and chips which can also be produced by RUM. It has been observed that the highest cutting force of RUM is 34.5% lower than that of conventional diamond side grinding process. 3D micro cavity was successfully made-up on silicon by a micro tungsten tool in cylindrical-shaped. Figure 2.25 shows the 150 μm hole on Soda glass. Figure 2.26 demonstrates the fabrication of 48 holes of 22 μm hole diameter by a single SD tool on Silicon workpiece of 20 μm tool diameter [28].

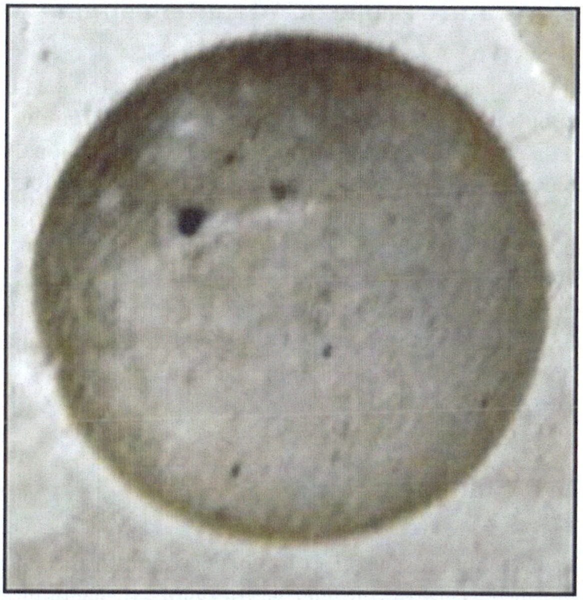

Fig. 2.24 Photographic view of hemispherical cavity on machined workpiece

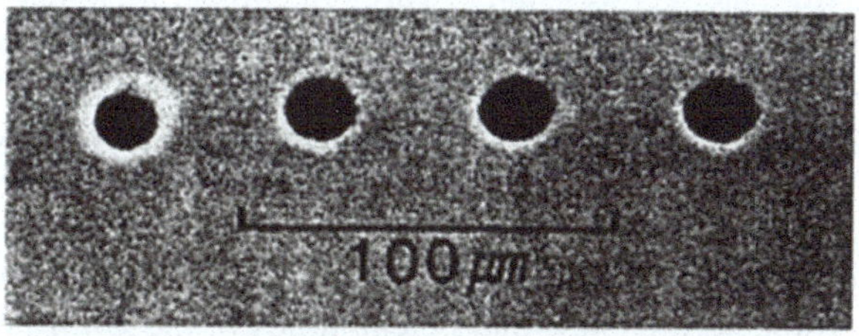

Fig. 2.25 150 μm hole on Soda glass [28]

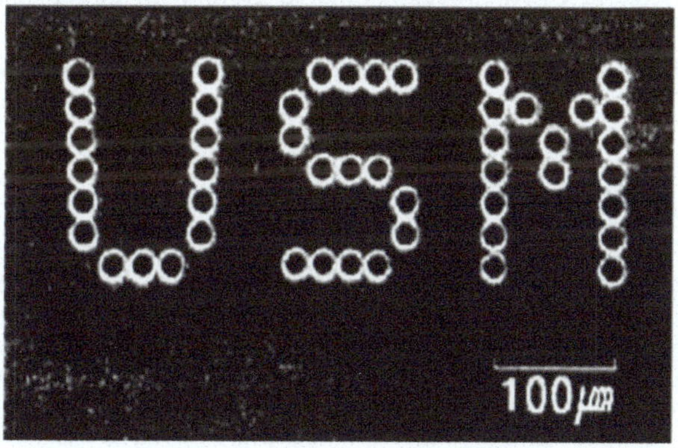

Fig. 2.26 Fabricated of 48 holes through single SD tool [28]

2.8 Future Scope of Advancement on USM

Process improvement for USM is one of the most essential matters. In USM mainly complicated task is to understand the material removal mechanism. A small number of discussions are existing till nowadays but additional analyse is of the mechanism of material removal is essential for this method. Ecological feature are one of the vital area which barely have been reported. This is one of the main topics of current industrialized process. However the present research work analysis will provide fruitful and technical information to the researchers, scientists and engineers who are working in the area of USM process. It can also provide direction for advancement of USM tool development and complex shaped profile machining of advanced ceramics those are highly demand in biomedical applications.

2.9 Summary

At first this present effort the background of the USM process is introduced. The significance of developing well-organized machining methodology for brittle and hard materials was stated. Through comparing a number of machining procedures, the potentiality of ultrasonic machining process in macro and micro-machining of different brittle and hard materials was pointed out. Within this book chapter the significant matters about various aspects of USM processes have been considered. In addition this section discusses on the working principle of different types of USM and improvement of USM process. Mechanism of material subtraction and influence of ultrasonic machining process parameters have been considered here. Profile accurateness and potentialities of the USM procedure have been discussed here. In USM process abrasive grain with larger sizes and larger abrasive slurry concentrations offer high value of MRR but reduced profile accuracy. In USM process high-quality surface finish can be attained using very smaller size of abrasive grain.

References

1. Pandey PC, Shan HS (1980) Modern machinig processes. Tata McGraw-Hill, New York
2. Doyle LE, Keyser CE, Leach JL, Schrader GF, Singer MS (1985) Manufacturing processes and materials for engineers. Prentice-Hall. Engle-wood Cliffs, NJ
3. Ratner BD, Hoffman AS, Schoen FJ, Lemons JE (2004) Biomaterials science. An introduction to materials in medicine, 2nd edn. Elsevier Academic Press, San Diego, p 162
4. Thoe TB, Aspinwall DK, Wise MLH (1998) Review on ultrasonic machining. Int J Mach Tools Manuf 38:239–255
5. Rozenberg LD (1973) Physical principles of ultrasonic technology. Ultrason Technol 1:7–20
6. Mishra PK (2005) Non conventional machining. Narosa Publishing House, New Delhi, pp 22–44
7. Benedict GF (1987) Non traditional manufacturing processes. Manufacturing engineering and materials processing, vol 19. Marcel Dekker. Inc., New York, pp 67–86

8. Kremer D, Saleh SM, Ghabrial SR, Moisan A (1981) The state of the art of ultrasonic machining. CIRP Ann—Manufacturing Technol 30:107–110
9. Evans AG (1974) Fracture mechanics determinations. Fracture mechanics of ceramics, vol 1. Plenum Press, New York, pp 17–48
10. Stevens R (1986) Zirconia and zirconia ceramics, 2nd edn. Magnesium Elektron Ltd.
11. Piconi C, Maccauro G (1999) Zirconia as a ceramic biomaterial. Biomaterials 20:1–25
12. Boutin P, Christel P, Dorlot JM, Meunier A, De Roquancourt A, Blanquaert D, Herman S, Sedel L, Witvoet J (1988) The use of dense alumina–alumina ceramic combination in total hip replacement. J Biomed Mater Res 22:1203–1232
13. Wang M (2003) Developing bioactive composite materials for tissue replacement. Biomaterials 24:2133–2151
14. Rodriguez-Lorenzo LM, Vallet-Regi M, Ferreira JMF (2001) Colloidal processing of hydroxyapatite. Biomaterials 22:1847–1852
15. Thamaraiselvi TV, Rajeswari S (2004) Biological evaluation of bioceramic materials—A review. Trends Biomater & Artif Organs 18:9–17
16. Shaw MC (1956) Ultrasonic grinding. Ann CIRP 5:25–53
17. Miller GE (1957) Special theory of ultrasonic machining. J Appl Phys 28:149–156
18. Rozenberg LD, Kazantsev VF, Makarov LO (1964) Ultrasonic cutting. Consultant Bureau, pp 97–102
19. Cook NH (1966) Manufacturing analysis. Addison-Wesley, pp 133–148
20. Kainth GS, Nandy A, Singh K (1979) On the mechanisms of material removal in ultrasonic machining. Int J Mach Tool Des 19:33–41
21. Nair EV, Ghosh A (1985) A fundamental approach to the study of mechanics of ultrasonic machining. Int J Prod Res 23:731–753
22. Rajurkar KP, Wang ZY, Kuppattan A (1999) Micro removal of ceramic material (Al2O3) in the precision ultrasonic machining. Precis Eng 23:73–78
23. Lee TC, Chan CW (1997) Mechanism of the ultrasonic machining of ceramic Composites. J Mater Process Technol 71:195–201
24. Wiercigroch M, Neilson RD, Player MA (1999) Material removal rate prediction for ultrasonic drilling of hard materials using an impact oscillator approach. Phys Lett A 259:91–96
25. Nath C, Lim GC, Zheng HY (2012) Influence of the material removal mechanisms on hole integrity in ultrasonic machining of structural ceramics. Ultrasonics 52:605–613
26. Ichida Y, Sato R, Morimoto Y, Kobayashi K (2005) Material removal mechanisms in non-contact ultrasonic abrasive machining. Wear 258:107–114
27. Lee BJ, Kim KE (2009) Characteristics of micro-hole machining of Al_2O_3 ceramics by ultrasonic longitudinal vibration. J Ceram Process Res 10(4):482–490
28. Egashira K, Masuzawa T (1999) Microultrasonic machining by the application of workpiece vibration. CIRP Ann—Manufacturing Technol 48(1):131–134

Chapter 3
Rotary Ultrasonic Machining–New Strategy of Cutting and Finishing

S. Kumar, S. Das, B. Doloi, and B. Bhattacharyya

Abstract Rotary ultrasonic machining (RUM) is an abrasive based advanced machining technique for cutting and finishing of various hard and fragile materials like ceramic, ceramics composite, glass, titanium and its alloy etc. RUM is the development over stationary ultrasonic machining for enhancement of MRR, geometrical accuracy and surface roughness. The basic mechanism of RUM is the combination of ultrasonic machining and conventional diamond grinding. In this chapter development, principle, mechanism, setup details of rotary ultrasonic machining has been discussed. It also highlights the effects of diverse input parameters on performance of RUM. The MRR always increases with spindle speed; tool feed rate and ultrasonic power. The surface roughness improved with spindle speed but worse with tool feed rate and ultrasonic power. The chipping size reduced with spindle speed but increase with tool feed rate and ultrasonic power. The cutting force reduces with spindle speed and ultrasonic power but increases with tool feed. Various advancements in RUM has also discussed. A new technique for drilling called robotic rotary ultrasonic drilling (RRUD) was developed. The application area of rotary ultrasonic machining (RUM) in micro domain has also been highlighted. It is extensively useful for generation of micro feature like micro hole, micro channel, complex micro cavity etc. on various materials such as quartz, glass, SiC, and Al_2O_3.

Keywords RUM · MRR · Surface finish · Edge chipping · Cutting force · Ceramics

S. Kumar (✉) · B. Doloi · B. Bhattacharyya
Production Engineering Department, Jadavpur University, Kolkata 700032, India
e-mail: santosh14fiem@gmail.com

B. Doloi
e-mail: bdoloionline@rediffmail.com

B. Bhattacharyya
e-mail: bb13@rediffmail.com

S. Das
Mechanical Engineering Department, Swami Vivekananda Institute of Science & Technology, Kolkata 700145, India
e-mail: somnath96@gmail.com

S. Das et al. (eds.), *Advances in Abrasive Based Machining and Finishing Processes*, Materials Forming, Machining and Tribology,
https://doi.org/10.1007/978-3-030-43312-3_3

3.1 Introduction

The recent up-gradation and innovation of various hard and brittle materials and advanced materials (ceramic, composite, optical etc.) have generated many challenging issues to machine these types of materials. Various advanced manufacturing techniques are used to machine these materials. These techniques are utilizing various energy sources like mechanical energy, thermal energy and electrochemical energy etc. to machine these materials [1, 2]. Utilizing mechanical energy for machining these materials is one of the best choices. Ultrasonic machining (USM) is relatively best technique for machining materials which are more hardness and brittleness without considering of their electrical and thermal properties [3–4]. For ultrasonic machining of alumina based ceramic composites fracture toughness plays vital role. The Increment of fracture toughness of ceramic composite means reduction in MRR and also surface roughness [5]. USM is used to fabricate a precise hole on soda lime glass. To fabricate precise hole coating was done on workpiece before machining, so crack formed on coating instead of workpiece [6]. An ultrasonic vibration holder designed and used to drill Inconel 738-LC. After experiment it is observed that all the circularity, roughness and hole accuracy enhanced by 60% [7]. Taguchi technique is applied to carry out experiment on titanium alloy and observed that hih quality product is produced [6]. After the collection of past experimental data, optimization techniques is used and observed that WPC and PCA-based TOPSIS techniques better performance than PCA based GRA technique. But, WPC technique is desirable as it is s easy computational process [8]. USM process has also been used for the manufacture of graphite electrodes [9]. Ultrasonic machining (USM) is a process which is also used for finishing ceramics material (Al_2O_3) [10].

Rotary ultrasonic machining (RUM) was developed for enhancement over ultrasonic machining (USM). In USM abrasive slurry (combination of abrasives and liquid media in appropriate quantity) is fed in top surface of workpiece throughout the operation. While in RUM, abrasives are closely affixed to the tool tip. Because of this, few drawback of the stationary ultrasonic machining are solved. In case of stationary USM abrasive and debris particles try to spoil the wall of the machined surface during the recalculation of abrasive slurry thus maintains to close tolerance is very difficult. The use of saturated diamond particles in RUM to improve the hole and geometrical accuracy and it is easy to drill deep holes. It is not possible to supply the abrasive slurry to the workpiece during thicker material and complex shape generation; so replaced of abrasive slurry with RUM for broad application is a best chaise. It has been reported that RUM is competent same work 10 times more rapidly than stationary USM in identical machining condition. The better quality surface obtained compared to stationary USM [11].

3.1.1 Ultrasonic Machining

The name "ultrasonic" is used to explain the vibratory waves have a frequency more than the audible range of healthy person i.e. far away 20 kHz. The audible range of vibration is the one which the human ear perceives (i.e. 20.0 Hz to 20.0 kHz). The ultrasonic range of vibration lies above this i.e. Frequency beyond 20 kHz. Ultrasonic vibration has been known not only as a feasible and reliable option to prophesy diseases in medical practices but also an established technique in non-destructive detection and manufacturing [12]. Ultrasonic energy is a type of mechanical energy. While the ultrasonic wave enters in the solid, it can create an extra stress field in the material. In manufacturing, ultrasonic machining has been extensively useful to difficult-to-machine materials. Traditional machining in the presence of ultrasonic can be delineating into ultrasonic machining (USM) and ultrasonic assisted machining (UAM). With the aid of ultrasonic it can extend to ultrasonic assisted turning, ultrasonic assisted drilling, ultrasonic assisted milling, ultrasonic assisted grinding, ultrasonic assisted EDM, and ultrasonic assisted LBM etc.

The use of ultrasonic vibration for material removal was first proposed in 1927 by Wood and Loomis [13]. For USM, 1st patent issued to Balamuth [14]. The first tool made for ultrasonic machining in 1954 [15]. However, ultrasonic machining (USM) paying attention into industry comparatively broad level and useful to the cutting of hard but brittle materials for example glass, silicon, ruby and ceramics, titanium, composite ceramic etc. Due to succeeding blow strokes of abrasives, material is removed from the work in the form of micro chip.

3.1.2 Development of RUM

To defeat the limitation of USM, RUM was first made-up by Percy Legge in 1964. He is a technical officer at UKAEA (United Kingdom Atomic Energy Authority). In this RUM device, abrasive slurry and tool are substituted by diamond bonded tool and with revolving workpiece. But, this device also some limitations, because of workpiece were fixed in rotating four-jaw chuck. So, relatively small workpiece could be fixing and only circular hole can be machined. Some improvement of RUM was done with tool rotation and stationary workpiece. The rotation of tool made it precisely machine with close tolerance. By using different shaped tools it extended to milling, slot cutting, screwing threading and grinding of internal and external surface. RUM had been considered as an upgrading mode from USM to defeat the limitations of USM. But by definition, it is a hybrid machining process, which combines two techniques, which is conventional diamond grinding and USM [16, 17]. It is also known by others name as ultrasonic drilling [18], ultrasonic grinding and ultrasonic twist drilling [16].

3.1.3 Basic Working Principle

Rotary ultrasonic machining system has basically three unit namely ultrasonic spindle unit, data acquisition unit and coolant supply unit [16, 19]. An ultrasonic spindle, a transducer, a power supply and an electric motor are main constitutes of ultrasonic spindle unit. The power supply unit change low-frequency (50–60 Hz) into high-frequency (20 kHz) [16]. Piezoelectric transducer is employ to change this high-frequency electrical energy into linear high frequency mechanical vibrations. In general, a piezoelectric transducer is preferred for ultrasonic machining processes as it work with more electromechanical conversion efficiency (capable of 96%) and as well low cooling is needed compare to a magnetostrictive transducer work with lower efficiency (capable of 20–35%). Piezoelectric transducer also consume smaller quantity of power for the equal output and also creates a less sound in working environments, which build easier for the machinist. The amplitude of the ultrasonic vibration is very low, so amplitude is boost my means of horn and transferred to the tool. At last tool vibrate at ultrasonic frequency. Figure 3.1 show the schematic view of working principle of RUM.

In stationary USM, the abrasive particles are free to move with liquid media so numbers of the acting abrasives are less during machining. So, the material removal rate for stationary USM is less than rotary USM where the abrasives are bonded on the tool and more abrasives are acting during machining. It increases as the machining depth increases. The free abrasives in stationary USM spoil the surface of the machined vertical wall as it comes back toward the wall surface, which restrictions geometrical accuracy, especially for small holes. In addition, the free abrasive particles; remove the tool material itself, which cause tool wear and, so it is not easy to found close tolerances in stationary USM. In case of rotary ultrasonic machining abrasive particles are bonded with tool itself so the positions of abrasives are fixed and the force at which the abrasive particles acting on the workpiece is higher compared to stationary USM and hence material removal rate is higher. Close tolerance

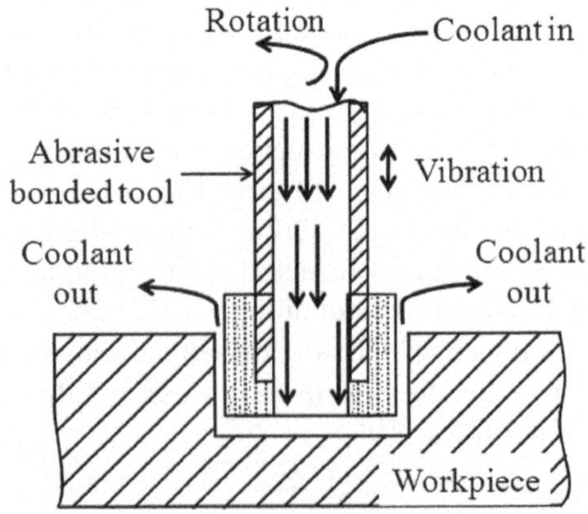

Fig. 3.1 Schematic view of working principle of RUM

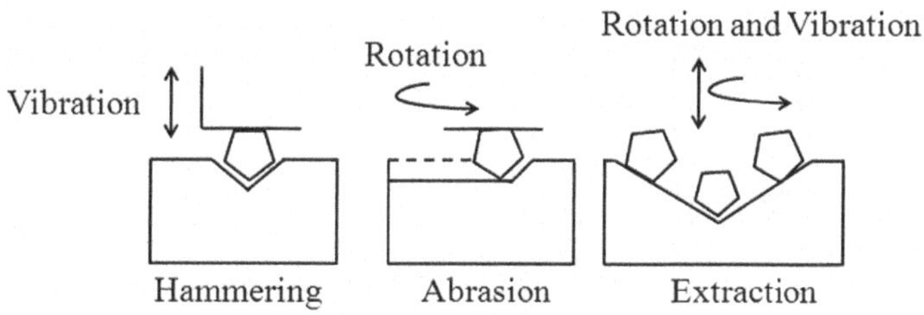

Fig. 3.2 Schematic view of material removal mechanism in RUM

is observed and also better surface finish is obtained during RUM by controlling size of the abrasive particles in the abrasive bonding rotary tool.

3.1.4 Material Removal Mechanism

RUM is the grouping of USM and conventional diamond grinding processes. So, during the machining; material removal mechanism involves of static USM and conventional grinding. Various methods such as hammering, extraction and abrasion actions are involved during rotary ultrasonic machining process [20].

(i) **Hammering action**: Indentation and crushing of workpiece under the shock of tool vibration more than 20 kHz is the source the material removal by "hammering action".
(ii) **Abrasion action**: Material removal by only rotation of abrasive bonded tool considers identical of conventional grinding and it is named as "abrasion action".
(iii) **Extraction action**: The vibration as well as rotation of tool simultaneously causes the material removal by "extraction action".

This is schematically represented in Fig. 3.2. Because of amalgamation of above mention mechanisms, RUM gives better performance than individuals achieved by ultrasonic machining or conventional diamond grinding.

3.2 Rotary Ultrasonic Machining System Details

A typical Rotary ultrasonic machining (RUM) system comprises of a machine and a tool. The major components of a RUM machine are ultrasonic spindle unit, and a coolant system and data accusation system. The details of all the components are discussed in following subsection. The schematic view of RUM system details is shown in Fig. 3.3.

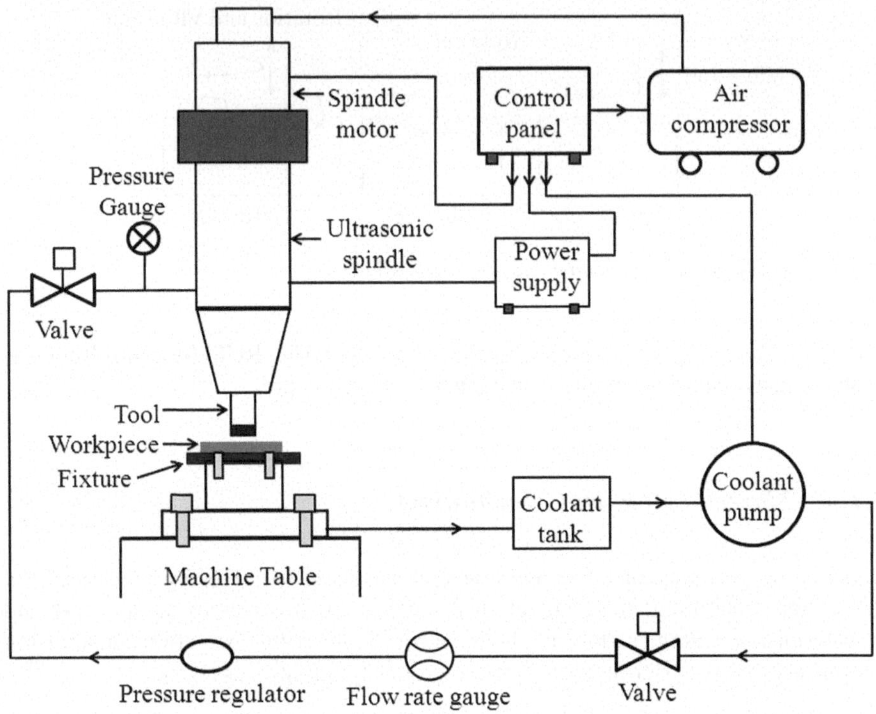

Fig. 3.3 Schematic views of RUM system details

(i) **Ultrasonic spindle unit**: In RUM system, ultrasonic spindle unit is uniquely design. It has an ultrasonic spindle, power supply, an electric motor, a pneumatic feed device and a control panel. The power supply changed low frequency electrical energy (60 Hz) to high frequency (20 kHz) electrical energy. By the use of piezoelectric transducer it changed to ultrasonic vibration ($\geq$20 kHz). The required amplitude of ultrasonic vibration is large so it is amplified through the horn located in the ultrasonic spindle and transferred to the cutting tool. The energy of ultrasonic vibration may possibly be changed by changing the output of the power supply. The motor at top of the ultrasonic spindle gives the revolving speed to the tool at diverse speed by regulating the speed of the tool.

Various types of shaped tool are used for RUM. Diamond-impregnated or electroplated tool are used for machining ceramic and glass. Diamond-impregnated tools are appropriate, but electroplated tools are inexpensive, so the choice of tool depends on the material to be machined.

(ii) **Data accusation system**: The main component of the system has a dynamometer, charge amplifier, an A/D convertor, a data acquisition card and a personal computer with software. The function of dynamometer is to measure force in the direction of ultrasonic vibration. Charge amplifier receive electrical signal from dynamometer. It amplifies and transferred into digital signal via A/D converter. After processing by signal conditioner, the digital signals received by data acquisition card on the computer by means of software.

(iii) **Coolant system**: The coolant system was used to clean the machining zone from the debris particle and also to give coolant for both the spindle and the tool-work interface. It contains coolant reservoir, coolant pump, valve, flow rate gauge, pressure regulator, and pressure gauge for the proper supply of coolant in the machining zone.

3.3 Effects of Process Parameters on Performance of RUM

The rotary ultrasonic machining (RUM) process is one of the faster and appropriate technique for machining of difficult to machine materials having brittleness like ceramic, glass, optical glass, ceramic matrix composite, bio-ceramic, titanium alloy etc. The MRR, accuracy, surface finish, tool wear rate and edge chipping have been improved in RUM compared to stationary USM. Different machining parameters of RUM process affect machining criteria like MRR, accuracy, average surface finish, tool wear and edge chipping. From the cause effect diagram of RUM represented in Fig. 3.4, the machining or input parameters and responses of RUM process are identified.

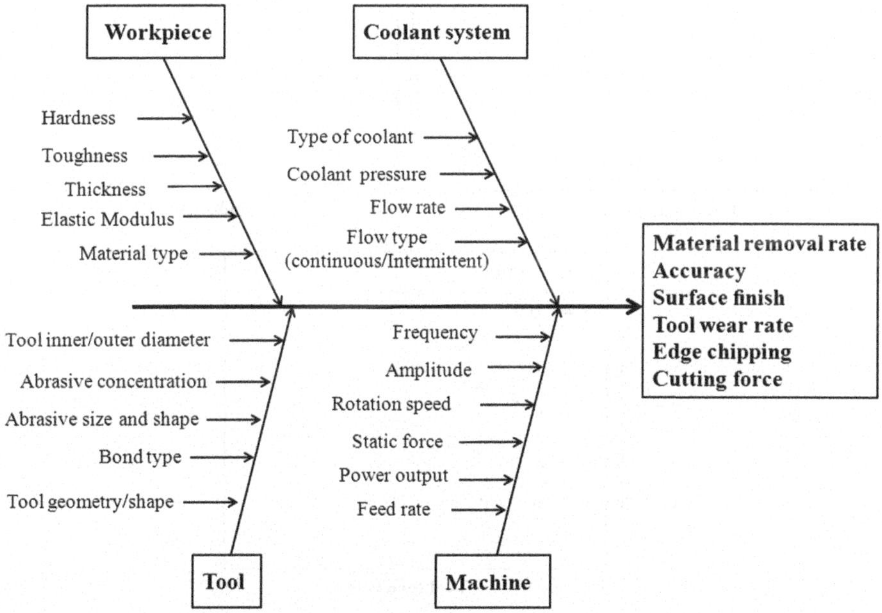

Fig. 3.4 Cause effect diagram of RUM process

3.3.1 Effect of Process Parameters on MRR

Figures 3.5a, b and c exhibit the effects of major process parameters such as ultrasonic power, spindle speed and feed rate on MRR respectively during machining of titanium alloy, alumina and ceramic matrix composites. The ultrasonic power has no major effect on MRR of titanium alloy and alumina. MRR rises with raise in ultrasonic power during RUM of ceramic matrix composite [21]. The variation of MRR during different material is different because of material properties.

The effect of spindle speed on MRR is shown in Fig. 3.5b. From the graph it can be observed that the spindle speed has less effect on MRR of titanium alloy. The effect of spindle speed on MRR for RUM of alumina has been reported by Jiao et al. [22]. Variation of MRR with spindle speed is very less. But in case of RUM of ceramic matrix composite it is different from other two materials such as titanium alloy and alumina. The MRR rises with raise in spindle speed for RUM of ceramic matrix composite materials [21]. When the feed rate increases MRR also increases. The increment of feed rate means, the tool movements quicker in the path of feed causing an increase in MRR. During RUM of ceramic matrix composite and alumina,

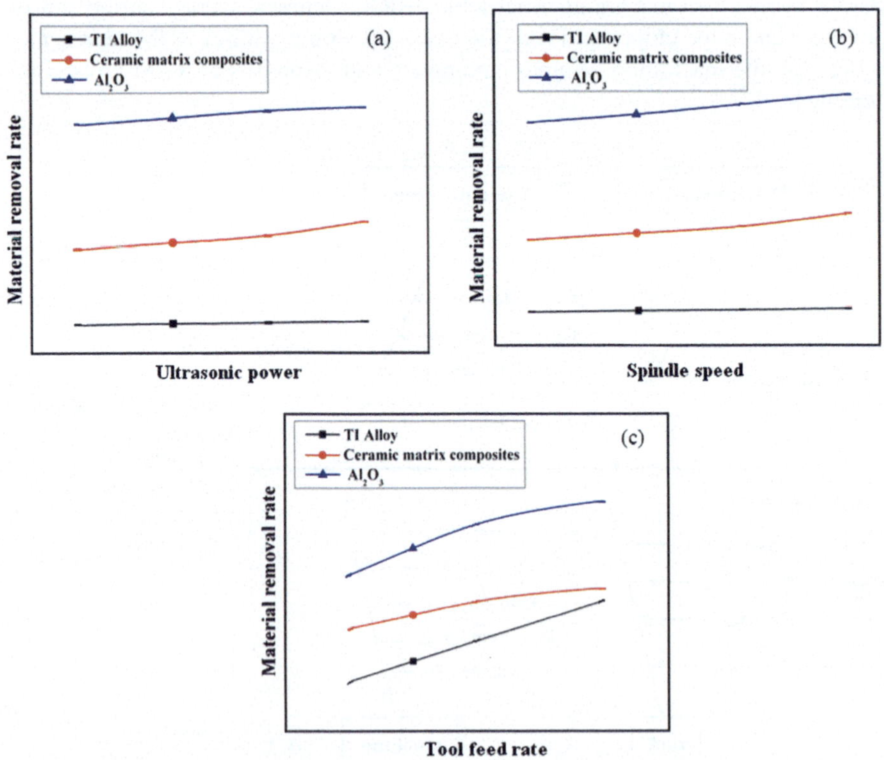

Fig. 3.5 Effect of process parameters on MRR

relationship is also similar between MRR and feed rate [21, 22]. MRR increases with feed rate for all material irrespective of its properties. During RUM, nature (brittle or ductile) of the material is more important for material removal rate. For all the process parameters MRR is higher during RUM of Alumina (Al_2O_3). It is followed by ceramic matrix composites and Ti alloy.

3.3.2 *Effect of Process Parameters on Surface Roughness*

Figure 3.6a, b and c illustrate the effects of the input process parameters such as power, revolving motion of tool and feed rate on the surface roughness of drilled hole respectively. Zhang et al. successfully performed drilling operation on optical K9 glass by utilizing RUM. Squeezed air acted as coolant during drilling operation. The result of ultrasonic power on roughness of machined surface has been revealed in Fig. 3.6a. For optical K9 glass the roughness of machined surface rises with raise in ultrasonic power. The results of the rotation speed of tool on the surface finish, is revealed in Fig. 3.6b. It indicates, with raise in spindle speed better surface finish has been obtained. Figure 3.6c illustrate the variation of feed rate and surface finish. The surface roughness increase as the feed rate increases.

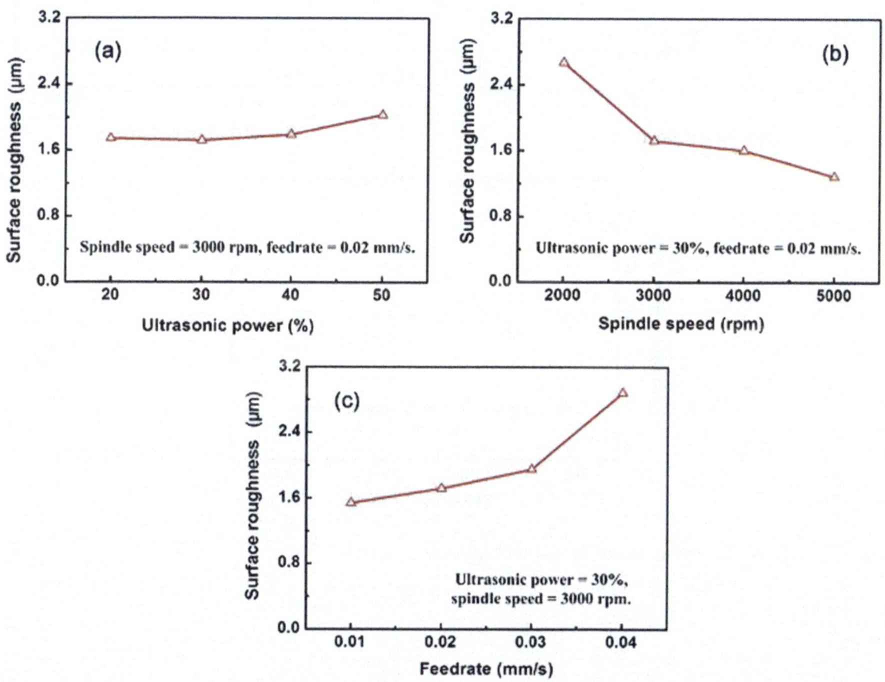

Fig. 3.6 Effect of process parameters on surface roughness [23]

3.3.3 Effect of Process Parameters on Edge Chipping

During hole generation on hard and brittle materials (ceramics and glass), edge chipping size (chipping thickness) is vital criterion to estimate the hole quality. Figure 3.7a, b and c demonstrates the results of the power, rotation speed of spindle, and feed rate on chipping thickness respectively. From Fig. 3.7a it observed that the chipping size reduces little while the ultrasonic power varies 'between' 20 and 40%, after that it drastically increases. From Fig. 3.7b, the chippings size reduces with the raise in the spindle speed. From the Fig. 3.7c it is noticed that, feed rate is more important process parameter for chipping thickness compared to other two process parameters. During drilling operation by using RUM of brittle materials edge chipping is induce due to fracture failure of machined workpiece. Stress distribution of interface area of tool and workpiece gives the mechanism of edge chipping.

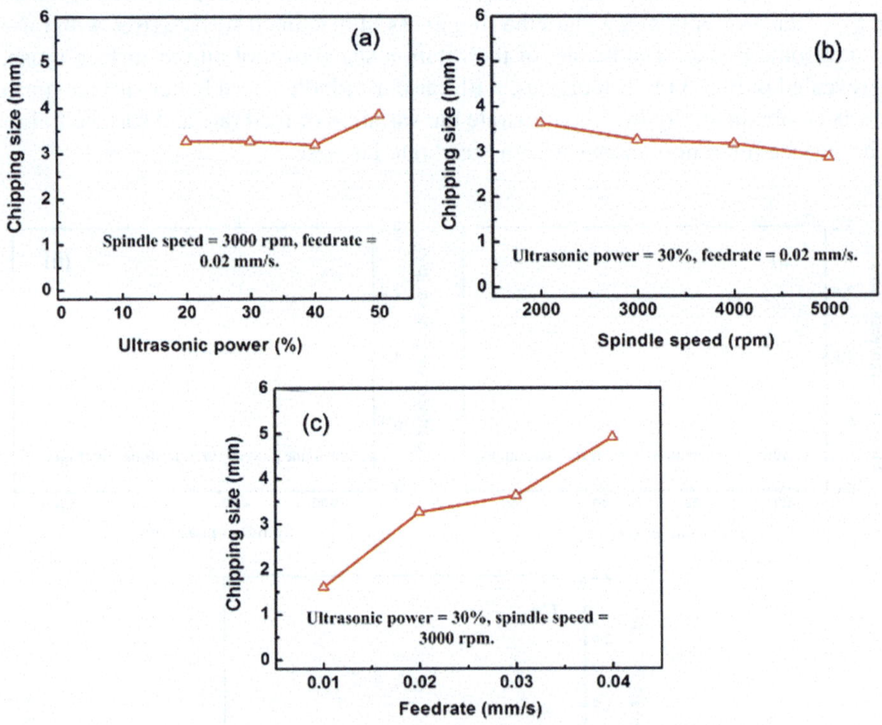

Fig. 3.7 Effect of process parameters on chippings size [23]

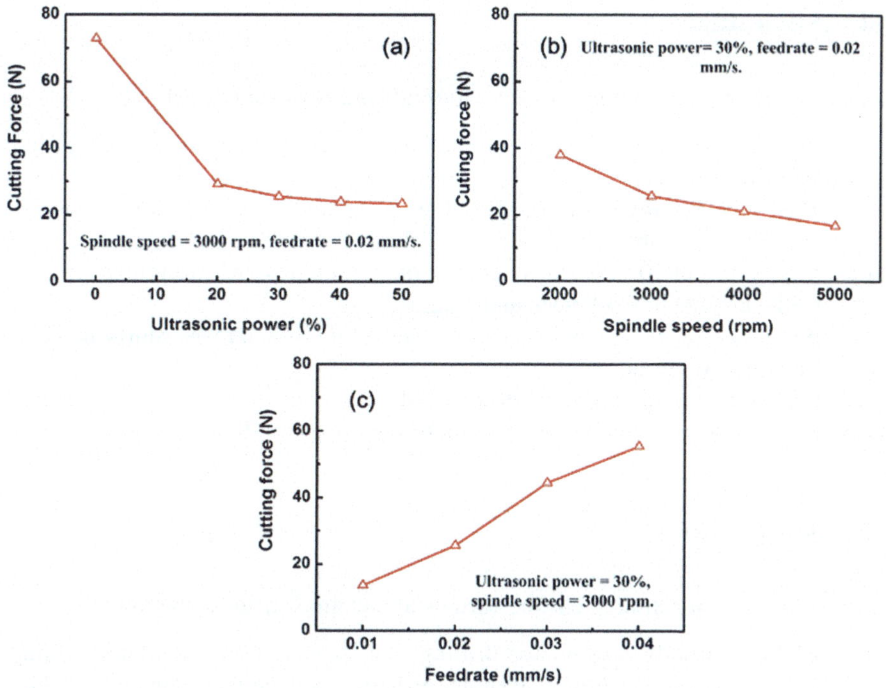

Fig. 3.8 Effect of process parameters on cutting force [23]

3.3.4 Effect of Process Parameters on Cutting Force

The effects of the ultrasonic power, spindle speed, and feed rate on the cutting force are revealed in Fig. 3.8a, b and c respectively. The cutting force reduces with rise in ultrasonic power. Up to the 20% ultrasonic power, cutting force suddenly reduced. Furthermore rise in the ultrasonic power, the decreasing speed of cutting force low. With increase in the spindle speed the cutting force reduces gradually as revealed in Fig. 3.8b. Figure 3.8c shows relation between feed rate and the cutting force. The cutting force rises with raise in feed rate. During RUM of CFRP, cutting force is consistent with increase in spindle speed and feed rate while cold air used as coolant. For RUM operation of optical K9 glass, cutting force is consistent with rising spindle speed and feed rate by using cutting fluid as coolant.

3.4 Advantages

Rotary ultrasonic machining has the various advantages listed as bellow:

(i) In rotary ultrasonic machining abrasives are fixed in tool tip instead of free abrasives.
(ii) The force required for RUM is low compared to conventional diamond grinding and stationary ultrasonic machining.
(iii) MRR is about 10 times faster comparing with ultrasonic machining.
(iv) Little damage to workpiece and long tool life.
(v) It is easy to drill deep hole because no need of loose abrasive slurry in RUM.
(vi) Low tool wear rate.
(vii) Higher machining accuracy is achieved.
(viii) RUM technique can be used for micro feature generation.

3.5 Applications

Rotary ultrasonic machining has the various applications listed as bellow:

(i) RUM is usually applied for drilling, but as well very useful for milling, complex shape machining, turning, grinding, surface texturing and finishing operations.
(ii) It can machine bio-ceramics like alumina, zirconia, hydroxyapatite which has great application in medical implants.
(iii) It can machine silicon, quartz, sapphire etc. which has wide application in semiconductor industry.
(iv) It can also produce deep holes in the borosilicate glass, ceramics, and ceramic composites wherever required good tolerance hole.
(v) RUM can also machine beryllium oxide, polycrystalline diamond, titanium and its alloy, bone, silicon nitride, stainless steel etc.
(vi) It can machine optical K9 glass which is used in optics, electronics, thermodynamics and fluidics appliances.
(vii) It can machine ceramic matrix composites (CMCs) which has application in fuel injector nozzles and gas turbine engine.
(viii) RUM is able to machine carbon fiber reinforced polymers (CFRP) which has been used in airspace, airplane and defense industries.

3.6 Recent Advancement

Continuous advancement in the existing machine setup always becomes essential for betterment of performances. It is compulsory to survive with regularly changing advanced techniques. The recent techniques come with additional benefit and wide application. Various researchers investigated in the area of RUM. The research areas of RUM are design and upgradation of main components such as sonotrode, spindle, ultrasonic vibrator, tool, measuring technique, frequency tracking, etc. of the RUM system for improving the machining performance.

An ultrasonic spindle is designed and developed for RUM and also validate. It is observed that amplitude of ultrasonic vibration is less than one micron generated on the spindle. It is also concluded that application of electromagnetic force shows the better execution [24]. The software namely ANSYS and FEM has been used to design ultrasonic transducer for maximize the amplitude and reduce the resonance frequency error [25]. A new tool designed for rotary ultrasonic milling process through utilizes the finite element method and ANSYS. It was concluded that increase in tool length means raise in mass as well as resonance amplitude but drop in frequency [26]. A novel approach was proposed for measuring vibration amplitude during RUM process. In this approach microscope was used for perceive the machined surface by RUM. The novel measurement approach give better performance than earlier used dial indicator method [27].

A new technique for drilling called robotic rotary ultrasonic drilling (RRUD) was developed. It consists of a robot with devoted end effector and a small rotary ultrasonic drilling (RUD) system [28]. The small RUD system is accumulated on the end effector. The weight of RUD system is one kilogram. It consists of ultrasonic generator, transducer, horn and drill bit. The schematic view of RRUD system is demonstrated in Fig. 3.9. In RRUD system, the ultrasonic generator gives the high-frequency electrical signal, after that the transducer changes it to mechanical vibration. Later than, amplitude of ultrasonic vibration is increased by horn and next transmits to the drilling tool. At last, the drilling tool obtains steady ultrasonic vibration.

The diameter variation of drilled holes during robotic conventional drilling (RCD) and robotic rotary ultrasonic drilling (RRUD) are illustrated in Fig. 3.10. It is clearly noticed that, the diameter variation of drilled hole of RRUD is lesser than RCD, particularly at higher spindle speed. It is also examined that machined hole quality of RRUD is better than RCD. Natural rock, which is not homogeneous materials, has been machined by RUM. RUM is used for drilling different kind of rocks such as basalt, marble and travertine. From the experimental investigation it was reported that RUM successfully applied for drill hard rocks. A higher feed rate and bigger diameter of drill bit gives high cutting force. Abrasive concentration of 100 (4.4 carat/cm^3) offer the small cutting force for basalt and marble and the cutting force reduces as tool rotation speed increases. Abrasive concentration is measure according to the weight of the diamond per cm^3. Rotation speed of tool, feed rate, and ultrasonic power had no significant outcome on quality of the machined surface [29].

Rotary ultrasonic micro machining (μ-RUM) is advancement of rotary ultrasonic machining (RUM) for micro domain application. It is extensively useful for generation of micro feature on various materials such as quartz, glass, SiC, and Al_2O_3 [30, 31]. A micro hole fabricated on quartz crystal with rotary tool in micro USM. The diameter of micro hole was 92 μm with high aspect ratio. Figure 3.11a and b show the fabricated micro hole and the micro tool after machining respectively [32].

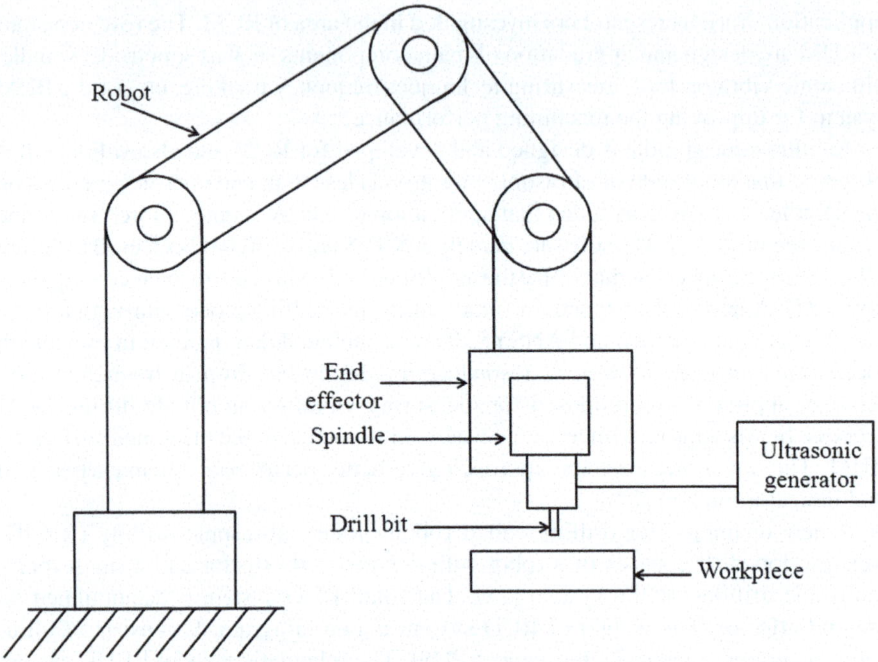

Fig. 3.9 Schematic view of RRUD system

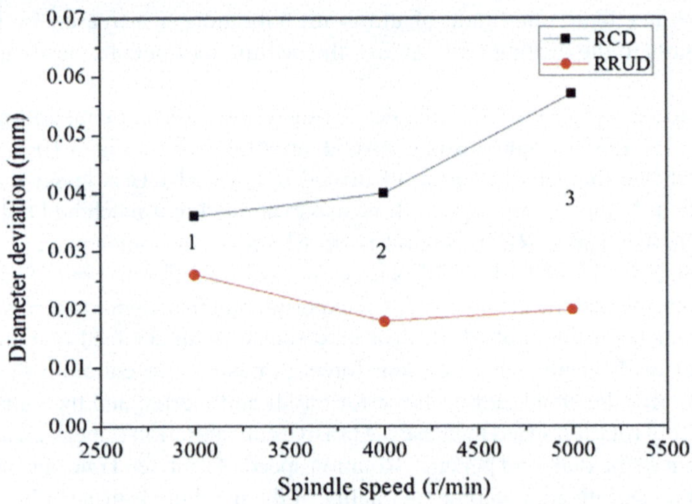

Fig. 3.10 The diameter deviations of holes by RCD and RRUD [28]

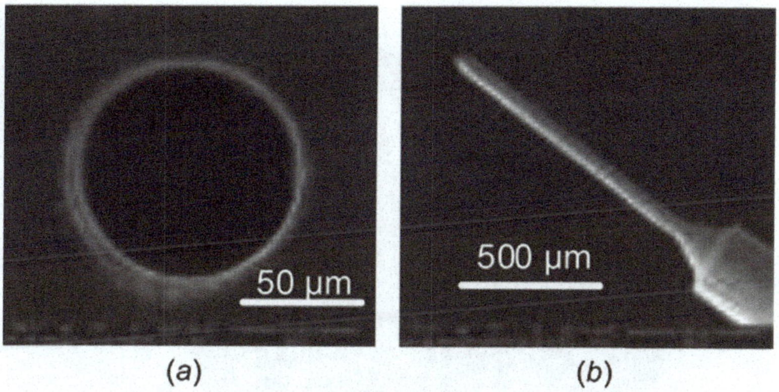

Fig. 3.11 **a** the fabricated micro hole on quartz and **b** The micro tool [32]

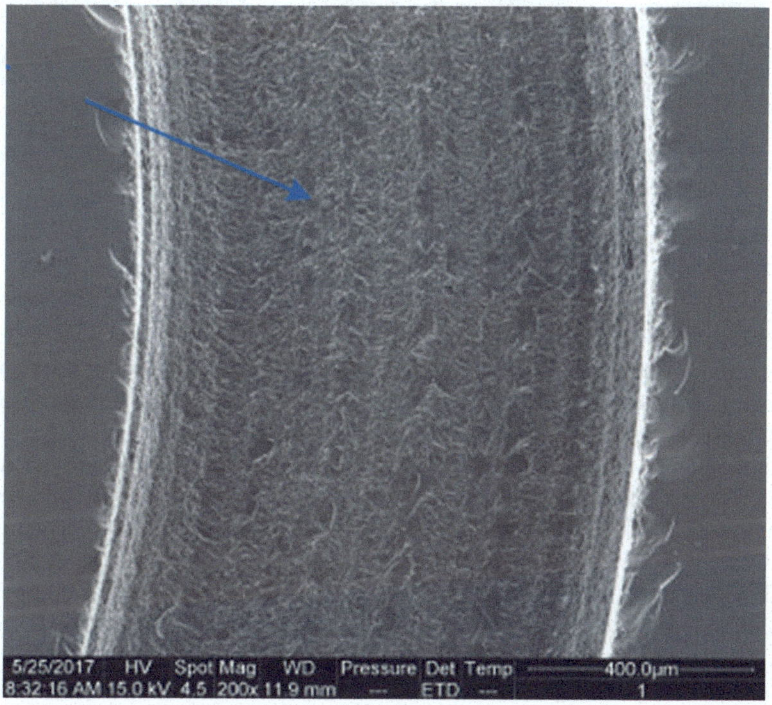

Fig. 3.12 SEM image of blind hole on Quartz [33]

Figure 3.12 shows scanning electron microscopy (SEM) image of blind hole on quartz. It represents the material removal characteristic of RUM. Figure 3.13 Shows the SEM image of hole quality of entrance side of alumina ceramic.

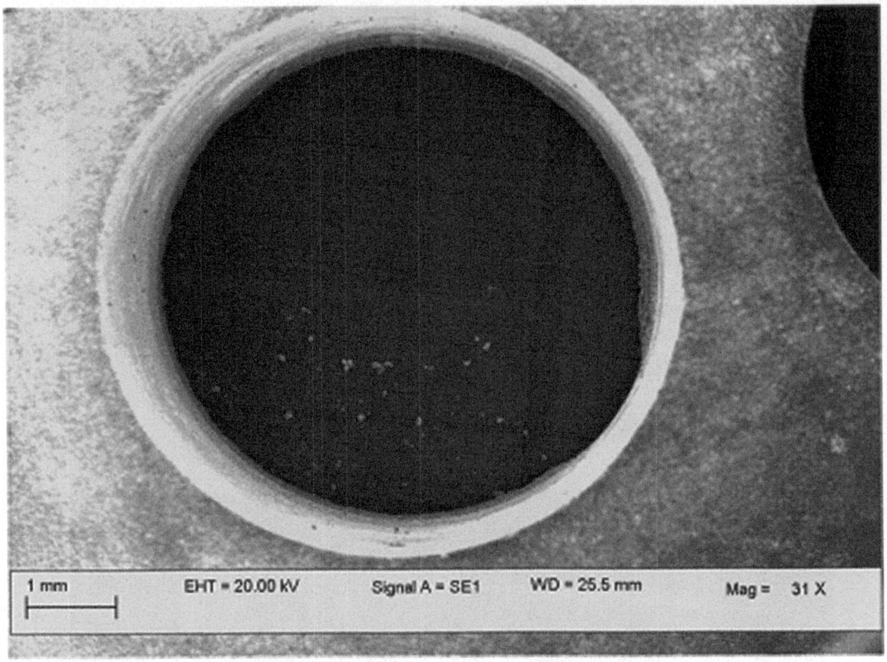

Fig. 3.13 SEM image of hole quality of entrance side of alumina ceramic [34]

3.7 Summary

The rotary ultrasonic machining process is probably faster and appropriate technique for machining materials having more hardness and brittleness such as ceramic, glass, quartz, optical glass, composite ceramics, bio-ceramic, titanium alloy etc. Rotary ultrasonic machining (RUM) is an upgrading over stationary ultrasonic machining (USM). The use of diamond impregnated tool in RUM improves the geometrical accuracy and it is effortless to produce holes in thicker workpiece. RUM is competent same work 10 times more rapidly than stationary USM in identical situation. A typical Rotary ultrasonic machining (RUM) system details with major components discussed in this chapter. RUM process provide enhanced performances in respect to MRR, surface finish, edge chipping and cutting force compared to stationary USM and conventional grinding. Different machining parameters of RUM process affect machining criteria like MRR, surface finish, edge chipping and cutting force has also been discussed. RUM can also be applied for micro machining domain for generation of micro feature like micro hole, micro channel, complex micro cavity etc.

References

1. Jain NK, Jain VK (2001) Modeling of material removal in mechanical type of advanced machining processes- A state of the art review. Int J Mach Tools Manuf 41:1573–1635
2. Jain VK (2013) Advanced machining processes. Allied publishers private limited, New Delhi, India
3. Kataria R, Kumar J, Pabla BS (2015) Experimental investigation into the hole quality in ultrasonic machining of WC-Co composite. Mater Manuf Process 30(7):921–933
4. Gilmore R (1998) Ultrasonic machining and orbital abrasion techniques. Soc Manuf Eng 419:1–20
5. Deng J, Lee T (2002) Ultrasonic machining of alumina based ceramic composites. J Eur Ceram Soc 22(8):1235–1241
6. Baek DK, Ko JT, Seung HY (2013) Enhancement of surface quality in ultrasonic machining of glass using a sacrificing coating. J Mater Process Technol 213:553–569
7. Azarhoushang B, Akbari J (2007) Ultrasonic assisted drilling of Inconel 738-LC. Int J Mach Tools Manuf 47:1027–1033
8. Gauri SK, Chakravorty R, Chakraborty S (2011) Optimization of correlated multiple responses of ultrasonic machining (USM) process. Int J Adv Manuf Technol 53:1115–1127
9. Gilmore R (1991) Ultrasonic machining: a case study. J Mater Process Technol 28(1–2):139–148
10. Ghahramani B, Wang ZY (2001) Precision ultrasonic machining process: a case study of stress analysis of ceramic (Al_2O_3). Int J Mach Tools Manuf 41(8):1189–1208
11. Pei ZJ, Khanna N, Ferreira PM (1995) Rotary ultrasonic machining of structural ceramics: a review. Ceram Eng Sci Proc 16:259–278
12. Shamoto E, Moriwaki T (1994) Study on elliptical vibration cutting. CIRP Ann Manuf Technol 43(1):35–38
13. Nishimura G, Jimbo Y, Shimakawa S (1954) Ultrasonic machining-Part I. J. Fac. Eng., Univ. Tokyo, 24(3): 65–100
14. Balamuth LA (1945) Method of abrading. British patent, 602801(1)
15. Rozenberg L, Kazantsev V, Makarov L, Yakhimovich D (1964) Ultrasonic cutting (translated from Russian). Consultants Bureau, New York
16. Churi N (2010) Rotary ultrasonic machining of hard-to-machine materials. Doctoral dissertation, Kansas State University, USA
17. Legge P (1966) Machining without abrasive slurry. Ultrasonics 4(3):157–163
18. Legge P (1964) Ultrasonic drilling of ceramics. Ind Diamond Rev 24(278):20–24
19. Neugebauer R, Stoll A (2004) Ultrasonic application in drilling. J Mater Process Technol 49(1):633–639
20. Pei ZJ, Ferreira PM, Kapoor SG, Haselkorn M (1995) Rotary ultrasonic machining for face milling of ceramics. Int J Mach Tools Manuf 35(7):1033–1046
21. Li Z, Jiao Y, Deines T, Pei ZJ, Treadwell C (2005) Rotary ultrasonic machining of ceramic matrix composites: feasibility study and designed experiments. Int J Mach Tools Manuf 45(12–13):1402–1411
22. Jiao Y, Hu P, Pei ZJ, Treadwell C (2005) Rotary ultrasonic machining of ceramics: design of experiments. Int J Manuf Technol Manage 7(2–4):192–206
23. Zhang C, Cong W, Feng P, Pei Z (2014) Rotary ultrasonic machining of optical K9 glass using compressed air as coolant: A feasibility study. Proc Inst Mech Eng, Part B J Eng Manuf 228(4):504–514
24. Wu Y, Yokoyama S, Sato T, Lin W, Tachibana T (2019) Development of a new rotary ultrasonic spindle for precision ultrasonically assisted grinding. Int J Mach Tools Manuf 49(12):933–938
25. Zeng FC, Zhang XH (2012) Optimal design of an ultrasonic transducer in rotary ultrasonic machining. Appl Mech Mater 184:352–355
26. Kuo KL (2008) Design of rotary ultrasonic milling tool using FEM simulation. J Mater Process Technol 201(1):48–53

27. Cong WL, Pei ZJ, Feng Q, Deines TW, Treadwell C (2012) Rotary ultrasonic machining of CFRP: A comparison with twist drilling. J Reinf Plast Compos 31(5):313–321
28. Dong S, Zheng K, Liao W (2018) Stability of lateral vibration in robotic rotary ultrasonic drilling. Int J Mech Sci 145:346–353
29. Fernando P, Zhang M, Pei Z (2018) Rotary ultrasonic machining of rocks: An experimental investigation. Adv Mech Eng 10(3):1–9
30. Hu P, Zhang JM, Pei ZJ (2003) Experimental investigation on coolant effects in rotary ultrasonic machining, NSF workshop on research needs in thermal aspects of material removal processes. Stillwater, UK, pp. 1–6
31. Kiran RS, Kolluru VSS (2005) Rotary ultrasonic machining a review. Proc Natl Aerosp Manuf Soc 22–23:105–114
32. Lei S, Yu Z, Zhou K, Li J, Kang R (2019) Influence of the planetary movement of tool on the aspect ratio of micro holes machined by micro-ultrasonic machining. J. Micro Nano-Manuf 7(1):010905
33. Wang J, Zhang J, Feng P, Guo P, Zhang Q (2018) Feasibility study of longitudinal–torsional-coupled rotary ultrasonic machining of brittle material. J Manuf Sci Eng 140(5):051008
34. Singh RP, Singhal S (2016) Experimental study on rotary ultrasonic machining of alumina ceramic: microstructure analysis and multi-response optimization. Proc Inst Mech Eng Part L: J Mater Design Appl, 146442071665737. https://doi.org/10.1177/1464420716657370

Chapter 4
Abrasive Water Jet Machining

Sachin Singh, Vishal Gupta, and M. R. Sankar

Abstract Abrasive water jet machining (AWJM) is a non-traditional technology widely used in industries for processing of materials. AWJM is a replacement for the other modern and conventional machining processes that have thermal or mechanical disadvantages for workpiece and tool material. AWJM is mainly used for cutting but few researchers successfully used AWJM process for controlled depth milling, drilling, turning, peening etc. Today's modern technology developed the AWJM process into a complete machining process that can produce precise and consistent results. AWJM can easily be employed for processing of intricate shape components made of brittle, hard, soft as well as ductile materials. A wide variety of materials ranging from stone, glass, titanium, steel, ceramics, composites, etc. can be easily machined by the AWJM.

Keywords Abrasive · Machining · Cutting · Modeling · Jet

4.1 Introduction

Machining of components used in high performance applications shares a considerable amount of its total cost. Also, there is a high demand for machining material with considerable speed and accuracy. Abrasive water jet machining (AWJM) is one of the modern machining process that came into existence during the last quarter of 20th century [1]. AWJM involves the high pressure (typically 200–350 MPa) water

S. Singh · V. Gupta
Department of Mechanical Engineering, Thapar Institute of Engineering and Technology, Patiala 147004, India
e-mail: sachin.singh@thapar.edu

V. Gupta
e-mail: vishal.gupta@thapar.edu

M. R. Sankar (✉)
Department of Mechanical Engineering, Indian Institute of Technology Tirupati, Tirupati 517506, India
e-mail: evmrs@iittp.ac.in; ravisankarm@gmail.com

S. Das et al. (eds.), *Advances in Abrasive Based Machining and Finishing Processes*, Materials Forming, Machining and Tribology,
https://doi.org/10.1007/978-3-030-43312-3_4

that is accelerated by a fine orifice (typically diameter 0.24–0.40 mm) containing abrasive particles (typically mesh size#50–#120) on the workpiece surface to be machined. AWJM machining is particularly useful during working with difficult to machine material. AWJM has number of advantages over conventional and advanced machining methods such as high flexibility, less cutting force, machining versatility, machined surfaces with minimal distortion, heat build-up and residual stresses [2]. Thus, AWJM is particularly useful in machining brittle materials and composites. Authors [3] experimentally showed the superior cutting performance of the AJM compared to the cutting of composites by plain water jet and laser. Also, compared to the electric discharge machining (EDM), chemical etching, plasma and flame cutting AWJM shows a higher damage free machining rate [4]. AWJM is primarily used for cutting operations but it can also be used for decoration/texturing of brittle materials, milling, turning, peening, forming and deep hole drilling [5]. Further, AWJM can be modified into several hybrid forms like hybrid machining with cutting fluids, abrasive cryogenic and ice jet machining etc. AWJM can be used to cut materials from thickness ranging between 4.6 to 300 mm with ±0.13 mm accuracy. It can drill holes up to a depth of 150 mm with tolerances below ±0.03 mm. Also, the minimum diameter hole that can be drilled by AWJM is 0.2 ± 0.03 mm [4].

4.2 Primary Components of the AWJM System

During AWJM more than 25 input parameters were identified by the researchers [6]. Major input parameters which affect the cutting performance of the AWJM process are water supply pressure (WP), nozzle diameter, stand-off distance (SOD), size and type of abrasive particles nozzle transverse speed (NTS) and abrasive particle flow rate (AFR) [7]. Relationship between various AWJM input parameters and corresponding output responses is shown in Fig. 4.1.

Authors in various literatures characterized the machining performance of the AWJM by studying the surface texture, surface integrity and material removal rate (*MRR*) of the machined workpiece surface as output responses (Fig. 4.2).

A typical system of AWJM setup is given in Fig. 4.3. The setup consists of following six major modules

(i) Hydraulic pump
(ii) Intensifier pump
(iii) Abrasive delivery system
(iv) Cutting head
(v) Position control system
(vi) Catcher unit.

The filter water without impurities is used in AWJM for the safe and proper working of the setup.

(i) **Hydraulic pump**: Hydraulic pump in combination with intensifier pump is heart of the AWJM system. It is a radial displacement pump that pumps the

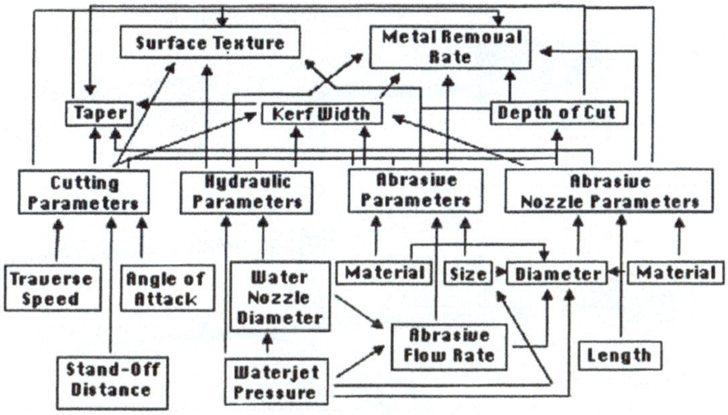

Fig. 4.1 Schematic showing the relationship between various input parameters of abrasive water jet machining and output responses [8]

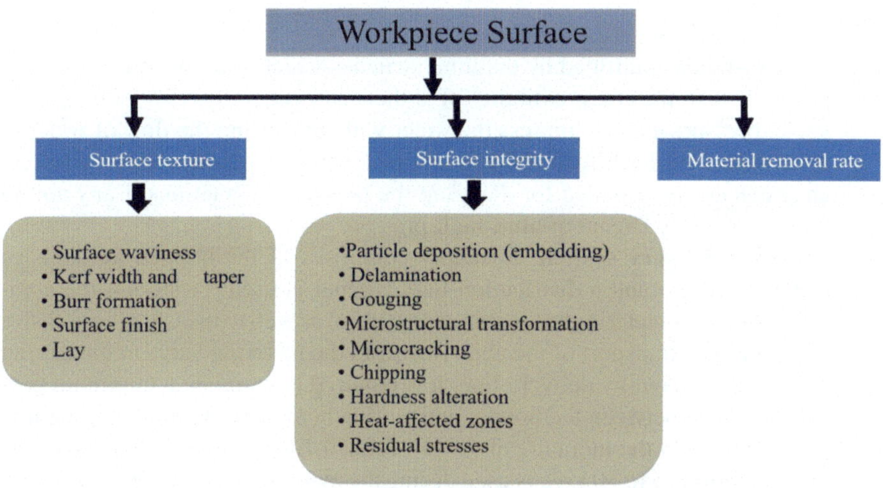

Fig. 4.2 Workpiece surface features associated with abrasive water jet machining

high pressure fluid to the system. Hydraulic pump generates the pressure of around 21 MPa in the fluid.

(ii) **Intensifier pump (IP)**: The primary function of the IP is to pump a high pressure water form a low pressure fluid input. The reciprocating piston of the IP is driven by the fluid pumped by the hydraulic pump. Intensifier contains a plunger on both ends of piston that produces pressure in two directions. Left side of the intensifier act as an inlet stroke, while the right side generates ultra-high pressure as output. During the first face of process pure water comes into the high-pressure cylinder. During the reverse motion of the piston, the water gets compressed and leaves the intensifier at ultra-high pressure. The motion

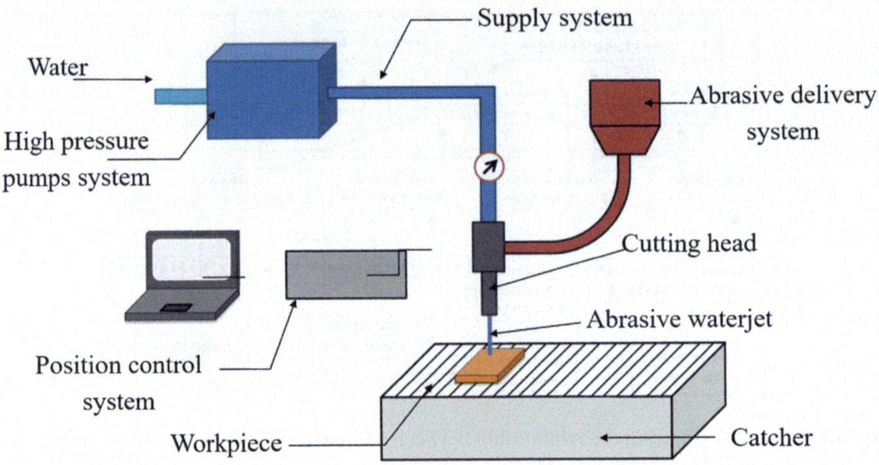

Fig. 4.3 Major components of the abrasive water jet machine

of the piston is controlled by the limit switches which were fixed at the end of piston. Due to the compressibility of at the water generally, 15% of the piston stroke pressurize and compress the water without causing the flow of water in the system. This leads to the fluctuations in the pressure of the water. Therefore, an attenuator is provided for damping the pressure fluctuations if any and to provide a steady stream of ultra-high-pressure water.

(iii) **Abrasive delivery system**: It mainly comprises of the hopper, an abrasive metering device and a distribution line. Hopper is used for storing the abrasives. Abrasive metering value is used to control as well to measure the abrasive flow rate. The transport of the abrasives from the metering valve to the cutting head is carried out by delivery line. It is required to maintain a minimum possible distance between the hopper and cutting head for obtaining high cutting performance. In the literature abrasive of size 0.1 to 0.3 mm with a mass flow rate of around 15 to 30 kg/h is used during the experimentation work. Focusing tubes with longer lengths results in coherent jets but also produces low velocity due to friction between jet and tubes inner wall [4].

(iv) **Cutting head**: The cutting head includes three components viz. orifice, nozzle and mixing chamber. Cutting head controls and provides the high pressure abrasive water jet. Primarily role of the orifice is to convert high pressure water into high velocity jet. Typically, the diameter of orifice varies from 0.08 to 0.8 mm and is made of hard material such as sapphire, diamond etc., the mixing chamber is located between the orifice and nozzle. Abrasive particles enter the mixing chamber by vacuum suction and are mixed with the water jet [9]. Thereafter, the water jet containing abrasive particles is accelerated by the nozzle. Nozzle length affects the velocity attained by the water jet. Therefore, an optimum length is required for attaining maximum velocity.

With tungsten carbide as the manufacturing material, nozzle length lays in the range of 50–100 mm.

(v) **Position control system**: Drive of the cutting head is controlled by the position control system. The device can be a mechanical manipulator or a multi-axis robot. Higher the controlled and multi-axis movement of the cutting head more accurately intricate shapes can be produced from the workpiece. Manufacturing software in combination with CNC controller can be employed for positioning the cutting head.

(vi) **Catcher**: During the AWJ process, a huge magnitude of kinetic energy generated at the nozzle outlet but nearly 75% of the contained energy is retained even after the cut of the workpiece by the water jet. Therefore, a component called catcher is required to collect the water, metal debris and abrasive particles. Catcher is broadly classified into two main categories tank catcher and point catcher. Point catcher is used in case of moving workpiece and fixed nozzle. While a tank catcher is required in case of a moving nozzle in combination with the fixed workpiece.

4.2.1 Various Cutting Stages During AWJM Process

There are mainly three cutting stages during the machining of the workpiece by the AWJM process [10]:

1. Initial stage: This is known as the development stage of the water jet. This stage extends for a length, X_O, thereafter the jet achieves maximum depth of cut, h.
2. Steady stage: During the steady stage cutting is performed in a cyclic manner. Steady stage cut extends up to a depth, h. Zone up to h_c is known as the cutting wear zone and material removal takes majorly by the particle impact on the workpiece surface at a shallow angle. Beyond h_c is the deformation zone which is marked by the impact of the jet at large angles causing jet instabilities and forming waviness in the machined surface.
3. Exit stage: When the jet advances to the workpiece exit edge, an uncut triangle is found with its top location at a depth of h_c. Therefore, complete separation of the opposite sides h_c should be greater than the thickness of the workpiece. This can be achieved by properly adjusting the AWJM input parameters (Fig. 4.4).

4.3 Experimental Study

Savrun and Taya [11] studied the performance of AWJM during machining of high temperature ceramics and composites. Authors studied the microstructure and surface finish of the machined workpiece. Machined surfaces are smooth free from

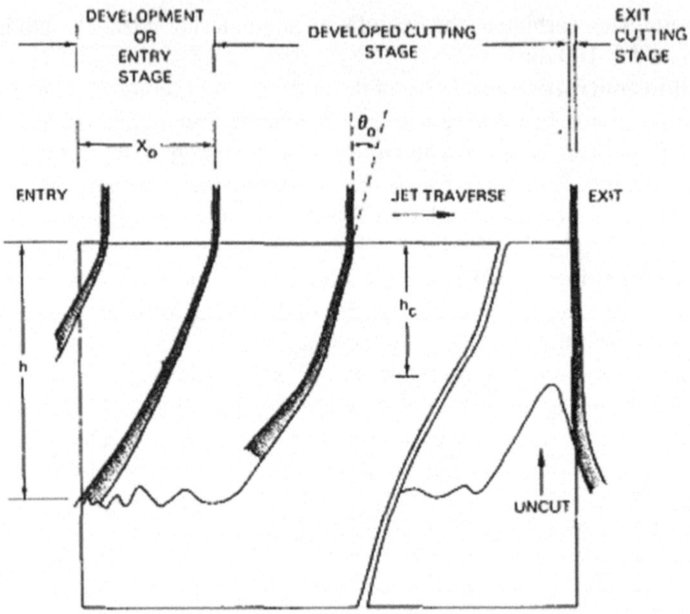

Fig. 4.4 Schematic showing the various cutting stage of an abrasive jet from entry to exit of the workpiece [10]

crack and don't show any microstructural changes. Similar study is carried out by Hamatani and Ramulu [7] during the piercing and slot cutting of ceramics and metal matrix composites. Authors evaluated the cutting performance in terms of hole taper produced during piercing and kerf width along with surface roughness during slotting operation. Authors [12] carried the AWJM of lexan, lucite and glass to study the cutting process dynamics. They observed that cutting process consists mainly of two erosion modes. First is cutting wear mode that occurs at a shallow angle of impact and results in steady cutting. While deformation wear mode results for a large angle of impact and causes unsteady penetration zone in the workpiece. Authors also studied the surface characteristics of the thin metal sheets [10], thick materials [13] amorphous material [14] and polymer matrix composite sheets [15] machined by the AWJM process. Chen and Siores [16] concluded that the striations on the workpiece surface are because of wavy nature of the abrasive particle kinetic energy, vibration of nozzle and fluctuations other AWJM input variable. Hascalik et al. [17] investigated the effect of jet traverse speed on the machined morphology. Authors identified three distinct cutting zones on the machined surface viz. initial damage zone (IDR) on top followed by a smooth cutting zone (SCR) and at last the rough cutting zone (RCR). The abrasive particles strike on the workpiece is getting decrease with increases in the water jet speed. This will lead to lesser depth of cut as well as the width of IDR and SCR also decreases (Fig. 4.5).

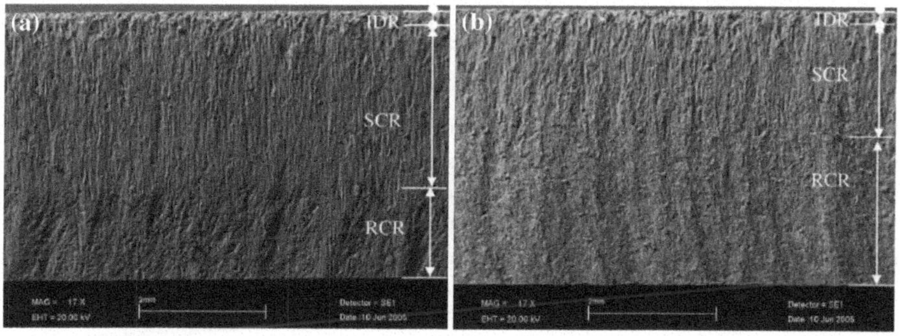

Fig. 4.5 Machined surface topograpgy at various traverse speeed **a** 60 mm/min **b** 200 mm/min [17]

Authors [18] also investigated the material to be used in making the mixing tube of the AWJM process. Performance of three materials viz. B4C, WC and composite carbide were investigated in terms of wear under abrasion and erosion. Authors concluded that composite carbide shows the least wear rate. Momber and Kovacevic [19] studied the effect of AWJM input parameter viz. NTS, AFR and WP during the AWJM of rocks. Authors showed that for every input parameter there exists a minimum and maximum threshold value above and below which machining is effectively performed. Later [20] authors also incorporated acoustic emission technique to study the material removal process of AWJM. Particle size and velocity affects the AWJM performance during cutting of rocks [21]. Further, hardness as well as the porosity also affects the mechanism by abrasive water jet process [22]. Gupta et al. [23] studied the machining performance of AWJM by cutting marble. Authors concluded that NTS is the major input parameter that affects the kerf width and taper. While, glazed vitreous tiles were machined by Santhana kumar et al. [24]. Authors reported the best combination of AWJM input parameters to achieve an optimum cutting surface quality. Effect of material properties on the surface integrity and texture after AWJM process is studied by Arola and Ramulu [25] by carrying out experiments on 6 different metals. Authors concluded that ductility, strain hardening and metal strength coefficient also decides the machined surface characteristics. Wand and Wong [26] carried out the experimental investigation for studying the kerf quality during the AWJM of coated metallic sheets. Based on the experimental study authors developed empirical relations for kerf width (W_t), kerf taper (T_a) and kerf surface roughness (K_a) as a function of input parameters which are given as:

$$W_t = -1.554 + 0.019P + (4.33S_d + 1.14VS_d - 0.01785VP + 4.33S_dP - 0.321P^2)10^{-4} \tag{4.1}$$

$$T_a = -1.067 + 0.008P - 0.212F_r - (0.0789VP + 3.887F_rS_d + 1.906VS_d - 1.363P^2 + 0.02016V^2)10^{-5} \tag{4.2}$$

$$K_a = 34.506 - 0.221P + (4V - 0.9242VS_d + 5S_dP - F_rS_d + 0.3717P^2)10^{-3} \quad (4.3)$$

where, P is WP, $V$is NTS, S_d is SOD and F_r is AFR. Similar study is carried out by Babu and Chetty [27] during the AWJM of trapezoidal aluminum work piece. Authors carried out the detailed experimental study and analyzed influence of input parameter on the AWJM output responses the same by performing analysis of variance (ANOVA). Authors [28] also used various advanced optimization techniques to optimize the AWJM during the cutting of aluminum workpiece.

Several authors [29, 30] performed detailed AWJM experimental study and developed regression equation for predicting the R_a of the machined glass/epoxy composite as a function of AWJM input parameters in combination with composite material properties.

$$\begin{aligned} R_a = 4.074 - 2.93 \times 10^{-4}x_1 - 1.9 \times 10^{-5}x_2 + 6.89 \times 10^{-2}x_3 - 6.67 \times 10^{-2}x_4 \\ + 5.66 \times 10^{-4}x_5 - 5.28 \times 10^{-3}x_6 + 9.6 \times 10^{-3}x_7 - 1.36 \times 10^{-2}x_8 \quad (4.4) \end{aligned}$$

where, x_1 is hardness of abrasive material, x_2 is pressure, x_3 is standoff distance, x_4 is abrasive flow rate, x_5 is traverse speed, $x_6$is tensile stress of fibre, x_7 is fibre volume fraction and x_8 is composite laminate thickness. Authors [31] also machined CFRP by using AWJM process. Authors concluded that lay up of the material doesn't influence the kerf width and taper. Also, to obtain minimum kerf taper AWJM should be carried out at high pressure, low SOD and high NTS. Parametric study of the AWJM machining process during the cutting of glass-vinyl ester composite was performed by Armağan and Arici [32]. During the cutting process, standoff distance contributes for 73.35% in determining the top kerf width. While, R_a of the machined surface is decided by SOD, WP, and AFR. Selvam et al. [33] machined hybrid composites (Reinforcement: E-Glass fabric and Carbon fabric; Matrix: Epoxy resin) with the AWJM process and developed the following empirical relation for the R_a of the composites:

$$\begin{aligned} R_a = {} & 2.62938 - 0.12158A + 0.032089\mathrm{B} + 0.022661\mathrm{C} + 1.70009\mathrm{D} \\ & +1.3675 \times 10^{-4}\mathrm{AB} + 5.625 \times 10^{-6}\mathrm{AC} + 3.4125 \times 10^{-3}\mathrm{AD} \\ & - 7.0875 \times 10^{-5}\mathrm{BC} - 4.8375 \times 10^{-3}\mathrm{BD} + 1.48125\mathrm{E} - 3\mathrm{CD} \\ & +2.73825 \times 10^{-4}\mathrm{A}^2 - 3.21754 \times 10^{-5}B^2 \\ & - 3.04386 \times 10^{-6}C^2 - 0.095439D^2 \quad (4.5) \end{aligned}$$

where, A to D are the magnitudes of NTS, WP, AFR and SOD. Kumar et al. [34] developed regression model for MRR and R_a during the machining of aluminium/tungsten carbide composites by AWJM as:

$$MRR = 1.00712 - (0.0235 \times SOD) + (0.0649 \times NTS) - (0.0298 \times WC)$$

$$+ (0.0126 \times TS \times SOD) - \left(0.0106 \times SOD^2\right)$$
$$- \left(0.0128 \times NTS^2\right) - \left(0.0116 \times WC^2\right) \quad (4.6)$$

$$R_a = 4.19385 - (0.2947 \times SOD) + (0.3442 \times NTS) + (0.6951 \times WC)$$
$$- (0.0637 \times NTS \times SOD) + (0.11555 \times SOD \times WC) + \left(0.1073 \times WC^2\right) \quad (4.7)$$

For achieving maximum *MRR* with minimum R_a authors carried out the optimization study and proposed *SOD* of 0.422 cm, N*TS* of 3.72 mm/s, and *WC*of 2.10% for obtaining the same.

Effect of AWJM input parameters and the MRR of the amorphous material during machining is studied by Aich et al. [35]. Authors used borosilicate glass as the workpiece material and developed the following linear model to map the relationship between depth of cut (*DOC*) with major input parameters:

$$DOC = e^{-3.9417} + P^{1.1441} + m_a^{0.5293} + V^{-0.7027} + F_r^{0.0542} \quad (4.8)$$

Kechagias et al. [36] used Taguchi method to design the experiments for determining the influence of input parameters on output responses during the cutting of steel sheets by the AWJM process. Authors concluded nozzle diameter contributes majorly i.e. 74.47 and 26.61%, in deciding the kerf width and R_a, followed by stand-off distance, thickness of the sheets and traverse speed. Authors [37] concluded from the extensive literature survey on machining of steel by AWJM that proper selection of nozzle diameter and orifice diameter improves the quality of kerf. Also, pressure and material of the abrasive particle are the significant parameters that influence the R_a. Axinte et al. [38] carried out experimental study on the AWJM process during the cutting of polycrystalline diamond (PCD). Authors used aluminum oxide, silicon carbide and diamond as abrasive particles. Machining with diamond abrasive gives highest *MRR* with a defect free machined surface having a R_a of less than 4.6 μm. Later, authors [39] extended the application of AJWM process to the dressing of grinding wheels. Figure 4.6 shows the developed abrasive water jet turning setup.

Paul et al. [40] extended the application of AWJM to pocket milling of the workpiece. Primarily concern during pocket milling by AWJM is change in the depth. Authors in their study controlled the depth of pocket within 0.04 mm with a fine surface finish. Shipway and Pashby [41] tuned various input parameters to perform milling of titanium alloys by abrasive water jet. Authors showed that waviness of the machined surface can be reduced by using high traverse speed, low pressure, small mesh size abrasive particles impacting the workpiece surface at low impact angle. However, such a combination of input parameters also reduces *MRR*. Effect of shape and hardness of abrasive particle on the Ti64 machined surface prepared by controlled depth milling using AWJM process is studied by Fowler et al. [42]. Authors concluded that the shape of the abrasive particle is non-significant during the machining process. However, as the hardness of the abrasive particle increases

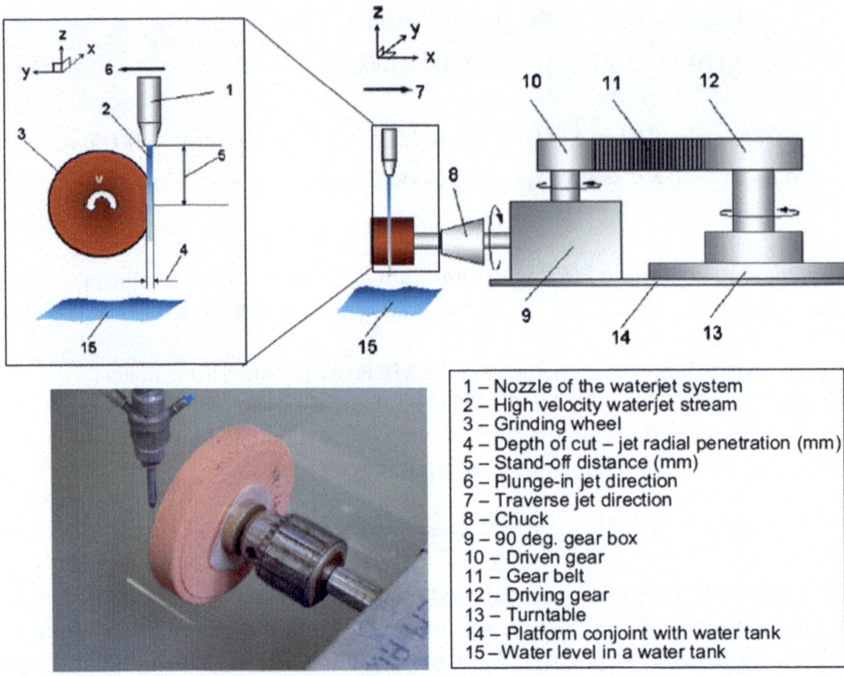

Fig. 4.6 Schematic of the modified abrasive water jet machining setup to be used for turning [39]

MMR and R_a on the workpiece surface increases. AWJM not only is effective for machining difficult to cut materials but can also be used for machining shape memory alloys. Kong et al. [43] carried out the milling of nickel–titanium (NiTi) alloys by AWJM process. Transformation temperature for shape memory alloy can be from room temperature to around 80 °C. A variation in temperature can cause a change in their mechanical and metallurgical properties. AWJM can be a viable alternative for machining shape memory alloys as thermal energy generated is least as compared to the other machining process. However, abrasive particle embedment on the workpiece surface at high jet velocities can result in reducing its fatigue life and still an area for improvement. Uthayakumar et al. [44] also performed machining of nickel based superalloys by AWJM process. Experimental study of the machining process shows that jet pressure plays a major role in determining machined surface quality.

Influence of feed rate during machining of ductile materials like aluminum alloy, steel and brass with varying thickness is studied by Akkurt et al. [45, 46]. Authors concluded that the material with higher strength has higher deformation and surface roughness after AWJM of thinner workpiece. Later, Akkurt [47] extended the use of AWJM process for the drilling of various metals such as steel, brass and aluminum alloy. Authors showed that AWJM process can successfully drill holes in metal workpiece. However, as the thickness of the workpiece increases the roughness of machined surface deteriorates. Also, machining time depends on the material of the

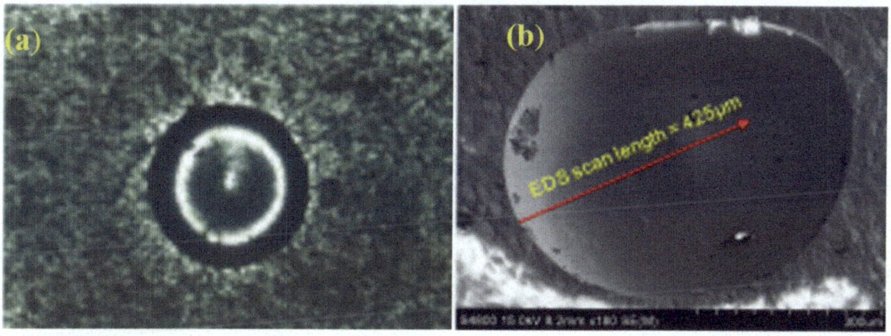

Fig. 4.7 Scan electron microscopic image showing hole drilled by **a** AWJM **b** EDM [49]

workpiece to be drilled. Authors [48] showed that AWJM process can also be used for drilling holes in glass. Ceramic thermal barrier coated (TBCs) components are used in the aerospace industry. It is highly challenging to make a micro hole in such components by electro-discharge machining (EDM) due to their non-conductive nature. To overcome such limitations of TBCs, authors [49] first used AWJM for removing the ceramic layer and drilling a blind hole. Later, EDM process is used to obtain the hole with the desired geometry (Fig. 4.7).

Park et al. [50] used the micro-AJWM process for micro-grooving of glass with 80 μm groove width. Authors concluded that with proper masking method in combination with tuned AWJM input parameters micromachining of semiconductors, electronic devices etc. can be performed. Later, authors [51] tried to fabricate complex features i.e., micro-pillars on the aluminum and steel alloy with the help of AWJM process. As shown in Fig. 4.8 for the fabrication of the micro-pillar the jet is moved along two cross raster paths with a distance of 4.2 times the jet diameter between two consecutive jet paths.

Cutting of bones with saw and milling machine can cause thermal damage to the bone tissues. To avoid such disadvantages Schwieger et al. [52] explored the potential of AWJM process by cutting of cancellous bone. AJWM of ductile sometimes results in the contamination of the surface due to the embedment of the fractured abrasives on the workpiece surface. So, to overcome such drawback Patel [53] proposed nozzle

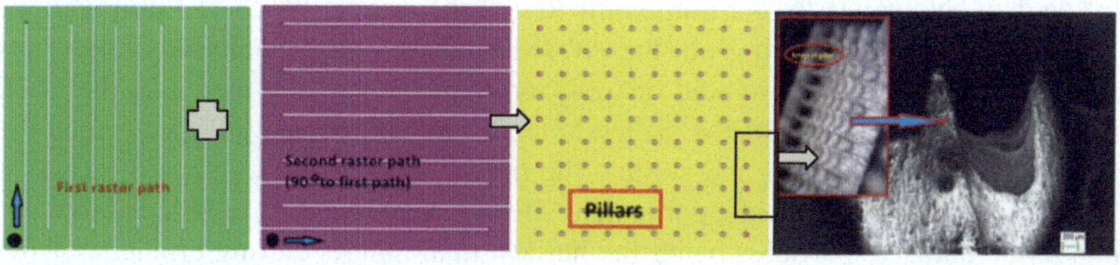

Fig. 4.8 Abrasive jet paths to fabricate micro-pillar with fabricated micro-pillar array with fabricated micro-pillars [51]

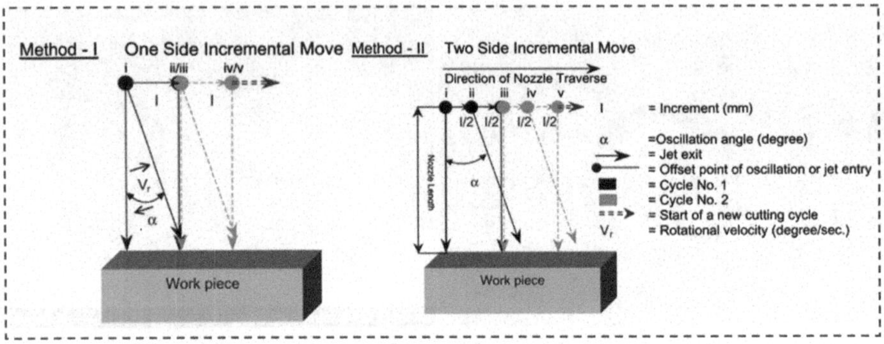

Fig. 4.9 Schematic showing the various nozzle oscillation methods [52]

oscillation technique. As shown in Fig. 4.9 authors used 2 modes of oscillation of the nozzle during the experimental runs. It was concluded that mode 1 results in decreasing the degree of contamination by 20%. While performing AWJM in mode generates the machined surface with a 50% reduction in the degree of contamination.

Wang and Guo [54] demonstrated the improved performance of the AWJM during multipass mode compared to the single pass. Later, authors [55] also combined jet forward impact angles with the multipass AWJM process for enhancing its performance in terms of the total depth of cut, surface roughness and reduced kerf taper. To make the AWJM more economical and environment friendly, recycling of abrasive particles is also carried out. Babu and Chetty [56] performed the experimental study on the performance of the AWJM during machining with recycled abrasives that are recharged with fresh abrasive particles. For precise micro-machining by AWJM, Miller [57] produced jet diameters less than 10 μm. To increase the cut area during the AWJM authors used a suspension system of passing abrasive particles in water rather than passing the abrasive particles carried by air in high velocity water jet. There is an increase in the orifice and nozzle diameter during the machining due to the erosive nature of high velocity abrasive particles. This affects the efficiency and quality of the AWJM process. Jegaraj and Babu [58] studied the influence of change in orifice and nozzle dimensions on the quality of the aluminum alloy machined surface. Authors concluded that ratio between nozzle to orifice size should be maintained in the ranges of 3–4.5 for a getting good quality machined surface. Desirable size of the nozzle is in between 0.76–4.2 mm, while that of the orifice is 0.25–0.3 mm. Palleda [59] showed that by carrying out the AWJM with the slurry containing chemicals such as acetone, phosphoric acid or (polyacrylamide) not only increases the *MRR* but also reduces the taper of the drilled holes. Khan and Haque [60] performed the cutting of glass by AWJM and studied the performance of three types of abrasive particles viz. aluminum oxide, garnet and silicon oxide. Authors concluded that silicon oxide abrasive particles being the hardest maintain its cutting efficiency from the top of the cut to its bottom. Thus, the cut slots produced are of minimum taper. Patel and Tandon [61] proposed to heat the workpiece material by an external heat source (oxy acetylene gas flame) during machining of difficult of cut material by the

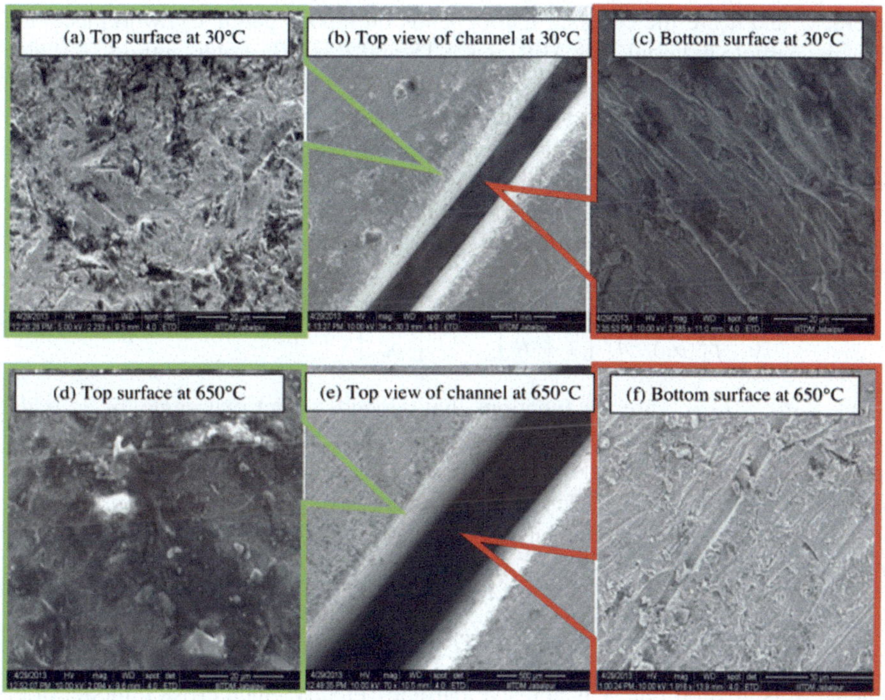

Fig. 4.10 Microstructure of the channels surface after machining by thermally enchanted abrasive water machining process at different temperatures [61]

AWJM process and termed the process as thermally enhanced AWJM (TEAWJM). Heating the workpiece locally reduces its yield strength, hardness and strain hardening. This results in an increase in *MRR* and improved machining time. As shown in Fig. 4.10 the microstructural study of the workpiece surface reveals that machining at high temperature changes the microstructure in machining region but the machined surface is free from any cracks.

4.4 Theoretical and Numerical Investigation

Chen et al. [62] during the machining of alumina-based ceramics by AWJM process introduced a mathematical model to find the depth of penetration of jet, *D* which is given as:-

$$D = C\left(\frac{qP}{u^a}\right) \tag{4.9}$$

where, C and a are the constant that depends on nozzle geometry, workpiece and abrasive particle material, q is AFR, u is NTS and P is WP. Later, authors [63] also proposed the use of providing forward oscillation to cutting head for increasing the efficiency of the AWJM process. Authors concluded that there is a 30% increase in kerf depth with fine finished surface in case of oscillation when compared to without oscillation. Abdel-Rahman and EI-Domiaty [64] also developed a cutting model during the AWJM of ceramics. Developed model gives a relation between depth of cut, H with the material properties and AWJM input parameters can be given as:-

$$H = \frac{3}{16} a \frac{f_1(\alpha_c)}{f_2^2(\alpha_c)} \frac{d_j R}{u} \left[\frac{1}{1+R}\right]^{19/6} \rho_w \left[\frac{2P}{\rho_w}\right] \left(r_p^{2/3} \rho_p^{7/12} K_c^{4/3} H_t^{1/4}\right) \tag{4.10}$$

where, $f_1(\alpha_c)$, $f_2(\alpha_c)$ are the functions that depends on jet exit angle, α_c, a is the constant, d_j is nozzle diameter, R is abrasive to water mass flow rate, u is traverse speed, P is pressure, H_t is vickers hardness, K_c is fracture toughness of the workpiece material, r_p is abrasive particle radius, ρ_p is abrasive particle density, ρ_w is water density. General erosion model during the impact of abrasive particles with the brittle materials is given as:

$$W \propto v^{e1} (2r_p^{e2}) \rho_p^{e3} K_c^{e4} H_t^{e5} \tag{4.11}$$

where, W is erosion rate, v is abrasive particle velocity. Such erosion model often results in generation of radial cracks on the workpiece surface which in turn degrades the material strength. Wakuda et al. [65] studied the feasibility of above erosion model during the AWJM of ceramics. Authors concluded that AWJM doesn't fit the erosion model and hence abrasive particles will not produce surface damages in the form of radial cracks during machining. Schwetz et al. [66] proposed the use of fine grained boron carbide or boron carbide/metal boride composites as nozzle material. Such material with increased fracture toughness improves the wear resistant and life time of the jet nozzle. Later, an extensive study about the effect of length of nozzle diameter and orifice diameter, AFR and WP on nozzle wear was performed by Nanduri et al. [67]. Authors developed mathematical models for weight loss rate of nozzle (W_n) and volume of nozzle material removed by single particle (V) which can be given as:-

$$W_n = (8.07E - 4) \frac{P^{0.9} d_0^{0.38} m_a}{d_n^{0.5} L^{0.8}} \tag{4.12}$$

$$V = (v \sin\alpha)^{1.9} (2r_p)^{3.28} \left(\frac{H_p}{H_t}\right)^{7.12} K_t^{-1.3} \tag{4.13}$$

where, d_n, d_o are the nozzle and orifice diameter, L is nozzle length, m_a is AFR, α is inlet angle, H_p, H_t are the abrasive particle and workpiece hardness respectively and K_t is workpiece material toughness. Vikram and Babu [68] presented a model

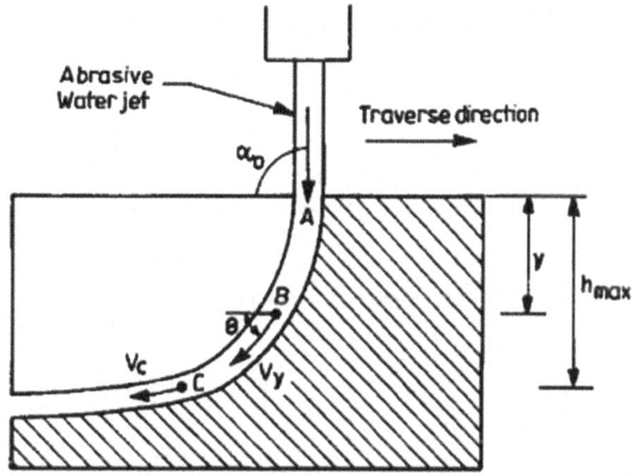

Fig. 4.11 Schematic showing the jet trajectory along the workpiece depth [68]

to predict the 2D topography of the workpiece surface machined by the AWJM. Trajectory of the jet during the AWJM is shown in Fig. 4.11.

Velocity of the abrasive jet, V_y at any given depth y can be given as:

$$V_y = ay^2 + by + c \tag{4.14}$$

Also,

$$a = K^2/4, \quad b = -KV_o^{1/2}, \quad c = V_0 \tag{4.15}$$

The trajectory of the jet is obtained by following differential equation:

$$\begin{aligned}\rho_m \Pi(d_j^2/4)\sqrt{(1+y'^2)} = {} & -m[y'y''(ay^2+by+c)^2/(\varepsilon(1+y'^2)) + (y'^2/\varepsilon + 1/\beta) \\ & \{(1+y'^2)(ay^2+by+c)(2ay+b) \\ & - y'y''(ay^2+by+c)^2\}/(1+y'^2)^2]\end{aligned} \tag{4.16}$$

where, K is the material and process constant, V_o is jet initial velocity, m, ρ_m is mass and density of abrasive particle, d_j is diameter of jet. Later, Liu et al. [69] performed computational fluid dynamics (CFD) simulation of high velocity water jet during the AWJM process. Authors primary motive of the study is to understand the dynamic characteristics of the jet such as distribution of jet velocity which in turn will help in understanding the cutting mechanism during the AWJM process. Cutting performance during the AWJM process depends on the AWJM input parameters and workpiece properties. Abrasive particle velocity and impact angle changes with the radius of curvature of the cutting surface along the depth of cut kerf which also changes the erosion process of the workpiece material. Ahmed et al. [70] did the numerical study to observe the impact characteristic of the abrasive particles as a function of curvature radius of the cutting surface. Equation 4.17 is solved for determine the motion of abrasive particles:

$$m_p \frac{\partial u}{\partial t} = F \tag{4.17}$$

where, F is the total force acting on the particle, which the sum of drag force (F_D) and buoyancy force (F_B) given as:

$$F_D = \frac{1}{8}\pi d^2 C_D |v_r| v_r \tag{4.18}$$

$$F_B = \frac{1}{6}\pi d^3 (\rho_g - \rho_f) g \tag{4.19}$$

where, m_p is mass of abrasive, C_D is drag coefficient, v_r relative velocity between abrasive and fluid, d is particle diameter, ρ_g is particle density, ρ_f is fluid density and g is gravitational acceleration. Matsumura and Fueki [71] machined micro channels with the AWJM process. Authors also used CFD to simulate the machining process (Fig. 4.12). Stagnation area generated under the jet during the machining of grooves develops horizontal flow of jet and abrasive particles with the workpiece surface at low impingement angles. Thus, producing a fine and crack free machine surface.

With proper understanding about the physics of the process it is easy to control and optimize the same. Wang [72] also modeled the jet penetration depth during the minor oscillation of the nozzle. Authors concluded that there is an improvement in depth of cut by 82% by using small oscillation to nozzle (4–6°) at high frequency (≈14 Hz). Depth of cut without nozzle oscillation (h_I) can be given as:

$$h_1 = \frac{k_1 m_a P}{\sigma_f u \rho_w d_j} \left(\frac{D}{S_d}\right)^x \left(\frac{P}{\sigma_f}\right)^y \left(\frac{H_d}{E}\right)^z \tag{4.20}$$

where, k_1, x, y, z, are the constant needed to be determined experimentally, u is traverse speed, P is WP, S_d is SOD, D is average abrasive diameter, H_d is dynamic hardness, E young's modulus, σ_f is material flow stress, m_a is AFR, d_j is nozzle diameter and ρ_w is density. While depth of cut considering nozzle oscillation (h_2) is:

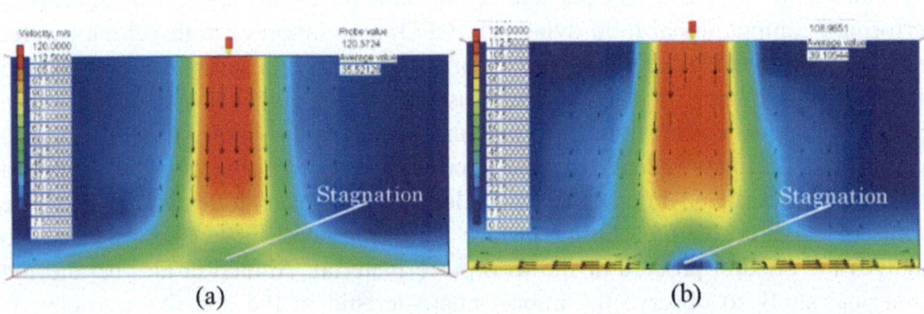

(a) (b)

Fig. 4.12 Velocity conoutrs during the mahcihing of **a** plate **b** groove (20 μm deep) [71]

$$h_2 = \frac{k_2 m_a P}{\sigma_f u \rho_w d_j}\theta^a \left(\frac{D}{S_d}\right)^a \left(\frac{P}{\sigma_f}\right)^b \left(\frac{H_d}{E}\right)^c \left(\frac{FH}{u}\right)^d \tag{4.21}$$

where, k_2, *a to d* are the constant needed to be determined experimentally, θ^a is nozzle oscillation angle and F is oscillation frequency. To further increase the efficiency of the AWJM process, Wang [73] extended their work by performing multi pass AWJM with nozzle oscillation. Authors modified their previously developed model for predicting depth of cut by incorporating new terms such as fracture toughness. Depth of cut for one pass h_1 is given as:

$$h_1 = \frac{C m_a}{\rho_p d_j u}\left(\frac{P}{\sigma_f}\right)^{r1}\left(\frac{H_d}{\sigma_f}\right)^{r2}\theta^{r3}\left(\frac{S_d}{d_p}\right)^{r4}\left(\frac{S_d m_a}{d_p^3 \rho_p u}\right)^{r5}\left(\frac{F d_p^3 \rho_p}{m_a}\right)^{r6}\left(\frac{\rho_p u^2}{P}\right)^{r7} \left(\frac{\rho_p}{\rho_w}\right)^{r8}\left(\frac{K_{IC}}{H_d d_p^{0.5}}\right)^{r9} \tag{4.22}$$

Also, depth of cutting for *j*th pass (h_j) considering nozzle oscillation is given as:-

$$h_j = \beta_j \frac{C m_a}{\rho_p d_p u_j}\left(\frac{P}{\sigma_f}\right)^{r1}\left(\frac{H_d}{\sigma_f}\right)^{r2}\theta_j^{r3}\left(\frac{S_{dj}}{d_p}\right)^{r4}\left(\frac{S_{dj} m_a}{d_p^3 \rho_p u_j}\right)^{r5}\left(\frac{F_j d_p^3 \rho_p}{m_a}\right)^{r6}\left(\frac{\rho_p u_j^2}{P}\right)^{r7} \left(\frac{\rho_p}{\rho_w}\right)^{r8}\left(\frac{K_{IC}}{H_d d_p^{0.5}}\right)^{r9} \tag{4.23}$$

where, β_j, C, *r1 to r9* are constants. Shanmugam et al. [74] proposed the incorporation of a taper compensation angle to the nozzle to reduce the amount of kerf taper produced during the machining of ceramics by AWJM process. Kerf angle, θ is given as:

$$\theta = k_1 \alpha \left(\frac{m_a u}{P}\right)^a \left(\frac{\rho_w u^2}{P}\right)^b \left(\frac{\rho_p u^2}{P}\right)^c \left(\frac{d_j}{S_d}\right)^d \left(\frac{d_p}{S_d}\right)^e \left(\frac{H_d}{P}\right)^f \left(\frac{\sigma_f}{P}\right)^g \left(\frac{K_{IC}}{P S_d^{1/2}}\right)^h \tag{4.24}$$

where, k_1, *a to h* are the constant needed to be determined experimentally, S_d is standoff distance, K_{IC} is fracture toughness, ρ_p is abrasive particle density, d_p is average abrasive particle diameter and α is kerf-taper compensation angle. Authors [75] concluded that minimum taper is produced during machining at low SOD, high WP and low NTS. Mathematical model to find the kerf taper angle for graphite/epoxy composites is given as:

$$\theta = k \left(\frac{S_d}{d_j}\right)^a \left(\frac{E d_j^2}{u m_a}\right)^b \left(\frac{P m_a^2}{\rho_w d_j^4 E^2}\right)^c \tag{4.25}$$

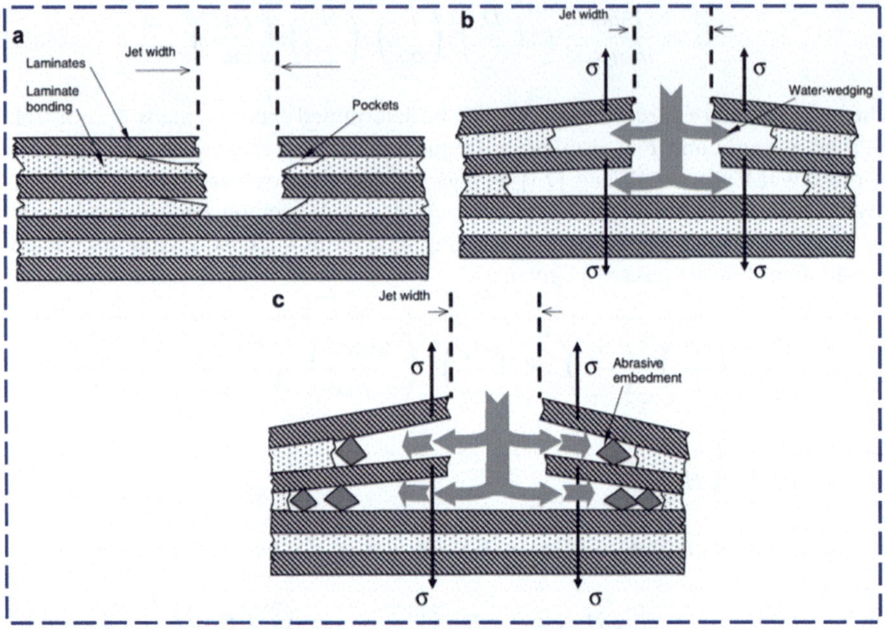

Fig. 4.13 Schematic showing various stages of graphite/epoxy composites delamination **a** initiation of facture **b** wedging of water **c** abrasive embedment [76]

Later, Shanmugam et al. [76] also examined the delamination behavior of the graphite composites during machining with AWJM process. Initially the cracks were generated on the workpiece surface due to the impact of the jet (Fig. 4.13a), later the crack propagates further due to the water wedge action (Fig. 4.13b) followed by the abrasive particle embedment into the cracks (Fig. 4.13c). Authors developed an analytical model for maximum crack length, *C* which is given as:

$$C = kd_j\left(\frac{d_j}{V_t}\right)^a\left(\frac{E * d_j}{m_s}\right)^a\left(\frac{P}{\rho_w}\right)^b\left(\frac{m_s^2}{E *^2 d_j^4}\right)^b \tag{4.26}$$

where, *K, a, b* are constants, *E** is equivalent young modulus and m_s mass flow rate of the jet.

The ease with which a material can be machined under a given input parameters is indicated by its machining index. Authors [77] tried to find out the machinability index of different composites experimentally. Later, same can be used to calculate traverse speed during the AWJM that will give the required quality of machined surface, Machinability model is given as:

$$V = \left(\frac{f_a \times N_m \times p^{1.594} \times d_j^{1.374} \times m_a^{0.343}}{C \times q \times H \times d_f^{0.618}}\right)^{1.15} \tag{4.27}$$

where, f_a is abrasive factor, N_m is machinability number, p is WP, C is constant, q is cutting quality and d_f is focusing tube diameter. Zohourkari and Zohoor [78] developed a cutting model to predict the final diameter of workpiece produced by turning using AWJM process. Authors found out a good agreement between the results predicted by model and the experimentally obtained results. For the kth revolution the reduction in radius, r_k is given as:

$$r_k = \frac{Q_k}{\pi D_k d_j} \tag{4.28}$$

where, Q_k is volume of material removed for kth revolution, D_k is starting diameter of workpiece for kth revolution and d_j is jet diameter. Retardation of abrasive jet is one of the major disadvantages of the AWJM process that causes declination in the kerf side walls especially at the entry and exit of the workpiece. Authors [79] developed theoretical model to predict the declination angle (θ_s) i.e.

$$\theta_s = \theta_{\text{lim}}\left(\frac{v_p}{v_{p\,\text{lim}}}\right)^{1.5} \tag{4.29}$$

By using above Eq. 4.29, traverse speed (v_p) for various declination angle can be calculated. Also, shift of the kerf side wall towards the jet axis ($\Delta\sigma$) is given as:

$$\Delta\sigma = \sqrt{R_1^2 + \left\{\frac{2}{5}Wtg\left[\theta_{\text{lim}}\left(\frac{v_p}{v_{\text{lim}}}\right)^{1.5}\right]\right\}^2} - R_1 + P_s \tag{4.30}$$

where, R_I is radius of curvature on the entry side surface of the cut, t is time, P_s is cut taper, θ_{lim}, v_{lim} is the limiting declination angle and limiting traverse velocity for cutting a material with thickness, Jerman et al. [80] also tried to model the striation formation phenomenon on the workpiece surface during the AWJM process. Authors used cellular automata (CA) model to study the influence of material property, abrasive jet intensity and its flow direction. Kumar and Shukla [81] did the finite element analysis for modeling the behavior of the stainless steel workpiece in terms of erosion rate, crater profileand depth of cut during the AWJM process as a function of number of impacting abrasive particles and their angle of impact. Figure 4.14 shows the top view of the crater profile at various traverse speed (V) and angle of impact (α).

Klocke et al. [82] carried out simulation of abrasive jet milling process. During the material removal process abrasive particle impacts the workpiece multiple times. Therefore, during the primary contact of the abrasive particle with the workpiece material removal (V_m) can be given as:

$$V_m = \frac{0.5mv^2\cos^2\alpha\sin n\alpha}{\varphi} + \frac{0.5m(v\sin\alpha - K)^2}{\varepsilon} \quad \alpha < \alpha_0 \tag{4.31}$$

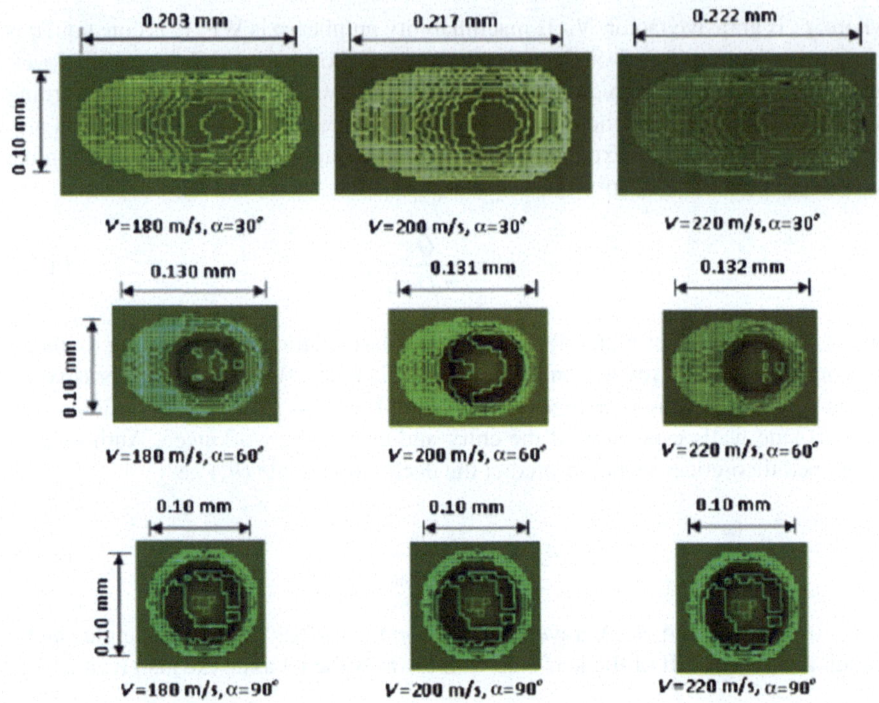

Fig. 4.14 Simulated profiles of the crater at various traverse speed and impact angle of jet [81]

$$V_m = \frac{0.5mv^2\cos^2\alpha}{\varphi} + \frac{0.5m(v\sin\alpha - K)^2}{\varepsilon} \qquad \alpha > \alpha_0 \tag{4.32}$$

While, material removal model during the secondary contact of the abrasive particle with the workpiece is:

$$V_m = \frac{0.5m(v\cos\alpha + \omega r)^2 \sin n\alpha}{\varphi} + \frac{0.5m(v\sin\alpha - K)^2}{\varepsilon} \qquad \alpha < \alpha_0 \tag{4.33}$$

$$V_m = \frac{0.5m(v\cos\alpha + \omega r)^2}{\varphi} + \frac{0.5m(v\sin\alpha - K)^2}{\varepsilon} \qquad \alpha > \alpha_0 \tag{4.34}$$

where, n is constant, α is the impact angle, α_0 is the critical impact angle, K is threshold velocity for material removal, v is velocity, ω is rotational velocity, φ, ε is erosion constant for cutting wear and deformation wear respectively. Authors [83] also developed mathematical model for simulating the water jet footprints during the milling process by AWJM process. Developed model predicts the depth of cut as a function of impact angle (θ), jet exposure time, traverse speed and jet trajectory of different orientation angle (β). Figure 4.15 shows predicted surface roughness and depth of cut on workpiece surface during forward and backward milling by abrasive jet.

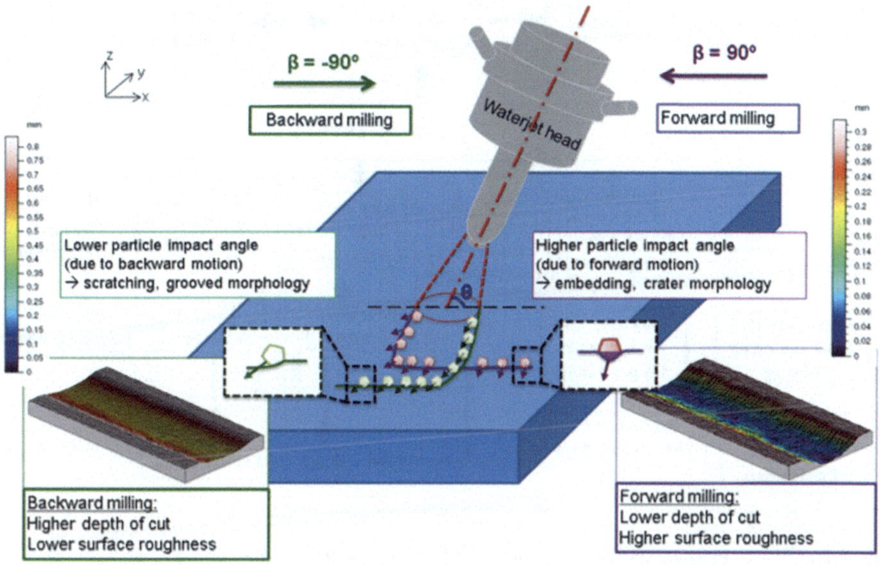

Fig. 4.15 Schematic showing surface roughness and depth of cut produced for upward and downward milling [83]

4.5 Summary

AWJM is well suited for cutting brittle materials such as ceramics, composites with minimum alteration in workpiece properties and distortion. Several techniques such as cutting with multi pass, locally heating the workpiece, forward angling the jet and providing small oscillation of the nozzle were proposed by the researchers to enhance the performance of the AWJM. Also, for the development of AWJM process authors proposed building an AWJM database that will map the relationship between various input parameters with the output responses. This will help to predict the nature of the machined geometric surface and simultaneously optimize the process. Some of the limitations of AWJM are jet lag, taper and cone effect during the machining of the workpiece. During the machining of the workpiece, a large number of AWJM input parameters affects the quality of the machined surface. Abrasive jet steadiness i.e. pressure, abrasive flow rate and traverse rate primarily affects the waviness, delamination; abrasive particle size significantly influences surface roughness, micro cracking, burr formation; while the machinability of the workpiece determines the kerf taper and wall straightness. Therefore, a proper selection of these parameters is necessary for obtaining a machined surface free from any defect. Modern technology and automation can change the abrasive jet from a viable to an overwhelmingly advantageous tool that can be used in wide variety of industrial applications. A smart system can be designed that predicts the set of input parameters for a particular material with the required surface quality and MRR. Figure 4.16 shows the AWJM that is at par with modern technology.

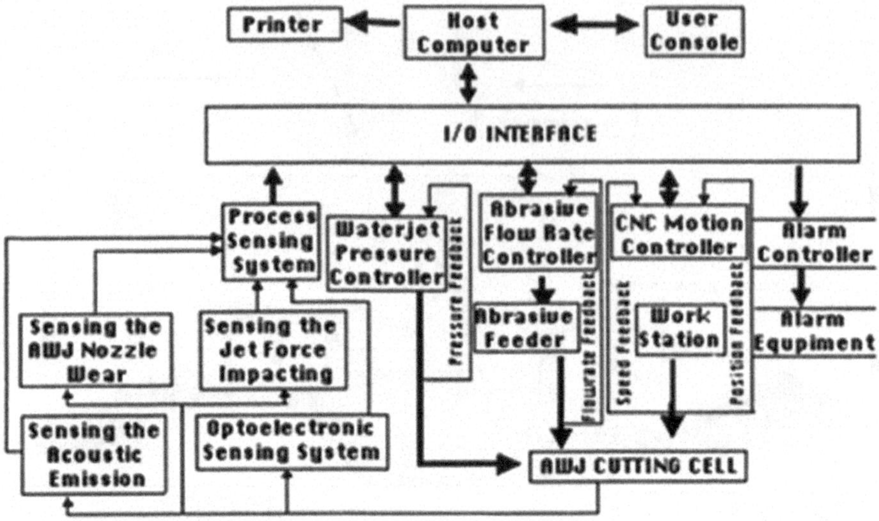

Fig. 4.16 Block diagram showing the automated abrasive water jet machining setup [84]

Some of the research areas that can be further explored include but not limited to the following points:

(i) Change in the orifice and nozzle diameter influences the cutting quality. Hence there is a need to explore material with exceptional fracture toughness and wear resistance properties for such equipment.
(ii) Elimination of the kerf taper and the striations generated on the machined surface.
(iii) More realistic numerical models need to be developed which should take into account the phenomenon such as interference, fragmentation and multiple impact of the abrasive particles during the AWJM process. A numerical model that can predict the combination of input parameters for a desired cutting quality and geometry on the given material.
(iv) Developing a multi-objective optimization of AWJM process with objective functions such as minimum power consumption, dimensional accuracy and minimum surface roughness etc.

References

1. Kovacevic R, Hashish M, Mohan R, Ramulu M, Kim TJ, Geskin ES (1997) State of the art of research and development in abrasive waterjet machining. J Manuf Sci Eng Trans ASME 119:776–785
2. Krajcarz D (2014) Comparison metal water jet cutting with laser and plasma cutting. Procedia Eng 69:838–843

3. Shanmugam DK, Chen FL, Siores E, Brandt M (2002) Comparative study of jetting machining technologies over laser machining technology for cutting composite materials. Compos Struct 57:289–296
4. Kulekci MK (2002) Processes and apparatus developments in industrial waterjet applications. Int J Mach Tools Manuf 42:1297–1306
5. Folkes J (2009) Waterjet-An innovative tool for manufacturing. J Mater Process Technol 209:6181–6189
6. Ramachandran N, Ramakrishnan N (1993) A review of abrasive jet machining. J Mater Process Tech 39:21–31
7. Hamatani G, Ramulu M (1990) Machinability of high temperature composites by abrasive waterjet. J Eng Mater Technol Trans ASME 112:381–386
8. Mazurkiewicz M (2000) Manufacturing tool for a new century. J Mater Process Technol 106:112–118
9. Hoogstrate AM, van Luttervelt CA (1997) Opportunities in AWJM—Hoogstrate A.pdf. CIRP Ann Manuf Technol 697–714
10. Hashish M (1991) Characteristics of surfaces machined with abrasive-waterjets. J Eng Mater Technol Trans ASME 113:354–362
11. Savrun E, Taya M (1988) Surface characterization of SiC whisker/2124 aluminium and Al2O3 composites machined by abrasive water jet. J Mater Sci 23:1453–1458
12. Mohamed H (1988) Visualization of the abrasive-waterjet cutting process. Exp Mech 28(2):159–169
13. Kovacevic R (1991) Surface texture in abrasive waterjet cutting. J Manuf Syst 10:32–40
14. Sano T, Takahashi M, Murakoshi Y, Suto S, Matsuno K (1992) Abrasive water-jet cutting of amorphous alloys. J Mater Process Tech 32:571–583
15. Wang J (1999) Machinability study of polymer matrix composites using abrasive waterjet cutting technology. J Mater Process Technol 94:30–35
16. Chen FL, Siores E (2003) The effect of cutting jet variation on surface striation formation in abrasive water jet cutting. J Mater Process Technol 135:1–5
17. Hascalik A, Çaydaş U, Gürün H (2007) Effect of traverse speed on abrasive waterjet machining of Ti-6Al-4 V alloy. Mater Des 28:1953–1957
18. Ness E, Zibbell R (1996) Abrasion and erosion of hard materials related to wear in the abrasive waterjet. Wear 196:120–125
19. Momber AW, Kovacevic R (1997) Test parameter analysis in abrasive water jet cutting of rocklike materials. Int J Rock Mech Min Sci 34:17–25
20. Momber AW, Mohan RS, Kovacevic R (1999) Acoustic emission study of cutting bauxite refractory ceramics by abrasive water jets. J Mater Eng Perform 8:450–454
21. Xiaohong L, Jiansheng W, Yiyu L, Lin Y, Huiming K, Jiajun S (2000) Experimental investigation of hard rock cutting with collimated abrasive water jets. Int J Rock Mech Min Sci 37:1143–1148
22. Miranda RM, Quintino L (2005) Microstructural study of material removal mechanisms observed in abrasive waterjet cutting of calcareous stones. Mater Charact 54:370–377
23. Gupta V, Pandey PM, Garg MP, Khanna R, Batra NK (2014) Minimization of Kerf Taper Angle and Kerf width using Taguchi's method in abrasive water jet machining of marble. Procedia Mater Sci 6:140–149
24. Santhanakumar M, Adalarasan R, Rajmohan M (2017) Analysis on abrasive waterjet cutting of glazed vitreous tiles using Taguchi-based RSM. Int J Manuf Technol Manag 31:472–489
25. Arola D, Ramulu M (1997) Material removal in abrasive waterjet machining of metals Surface integrity and texture. Wear 210:50–58
26. Wang J, Wong WCK (1999) A study of abrasive waterjet cutting of metallic coated sheet steels. Int J Mach Tools Manuf 39:855–870
27. Babu MK, Chetty OVK (2006) A study on the use of single mesh size abrasives in abrasive waterjet machining. Int J Adv Manuf Technol 29:532–540
28. Shukla R, Singh D (2017) Experimentation investigation of abrasive water jet machining parameters using Taguchi and evolutionary optimization techniques. Swarm Evol Comput 32:167–183

29. Azmir MA, Ahsan AK (2008) Investigation on glass/epoxy composite surfaces machined by abrasive water jet machining. J Mater Process Technol 198:122–128
30. Azmir MA, Ahsan AK (2009) A study of abrasive water jet machining process on glass/epoxy composite laminate. J Mater Process Technol 209:6168–6173
31. El-Hofy M, Helmy MO, Escobar-Palafox G, Kerrigan K, Scaife R, El-Hofy H (2018) Abrasive water jet machining of multidirectional CFRP laminates. Procedia CIRP 68:535–540
32. Armağan M, Arici AA (2017) Cutting performance of glass-vinyl ester composite by abrasive water jet. Mater Manuf Process 32:1715–1722
33. Selvam R, Karunamoorthy L, Arunkumar N (2017) Investigation on performance of abrasive water jet in machining hybrid composites. Mater Manuf Process 32:700–706
34. Ravi Kumar K, Sreebalaji VS, Pridhar T (2018) Characterization and optimization of abrasive water jet machining parameters of aluminium/tungsten carbide composites. Meas J Int Meas Confed 117:57–66
35. Aich U, Banerjee S, Bandyopadhyay A, Das PK (2014) Abrasive water jet cutting of borosilicate glass. Procedia Mater Sci 6:775–785
36. Kechagias J, Petropoulos G, Vaxevanidis N (2012) Application of Taguchi design for quality characterization of abrasive water jet machining of TRIP sheet steels. Int J Adv Manuf Technol 62:635–643
37. Supriya SB, Srinivas S (2018) Machinability studies on stainless steel by abrasive water jet -review. Mater Today Proc 5:2871–2876
38. Axinte DA, Srinivasu DS, Kong MC, Butler-Smith PW (2009) Abrasive waterjet cutting of polycrystalline diamond: A preliminary investigation. Int J Mach Tools Manuf 49:797–803
39. Axinte DA, Stepanian JP, Kong MC, McGourlay J (2009) Abrasive waterjet turning-An efficient method to profile and dress grinding wheels. Int J Mach Tools Manuf 49:351–356
40. Paul S, Hoogstrate AM, Van Luttervelt CA, Kals HJJ (1998) An experimental investigation of rectangular pocket milling with abrasive water jet. J Mater Process Technol 73:179–188
41. Shipway PH, Fowler G, Pashby IR (2005) Characteristics of the surface of a titanium alloy following milling with abrasive waterjets. Wear 258:123–132
42. Fowler G, Pashby IR, Shipway PH (2009) The effect of particle hardness and shape when abrasive water jet milling titanium alloy Ti6Al4V. Wear 266:613–620
43. Kong MC, Axinte D, Voice W (2011) Challenges in using waterjet machining of NiTi shape memory alloys: An analysis of controlled-depth milling. J Mater Process Technol 211:959–971
44. Uthayakumar M, Khan MA, Kumaran ST, Slota A, Zajac J (2016) Machinability of Nickel-based superalloy by abrasive water jet machining. Mater Manuf Process 31:1733–1739
45. Akkurt A, Kulekci MK, Seker U, Ercan F (2004) Effect of feed rate on surface roughness in abrasive waterjet cutting applications. J Mater Process Technol 147:389–396
46. Akkurt A (2010) Cut front geometry characterization in cutting applications of brass with abrasive water jet. J Mater Eng Perform 19:599–606
47. Akkurt A (2009) The effect of material type and plate thickness on drilling time of abrasive water jet drilling process. Mater Des 30:810–815
48. Srikanth DV, SreenivasaRao M (2014) Metal removal and kerf analysis in abrasive jet drilling of glass sheets. Procedia Mater Sci 6:1303–1311
49. Gao C, Liu Z, Zhao K, Guo C (2018) Abrasive water jet drilling of ceramic thermal barrier coatings. Procedia CIRP 68:517–522
50. Park DS, Cho MW, Lee H, Cho WS (2004) Micro-grooving of glass using micro-abrasive jet machining. J Mater Process Technol 146:234–240
51. Kumar Pal V, Choudhury SK (2014) Fabrication and analysis of micro-pillars by abrasive water jet machining. Procedia Mater Sci 6:61–71
52. Schwieger K, Carrero V, Rentzsch R, Becker A, Bishop N, Hille E, Louis H, Morlock M, Honl M (2004) Abrasive water jet cutting as a new procedure for cutting cancellous bone—In vitro testing in comparison with the oscillating saw. J Biomed Mater Res Part B Appl Biomater 71:223–228
53. Patel KJ (2004) Quantitative evaluation of abrasive contamination in ductile material during abrasive water jet machining and minimising with a nozzle head oscillation technique. Int J Mach Tools Manuf 44:1125–1132

54. Wang J, Guo DM (2003) The cutting performance in multipass abrasive waterjet machining of industrial ceramics. J Mater Process Technol 133:371–377
55. Wang J, Kuriyagawa T, Huang CZ (2003) An experimental study to enhance the cutting performance in abrasive waterjet machining. Mach Sci Technol 7:191–207
56. Kantha Babu M, Krishnaiah Chetty OV (2003) A study on recycling of abrasives in abrasive water jet machining. Wear 254:763–773
57. Miller DS (2004) Micromachining with abrasive waterjets. J Mater Process Technol 149:37–42
58. Jegaraj JJR, Babu NR (2005) A strategy for efficient and quality cutting of materials with abrasive waterjets considering the variation in orifice and focusing nozzle diameter. Int J Mach Tools Manuf 45:1443–1450
59. Palleda M (2007) A study of taper angles and material removal rates of drilled holes in the abrasive water jet machining process. J Mater Process Technol 189:292–295
60. Khan AA, Haque MM (2007) Performance of different abrasive materials during abrasive water jet machining of glass. J Mater Process Technol 191:404–407
61. Patel D, Tandon P (2015) Experimental investigations of thermally enhanced abrasive water jet machining of hard-to-machine metals. CIRP J Manuf Sci Technol 10:92–101
62. Chen L (1996) Kerf characteristics in abrasive waterjet cutting of ceramic materials. Int J Mach Tools Manuf 36:1201–1206
63. Chen L, Siores E, Wong WCK (1998) Optimising abrasive waterjet cutting of ceramic materials. J Mater Process Technol 74:251–254
64. Abdel-Rahman AA, El-Domiaty AA (1998) Maximum depth of cut for ceramics using abrasive waterjet technique. Wear 218:216–222
65. Wakuda M, Yamauchi Y, Kanzaki S (2002) Effect of workpiece properties on machinability in abrasive jet machining of ceramic materials. Precis Eng 26:193–198
66. Schwetz KA, Sigl LS, Greim J, Knoch H (1995) Wear of boron carbide ceramics by abrasive waterjets. Wear 181–183:148–155
67. Nanduri M, Taggart DG, Kim TJ (2002) The effects of system and geometric parameters on abrasive water jet nozzle wear. Int J Mach Tools Manuf 42:615–623
68. Vikram G, Babu NR (2002) Modelling and analysis of abrasive water jet cut surface topography. Int J Mach Tools Manuf 42:1345–1354
69. Liu H, Wang J, Kelson N, Brown RJ (2004) A study of abrasive waterjet characteristics by CFD simulation. J Mater Process Technol 153–154:488–493
70. Ahmed DH, Naser J, Deam RT (2016) Particles impact characteristics on cutting surface during the abrasive water jet machining: Numerical study. J Mater Process Technol 232:116–130
71. Matsumura T, Muramatsu T, Fueki S (2011) Abrasive water jet machining of glass with stagnation effect. CIRP Ann Manuf Technol 60:355–358
72. Wang JÃ (2007) Predictive depth of jet penetration models for abrasive waterjet cutting of alumina ceramics. Int J Mech Sci 49:306–316
73. Wang J (2010) Depth of cut models for multipass abrasive waterjet cutting of alumina ceramics with nozzle oscillation. Front Mech Eng China 5:19–32
74. Shanmugam DK, Wang J, Liu H (2008) Minimisation of kerf tapers in abrasive waterjet machining of alumina ceramics using a compensation technique. Int J Mach Tools Manuf 48:1527–1534
75. Shanmugam DK, Masood SH (2009) An investigation on kerf characteristics in abrasive waterjet cutting of layered composites. J Mater Process Technol 209:3887–3893
76. Shanmugam DK, Nguyen T, Wang J (2008) A study of delamination on graphite/epoxy composites in abrasive waterjet machining. Compos Part A Appl Sci Manuf 39:923–929
77. Alberdi A, Suárez A, Artaza T, Escobar-Palafox GA, Ridgway K (2013) Composite cutting with abrasive water jet. Procedia Eng 63:421–429
78. Zohoor M, Zohourkari I (2011) Modeling of abrasive waterjet turning. Aust J Basic Appl Sci 5:70–79
79. Hlaváč LM, Strnadel B, Kaličinský J, Gembalová L (2012) The model of product distortion in AWJ cutting. Int J Adv Manuf Technol 62:157–166

80. Jerman M, Valentinčič J, Lebar A, Orbanić H (2015) The study of abrasive water jet cutting front development using a two-dimensional cellular automata model. Stroj Vestnik/J Mech Eng 61:292–302
81. Kumar N, Shukla M (2012) Finite element analysis of multi-particle impact on erosion in abrasive water jet machining of titanium alloy. J Comput Appl Math 236:4600–4610
82. Klocke F, Schreiner T, Schüler M, Zeis M (2018) Material removal simulation for abrasive water jet milling. Procedia CIRP 68:541–546
83. Kong MC, Anwar S, Billingham J, Axinte DA (2012) Mathematical modelling of abrasive waterjet footprints for arbitrarily moving jets: Part i—Single straight paths. Int J Mach Tools Manuf 53:58–68

Chapter 5
Advances in Abrasive Assisted EDM and Micro-EDM

Asif Rashid and Muhammad P. Jahan

Abstract This chapter discusses the research trends and developments in the field of abrasive assisted EDM and micro-EDM. The parameters that control the performance of the abrasives in the hybrid system of abrasive mixed EDM has been explored. The actual improvement of EDM process has been studied by means of looking into the performance parameters. This chapter also focuses on materials that have been the most challenging to machine through EDM, and how the abrasive-mixed process has changed the scenario. This chapter will provide a comprehensive overview of this highly-efficient and productive manufacturing technique from the mechanism and application perspective.

Keywords EDM · Micro-EDM · PMEDM · APMEDM · Hybrid machining · Abrasive · Powder

5.1 Introduction

Electrical Discharge Machining (EDM) is a non-conventional machining process where voltage difference is applied between tool and workpiece, and material is removed by means of precisely controlled electrical sparks that pass through a channel of dielectric. Any electrically conductive material can be machined by this non-conventional process irrespective of its hardness. So EDM is quite a comprehensive process in terms of the materials that it can cover. But EDM process has its own drawbacks. The material removal rate of this non-conventional manufacturing process is low. Depending on the machining conditions, the surface finish may become inferior and can also lead to high tool wear ratio. EDM process also has limitations

A. Rashid · M. P. Jahan (✉)
Departmental of Mechanical and Manufacturing Engineering, Miami University, Oxford, OH 45056, USA
e-mail: jahanmp@miamioh.edu

A. Rashid
e-mail: rashidm@miamioh.edu

S. Das et al. (eds.), *Advances in Abrasive Based Machining and Finishing Processes*, Materials Forming, Machining and Tribology,
https://doi.org/10.1007/978-3-030-43312-3_5

about the conductivity of the workpiece that it can machine. Any material with a higher resistivity than a preset threshold cannot be machined in this process.

In order to solve the problems associated with the EDM process, compound, assisted, and hybrid machining processes have been approached in recent years. Sometime, all these processes are broadly categorized as a single type of machining process named hybrid machining. Hybrid machining is the process of combining two machining processes in the same machining zone concurrently to get a better performance than the output that could have been achieved by deploying the involved processes one after another, concurrently, or with assistance. Combining abrasives in the EDM machining system creates a hybrid effect that leads to the betterment of the overall process. Many limitations of the EDM process can be removed in this manner. Adding powder or abrasives into the EDM dielectric changes the dielectric system properties and leads to altered performance and better productivity. This process is also known as powder-mixed electro-discharge machining (PMEDM). The inherent drawbacks of the EDM process can be eliminated by using the PMEDM process in a calculated way. So this hybrid manufacturing process leads to better material removal rate, lesser tool wear, and greater surface integrity. Adding powder to the system can also broaden the scope of EDM process as many non-conductive and low-conductive materials can be allowed to be machined in this system. This chapter provides a comprehensive overview of the abrasive powder-mixed EDM process and its application in the improvement of machining performance and in the machining of difficult-to-machine materials.

5.2 Electrical Discharge Machining (EDM)

Electrical discharge machining (EDM) is an effective non-conventional manufacturing approach that can be used to machine hard materials. This process can be applied on any workpiece that is electrically conductive. Materials used in high-performance applications that are deployed in semiconductor, aerospace, automotive, and power generation industry can be shaped using this process. Complex three-dimensional geometries can be generated in this manufacturing process. The recent trend of miniaturization of components and demand for high-tolerance parts has led to the growth of this EDM method. Advanced engineering materials can be machined with precision at micro-level in this process.

In EDM process, a voltage difference is applied between the workpiece and a tool electrode. The workpiece and the electrode are both submerged in dielectric oil. When the voltage is over a threshold value and the electrode comes close to the workpiece, the dielectric breaks down. This breakdown of dielectric provides a smooth channel for the electrical energy to pass through. As a result, material is removed from the workpiece in the form of crater [1]. The dielectric also serves as the flushing agent and helps remove the debris from the gap between the workpiece and electrode. There is no physical contact between the workpiece and the electrode. This non-contact approach makes this process universal to all electrically-conductive

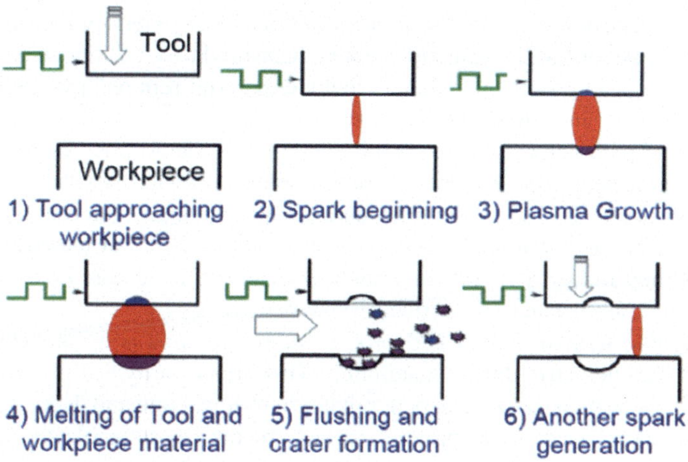

Fig. 5.1 Physical mechanism of EDM process [2]

Fig. 5.2 Interaction of tool and electrode in EDM process [3]

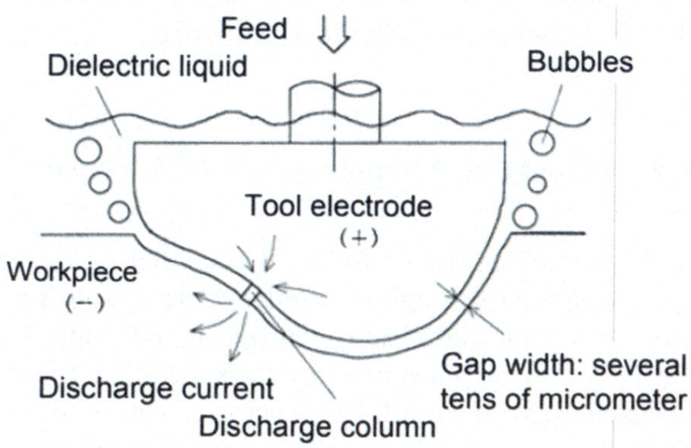

materials. Figure 5.1 shows the physical mechanism of the EDM process indicating each steps during the material removal process. Figure 5.2 indicates the interaction between the tool electrode and the workpiece at the spark gap during the EDM process.

5.3 Abrasive Assisted Electrical Discharge Machining

For electrical discharge machining (EDM) process, there are a few variables that determine the state of the machining system. These process parameters can be electrical or non-electrical quantities. Electrical process parameters, like voltage, current, capacitance determine the energy input to the system. Non-electrical parameters, like flushing and electrode rotation also control the machining environment of the system. There are quantifiable parameters that be measured to determine the machining performance. These process parameters include material removal rate (MRR), tool wear ratio (TWR), and surface roughness (SR).

The process parameters involved in the EDM process determines the performance parameters of a particular system. There are some intrinsic drawbacks that come with EDM process. The material removal rate is not that high. The roughness of the machined surface is not smooth. Tool wear is another issue. But the EDM process can be modified in a way that those drawbacks can be encountered by making the process more effective and productive.

One possible way to make EDM process more effective is by introducing fine particles to the dielectric during machining. This creates a hybrid machining phenomenon. This process can lead to better material removal rate and decreases surface roughness and tool wear. So the process can become more sustainable and productive at the same time. Addition of external fine particles can alter the process parameters in a positive way that it produces a better machining outcome. This modified EDM process that is carried out by adding fine particles to the dielectric is known as abrasive powder mixed electrical discharge machining (APMEDM) or powder mixed electrical discharge machining (PMEDM).

5.4 Effects of Abrasives in EDM Process

Adding abrasive particles to the dielectric during electrical discharge machining process changes the dynamics of the dielectric system. Various electrical and mechanical properties of the dielectric system are altered in this process. The process parameters involved with the machining becomes modified. Figure 5.3 shows the list of process parameters involved in EDM process. Changes to process parameters also modify the performance parameters. The changes in the result of machining in this process can vary from little to drastic effects.

5.4.1 Influence on the Machining System and Parameters

Adding a powder to the dielectric can alter various machining properties. The intrinsic properties of the powder materials are transferred onto the dielectric itself. Adding a conductive powder increases the conductivity of the dielectric. As a result, an increase in the number of sparks can be observed. This phenomenon can lead to a bridging effect, which contributes to the mechanism of dielectric breakdown in the machining system. Addition of abrasives creates a smooth path for the discharge to reach the machined surface. The resulting plasma channel can allow more energy to pass through. These changes to the system can increase the spark gap because of the increment of spark energy. So the flushing becomes

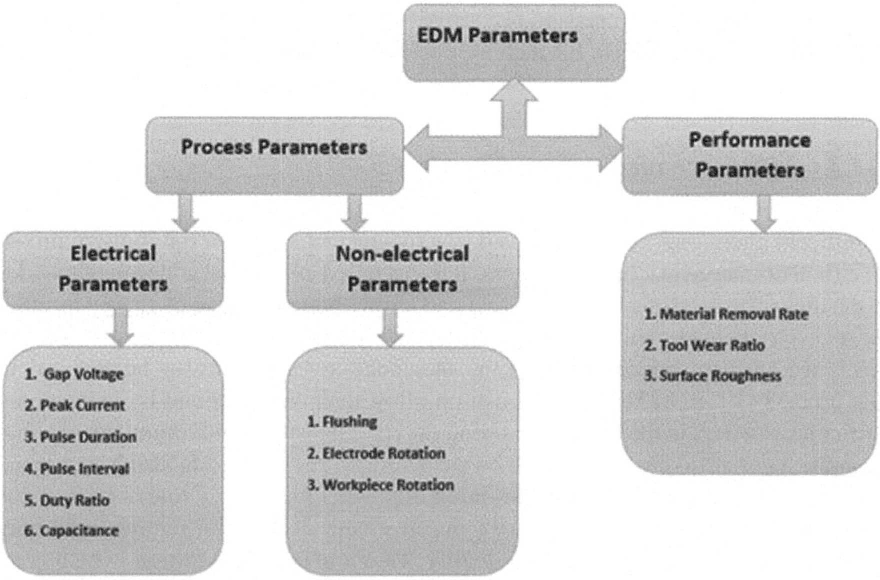

Fig. 5.3 Parameters involved in electrical discharge machining process [4]

better and better surface roughness can be obtained. Better machining efficiency and higher productivity are observed.

5.4.2 Influence on the Material Removal Rate

As discussed in the previous section, the addition of powder to the dielectric modifies the spark gap and flushing capabilities of the system. The plasma channel can get concentrated and more energy can be transferred in an organized manner. These changes to the system can also have an effect on the material removal rate (MRR) of the system as well. Increment of peak current results in greater material removal rate in powder mixed electrical discharge machining process. Aluminum powder improves the MRR of Inconel alloy for the increased amount of energy transfer through the plasma channel [5]. The size of the discharge channels can also get reduced as a result [6]. Addition of powder can lead to the reduction of breakdown voltage and the increment of discharge frequency. Such a mechanism is generated by adding graphite while machining steel alloys. This process can improve the material removal rate by 60% [7]. Titanium and almost all other conductive powders also work in the same mechanism [8]. Combining aluminum and graphite together can lead to even better results in terms of material removal rate [9]. The concentration of added powder has an influence on the enhancement of material removal rate. In some cases, optimization needs to be carried out to determine the best concentration of abrasive powder particles for increased material removal rate.

5.4.3 Influence on Surface Finish

Significant alterations can be achieved in terms of surface finish for abrasive mixed electrical discharge machining process. It is discussed before that addition of powder to the dielectric creates a bridging effect and allows better passage of energy through the plasma channel. This bridging effect also reduces the size of craters on the machined surface. Microcracks on the machined surface can also be reduced in this process [10, p. 825]. Mirror finish on alloy steel can be found by using some particular powders in the dielectric. Using the right density of aluminum powder can generate a mirror-like surface finish on steel surface [11]. This can also be achieved by using silicon powder. The concentration of powder plays a key role in getting this mirror-like finish [12]. In some cases, improvement of material removal rate may come at the cost of reduced surface quality. Titanium carbide addition, which is an well-known abrasive, improves material removal rate for alloy steel, but increases the surface roughness in the process [13]. Improved surface roughness leads to increased biocompatibility for different alloys, like titanium alloys [14]. By adding graphene, materials of lower conductivity can be machined with very few surface defects [15]. Inconel alloys can be machined in this process. Carbon nanotube is another powder material that performs very well in terms of surface finish for hard materials [13].

5.4.4 Influence on Tool Wear

The tool wear is another important factor that can be affected or modified in the abrasive or conductive powder mixed electrical discharge machining process. Improvement of material removal rate or surface roughness does not guarantee the reduction of tool wear. Graphene and carbon nanotube (CNT) performs very well in terms of material removal, such may not be the case for tool wear. Aluminum powder performs very well in terms of reducing the tool wear ratio during the machining of steel alloys [12]. Graphite powder is also effective in reducing the tool wear by generating a small discharge channel [6]. Boron carbide powder can also reduce the tool wear ratio [16]. Sometimes, the increment of material removal rate has an adverse effect on tool wear. So proper optimization needs to be carried out for reducing the tool wear ratio, while improving the surface finish and/or material removal rate.

5.4.5 Influence on Machining Accuracy

Machining accuracy can be affected in the abrasive mixed electrical discharge machining process. Focus on the increment of material removal rate and decrement of surface roughness and tool wear can have an adverse effect on the accuracy of machining. Boron carbide powder is known to increase the material removal rate and decrease the tool wear during machining titanium alloys. But the addition of this powder reduces the machining accuracy. The machined micro-holes can become non-uniform and form taper [17]. Aluminum powder is known to increase the machining accuracy [13].

5.5 Abrasive Parameters for Abrasive-Assisted EDM Process

Various parameters affect the outcome of the abrasive powder mixed electrical discharge machining (APMEDM) process. Suitable powder material has to be selected for a particular machining system. The concentration of the powder has to be optimized in a way that material removal rate and surface roughness can have acceptable values. Particle size plays an important role in the machining outcome as well. The electrode material is also vital for this PMEDM process. The tool polarity and flushing conditions can have a say on the performance parameters as well.

5.5.1 Abrasive Material

The selection of abrasive material in powder mixed electrical discharge machining process is the most important factor in successfully implementing this procedure. Past researches have been carried out in finding out the suitable abrasive material for machining hard materials in EDM process. The output of machining depends on the selection of the right abrasive or powder material.

Aluminum powder has been used in machining hard Inconel alloys. The material removal rate and surface finish both improve as a result [5]. Addition of aluminum powder reduces the tool wear ratio while machining steel alloys. Although other powders can perform better on steel when it comes to material removal rate and surface roughness, aluminum powder was found to be more effective [13]. Aluminum powder is also effective in machining some metal matrix composites.

Graphite powder is effective in terms of increasing the material removal rate for alloy steels [7]. Graphite is also useful for tungsten carbide-cobalt superalloys [18]. Titanium powder is a good alternative to aluminum or graphite for machining alloy steel compounds [13]. Titanium powder can improve the biocompatibility of titanium alloys [14]. Boron carbide powder can improve the material removal rate and decrease tool wear during the machining of titanium alloys [16].

Silicon powder, a known abrasive, is found to improve the material removal rate during EDM of alloys [19]. Surface finish can be improved significantly by adding silicon carbide into the dielectric. Near mirror finish and glossy surface can be generated by this process [12].

Addition of chromium powder to the dielectric reduces the size of the recast layer [12]. Highly conductive graphene or carbon nanotubes can also be used as powder materials in the dielectric. Addition of graphene results in a lower amount of thermal cracks on the machined surface. The hard materials, like Inconel 825 can be machined with ease [20]. Carbon nanotubes can perform better than aluminum or graphite in terms of both material removal rate and surface roughness [13]. Metal matrix composites can be successfully machined using carbon nanotubes as powder in the dielectric.

5.5.2 Abrasive Particle Size

Abrasive particle size plays an important role in determining the outcome of powder mixed electrical discharge machining process. The particle size can indirectly determine the spark gap associated with machining and control the flushing quality. Aluminum powder allows more spark gap than graphite, silicon or molybdenum disulfide. Removal of debris is easier for higher spark gap. Easier breakdown of powder mixed dielectric contributes to the increment of the spark gap. Molybdenum disulfide (MoS_2) particles are the smallest compared to aluminum or silicon carbide. MoS_2 powder allows mirror-like surface finish on steel alloys. Achieving the mirror like finish also comes at a cost of reduced machining efficiency [12].

Smaller particle size attributes to greater surface finish in powder mixed EDM process. But the smallest particles also generate the thickest of recast layers. Improvement of surface finish is not dependent on the concentration of powder materials, but only on the powder size itself. Smallest particles perform better as they produce fine cutting effects. The recast layer size decreases with the increment of particle size. This happens due to the fact that electrical power density reduces with increasing spark size [21, p. 11].

The spark gap increases with an increment in particle size. This happens because large particles enhance the connection between electrode and workpiece by occupying more space. Increase in powder size can result in the reduction of improvement of material removal rate. The machining efficiency can also become lower for higher particle size. The tool wear ratio does not follow any particular trend in terms of particle size [22].

5.5.3 *Abrasive Particle Concentration*

Concentration of abrasive particle has to be optimized in a way that the material removal rate and surface roughness has an acceptable value. High powder concentration can lead to unstable machining. The spark gap is determined by the concentration. Increment of powder concentration increases the spark gap [12]. An optimized amount of powder concentration exists for the best possible surface finish. This happens because only a limited number of particle contributes to the melted machined zone in an effective manner. Powder concentration also controls the thickness of the recast layer [21]. Because of the presence of an optimal spark gap that determines the maximum material removal rate, an optimum powder concentration also exists for maximum productivity. Powder concentration has very less effect on tool wear ratio [22]. Excessive powder concentration leads to short-circuit in the system, which results in a higher number and depth of craters developed on the machined surface [23].

5.6 Various Materials Used in Powder or Abrasive-Assisted EDM

5.6.1 *Aluminum*

Aluminum and its compounds are very often used in powder mixed electrical discharge machining (PMEDM) process. Inconel is a very hard alloy that is used in combustion chambers and thrust chambers of rocket engines. This material is difficult to machine due to its fast work hardening properties. So EDM is a viable option in this case as there is no contact between tool and workpiece. Adding aluminum powder in dielectric can increase the efficiency of the EDM process during EDM of Inconel alloy. Addition of abrasive aluminum oxide powder can enhance the process. In this case, the material removal rate improves with increased peak current. This is observed because the powder in the dielectric creates a bridging effect during machining. The number of sparks increase as aluminum powder is added to the dielectric. The bridging effect also helps reduce the size of holes and craters on the machined surface, and subsequently improving the surface finish [5]. The amount of microcracks on the surface also reduces in this process [10]. Figure 5.4 demonstrates the improvement of machined surface by using this aluminum powder mixed approach. So aluminum powder improves both MRR and SR while machining Inconel.

Steel is another alloy which can be machined by powder mixed EDM process. Aluminum powder is often used in machining EN19 alloy. Aluminum may not be the best choice as powder while machining steel alloys in terms of material removal rate or surface roughness, but it is the most effective in terms of tool wear ratio (TWR) and radial overcut (ROC). Aluminum powder performs better than carbon nanotubes and graphene when it comes to improving the tool life and increasing the machining

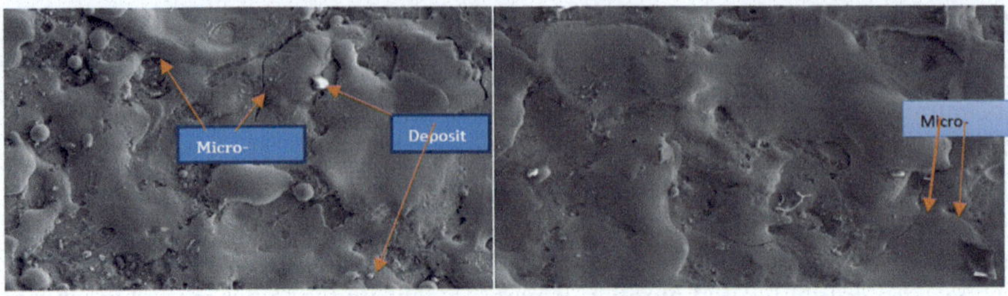

Fig. 5.4 Improvement of surface roughness of EDMed surface by aluminum powder mixed EDM process (right) over conventional EDM process (left) for machining Inconel 825 [10]

Fig. 5.5 Tool wear ration for various powder in dielectric while machining EN19 by EDM process [13]

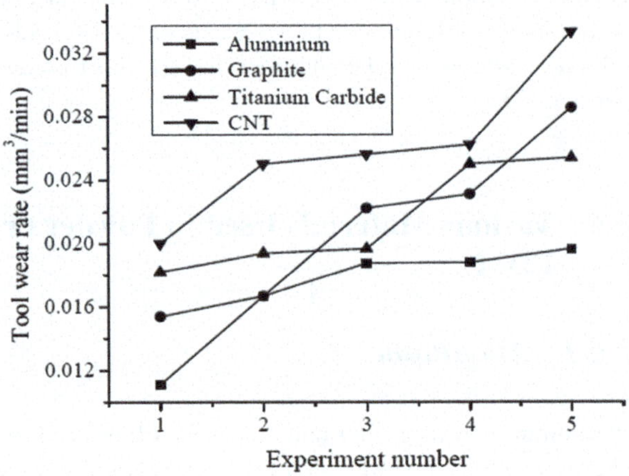

accuracy [13]. Figure 5.5 compares the performance of various powder materials for machining EN19 alloy in PMEDM process and shows that aluminum generates the least amount of TWR. Aluminum powder can also create a mirror-like surface finish for SKH-51 alloy [12]. But aluminum powders have a tendency to agglomerate in the dielectric. A surfactant can be deployed to separate the powder materials and creating a homogeneous mixture. So the density of aluminum powder and the density of the surfactant are both very important parameters for obtaining a mirror-like finish in steel alloys [11]. Figure 5.6 demonstrates the importance of a surfactant for attaining mirror-like surface finish by reduction of recast layer in aluminum powder mixed EDM process.

Aluminum powder is also used in the PMEDM process of some metal matrix composites. Silicon carbide reinforced aluminum matrix composites is a material that is used extensively for its extreme properties. The machining of this material can be enhanced by adding aluminum powder to the dielectric in EDM process. The surface roughness decreases in this method. The hardness is increased. The wear resistance is almost doubled [24]. Figure 5.7 depicts a comparison of surface finish for conventional and aluminum powder mixed EDM of silicon carbide reinforced

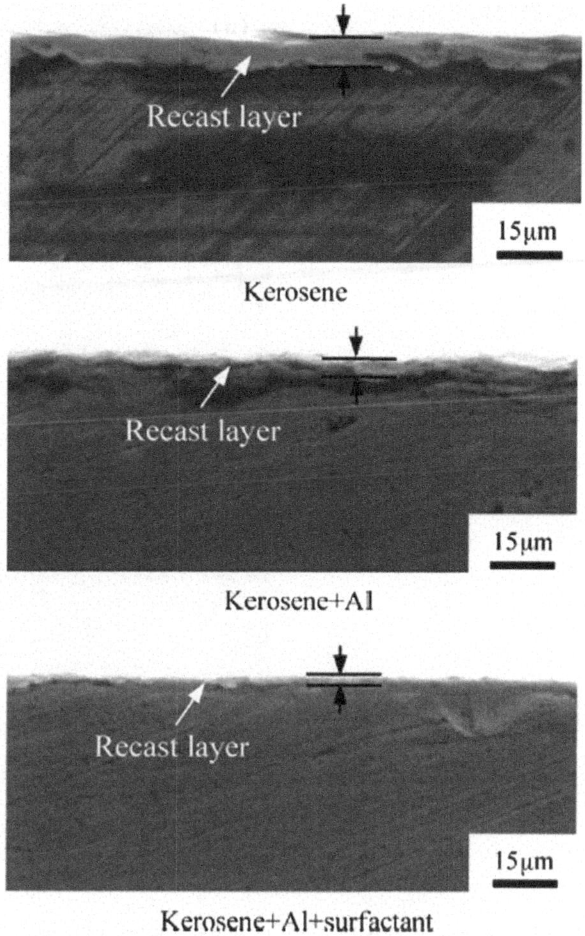

Fig. 5.6 Recast layer thickness for various configuration of aluminum powder mixed EDM [11]

aluminum matrix composite (SiCp-Al). It can be seen that aluminum powder mixed EDM generates smoother machined surface with fewer defects.

5.6.2 Graphite

Graphite can be a useful powder to use in the EDM process of hard materials. Graphite is mostly used in the PMEDM of steel and its various alloys. Adding fine graphite powder to kerosene dielectric for EDM of mild steel, the material removal rate can be increased by 60%. Adding graphite powder reduces the breakdown voltage and results in higher discharge frequency, which subsequently increases the MRR [7]. Adding graphite powder to kerosene dielectric while machining EN19 alloy steel

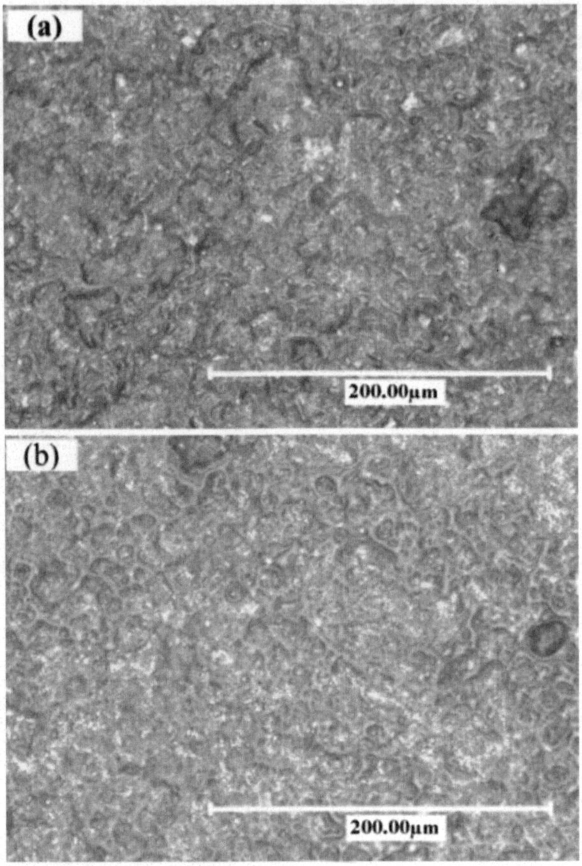

Fig. 5.7 Micro-surface texture for SiCp-Al composite after **a** conventional EDM, **b** aluminum powder mixed EDM [24]

can bring about positive influences. Increment of graphite powder concentration increases the material removal rate. The material removal rate also increases with the increment of peak current and gap voltage in this case [25]. Addition of graphite can alter the surface roughness and microstructure of the machined surface of EN19 alloy steel. Increasing peak current and gap voltage can increase the hardness of the machined surface [26]. Graphite and aluminum can be used together in the same system while machining steel alloys. Combining graphite and aluminum produces a better result than using the individual powders in terms of material removal rate. But the increased material removal rate comes at the cost of higher tool wear ratio [9]. Figure 5.8 shows the combined effect of using graphite and aluminum as powder while machining EN19 steel alloy by EDM process. Aluminum and graphite used together seem to generate better material removal rate.

Graphite powder has also been used in machining very hard materials, like tungsten carbide-cobalt (WC-Co) superalloy. Increasing the concentration of graphite powder during the machining of WC-Co alloy increases the material removal rate

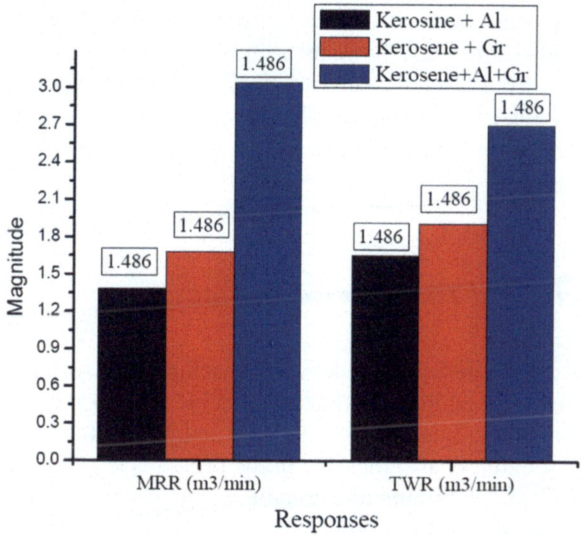

Fig. 5.8 Effect of using various configuration of aluminum and graphite powder while machining EN19 alloy steel by powder mixed EDM process [9]

and decreases the electrode wear ratio. But there exists an optimum point for maximum material removal rate and minimum electrode wear ratio. Such an optimized point also exists for surface roughness as well. For concentration of less than or greater than the optimized value, the material removal is not superior and the surface roughness increases. The optimized concentration is not the same for both material removal rate and surface roughness [18]. Figure 5.9 demonstrates this fact. So machining has to be carried out in a way that it only prioritizes material removal rate or surface roughness. Adding graphite powder in machining Cu 605 superalloy

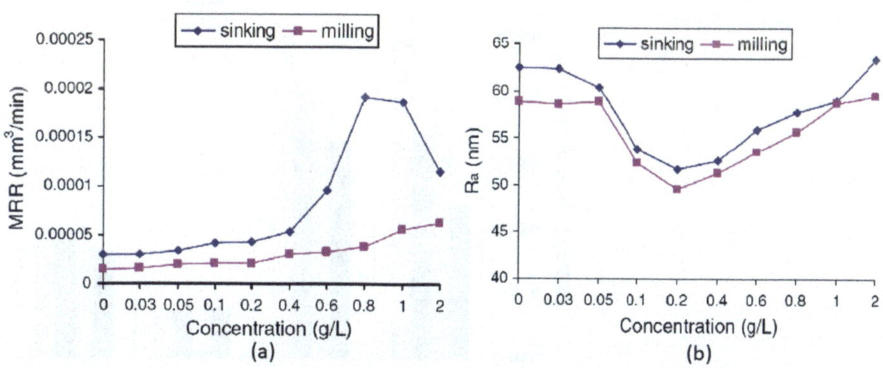

Fig. 5.9 Effect of graphite powder concentration on **a** material removal rate and **b** average surface roughness while machining WC-Co in EDM process [18]

improves the micro-hardness and surface finish of the alloy. The surface generated in this process is also smoother. To maintain this enhanced surface finish, the peak current is prevented from reaching a high value [27]. Mixture of graphite in dielectric fluid also leads to reduced tool wear. Graphite powder accomplishes this by an energy dissipation mechanism which is generated as a result of small discharge channels [6].

5.6.3 Titanium and Titanium Compounds

Titanium carbide can be used as a powder while machining alloy steel. While machining EN19 alloy steel, the addition of titanium carbide powder improves the material removal rate with the increment of peak current and gap voltage. But the tool wear ratio and surface roughness also increase in the process [13]. Titanium carbide can be a good alternative to aluminum and graphite powder in the machining process of EN19 alloys steel as seen in Fig. 5.10.

Titanium compounds as powder in the dielectric have been used for surface modification process. The quality of the machined surface of AISI D2 steel can be enhanced by adding titanium nanopowder. EDM process usually results in higher surface roughness if the material removal rate is enhanced. The increment of spark energy is responsible for this phenomenon. Addition of titanium nanopowder in hydrocarbon oil dielectric while machining AISI D2 steel can enhance the MRR and decrease the surface roughness at the same time. Other surface micro-defects, like voids, cracks and droplets also diminish [8]. Figure 5.11 reveal the improvement of MRR and decrement of average surface roughness with Ti nano powder mixed

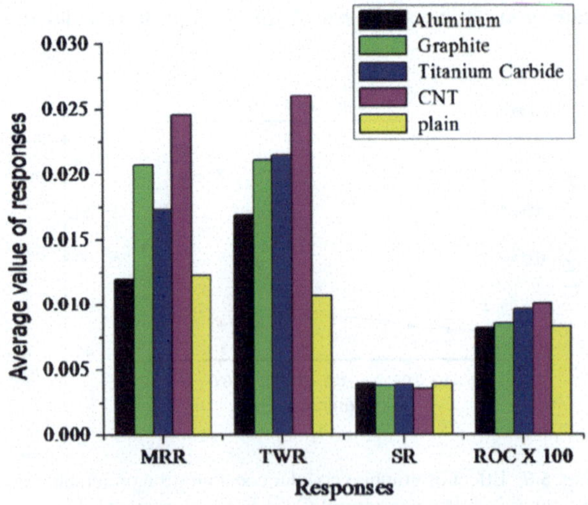

Fig. 5.10 Machining responses for different powder materials for machining of EN19 alloy steel in electrical discharge machining process [13]

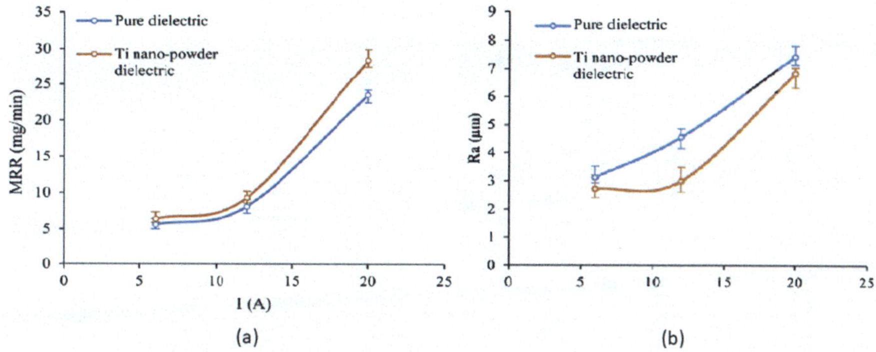

Fig. 5.11 Improvement of **a** material removal rate (MRR), and **b** average surface roughness (Ra) for machining D2 steel in EDM process by adding Ti nanopowder [pulse duration 210 μs, pulse interval 140 μs] [8]

dielectric compared to those of pure dielectric. Accumulation of titanium atoms on crack surface leads to this incredible improvement.

Titanium alloys are used in the implant industry. EDM is a viable option for machining titanium alloys. But this process leads to an increased amount of recast layer and microcracks. Adding titanium powder to deionized water dielectric reduces the formation of surface microcracks on titanium workpiece surface. An optimum powder concentration exists for which the best results are found [14]. Figure 5.12 shows the effect of different powder concentration indicating that the thickness of the recast layer increased with higher powder concentration.

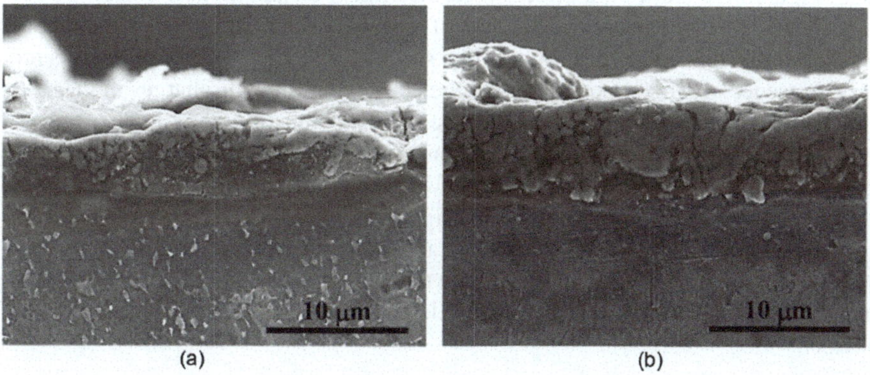

Fig. 5.12 Cross-sectional SEM micrographs of titanium surface after machining with EDM by titanium powder mixed dielectric for **a** 3 g/l, **b** 6 g/l [14]

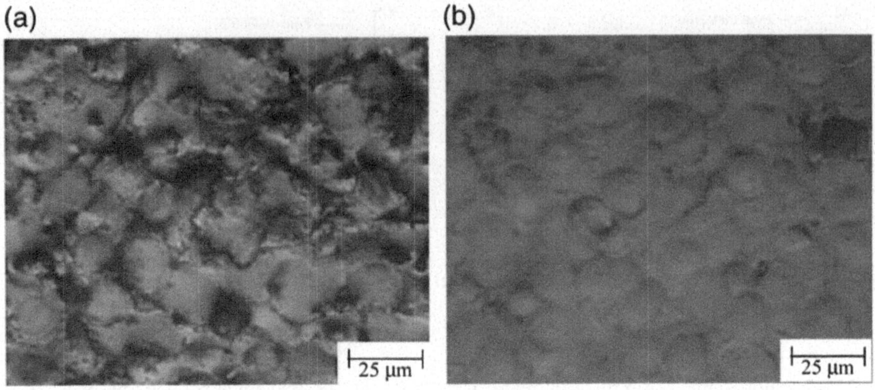

Fig. 5.13 Surface texture from EDM of steel **a** without silicon powder, **b** with silicon powder [30]

5.6.4 Silicon and Silicon Compounds

Silicon powder can be used to increase the machining efficiency in electrical discharge machining process. Researchers have primarily used silicon powder during the machining of various steel alloys. In this abrasive powder mixed electrodischarge machining (PMEDM) process, peak current, pulse-on time, concentration of silicon powder determines the outcome [28]. Peak current and concentration of silicon powder is the most influential factor in the outcome of the process [29]. Addition of silicon powder improves the material removal rate of EDM process [19]. Presence of silicon powder significantly reduces the surface roughness of machined steel surface. Figure 5.13 shows the effect on surface finish for addition of silicon powder in dielectric of EDM process. Presence of silicon powder also reduces the amount of undesirable discharge [30]. Mirror-like surface finish can be achieved in this process. Presence of silicon powder in the dielectric allows the discharges in the spark gap to create a glossy mirror-like surface. The amount of craters is also reduced [12]. Concentration of silicon powder plays a vital role in the surface roughness and achieving a mirror-like finish. The powder concentration has to be optimized to achieve the best-possible surface finish. Figure 5.14 depicts the fact that there exists a right amount of powder concentration, above or below which the surface roughness is not optimized. It is found that with the increase of powder concentration, the surface finish improves first, then again deteriorates at very high concentration of powder.

5.6.5 Chromium

Chromium powder is another element that can be used to improve the machining efficiency of electrical discharge machining process. Addition of chromium powder

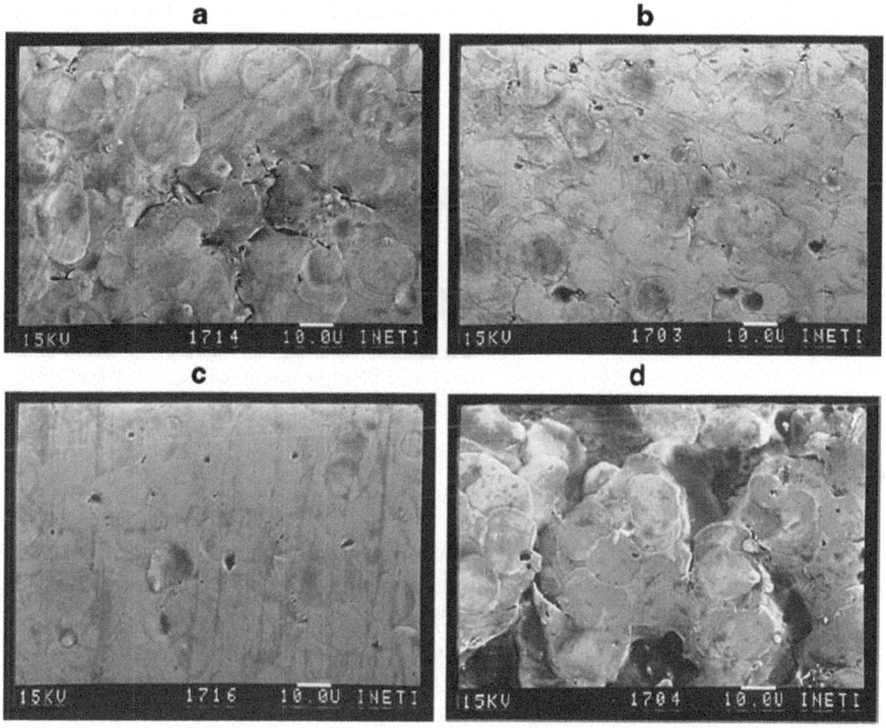

Fig. 5.14 Surface topography of machined surface by silicon powder mixed EDM **a** 0 g/l, **b** 1 g/l, **c** 2 g/l, **d** 20 g/l [23]

induces a large spark gap. Chromium powder allows more spark gap than silicon carbide or copper powders. Addition of chromium powder can significantly improve the material removal rate of the process. In terms of productivity and MRR, chromium powder performs better than aluminum or silicon carbide powders [22]. Chromium powder addition can also reduce the thickness of recast layer [21]. Proper optimization needs to be carried out to get the optimum combination of material removal rate and tool wear ratio [31].

Addition of chromium powder generates a layer of chromium on top of the machined surface. This phenomenon reduces the surface roughness. Deposition of chromium powder increases with the concentration of chromium in the dielectric and decreases with the increment of peak current. The thickness of this chromium layer increases with the increment of chromium powder concentration and discharge current. Addition of chromium powder not only improve the material removal rate and surface finish, but also enhances micro-hardness and corrosion resistance of the machined surface. Water repellency also increases with addition of chromium powder to the machined surface [32]. Current and pulse-on time are significant parameters in the process [33]. Figure 5.15 compares the Vickers hardness of machined surface for various dielectric configurations.

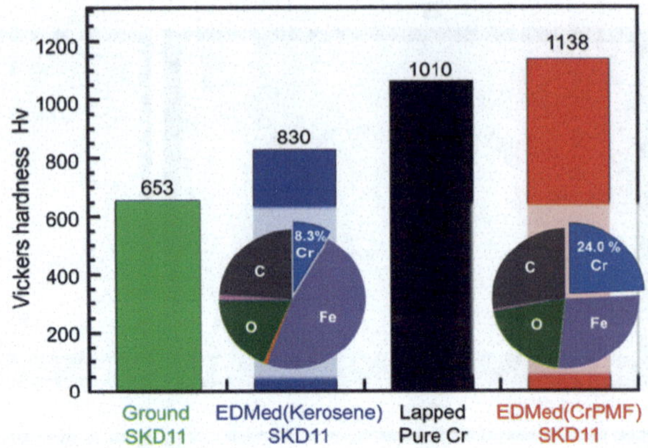

Fig. 5.15 Vickers hardness for various dielectric configurations [32]

5.6.6 Graphene

Graphene and graphene oxide (GO) has been used as suspended powder for dielectric during electrical discharge machining. Titanium and its alloys are one of the most difficult materials to machine, even by EDM process. This is because of their low thermal conductivity and high chemical reactivity. So machining high-quality holes in titanium alloys is a challenge. Suspending graphene oxide in the dielectric improves the machinability of titanium alloys. The machined surface generated in this process contains less amount of thermal cracks than the surface produced by conventional process. So this powder suspended process results in reduced surface roughness than conventional process [15]. Figure 5.16 compares the machined surface in conventional and GO suspended EDM process. Figure 5.16 shows the reduction of cracks in machined surface by this process. Addition of GO particles also improves the machining process of Inconel 825 alloy [20].

Electrical discharge machining is an effective process for machining any workpiece irrespective of its hardness as long as it is electrically conductive. Nonconductive materials are usually not machinable in this process. Graphene has opened up possibilities for important non-conductive materials that can be machined by EDM process. Ceramic materials have applications in semiconductor, biomedical, transportation, and aerospace industries. These materials are also brittle and hard. So conventional machining process cannot be applied to ceramic materials. Addition of graphene to the dielectric can increase the electrical conductivity of the system and raise the value upwards of the threshold conductivity that exists for EDM process. So addition of graphene powder to the dielectric can allow engineering ceramic materials to be machined by electrical discharge machining process.

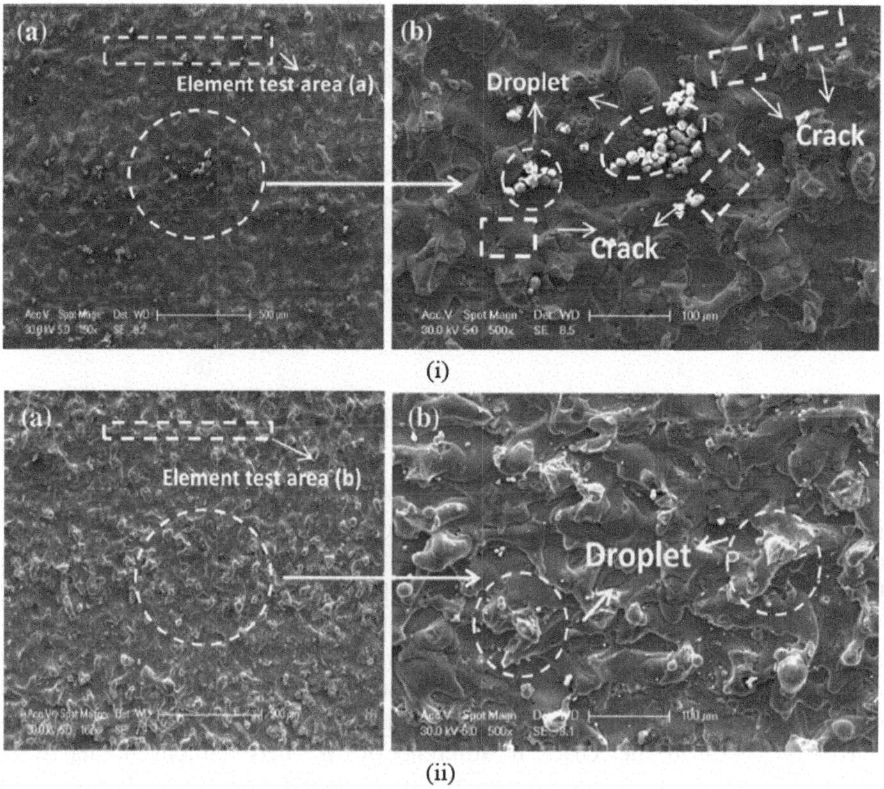

Fig. 5.16 SEM images of machined surface of **i** conventional EDM, and **ii** with graphene oxide suspended dielectric in EDM **a** 150x magnification, **b** 500x magnification [15]

5.6.7 Carbon Nanotube

Using carbon nanotube (CNT) as a powder in the dielectric has become a recent trend in the abrasive assisted EDM. Researchers have used CNT during machining EN19 alloy steel. Addition of CNT to the dielectric results in the increment of material removal rate and the decrement of surface roughness. CNT performs better than aluminum, graphite or titanium as powder while machining EN19 alloy steel in terms of both material removal rate and surface roughness [13]. In terms of tool wear ratio and tool life, CNT does not perform as well as aluminum powder. Peak current is the key factor that determines the output of material removal rate [34]. Figure 5.17a depicts the fact that CNT results in better material removal rate compared to aluminum, graphite, and titanium powder. Figure 5.17b shows that addition of CNT allows better surface finish compared to aluminum, graphite, and titanium powder.

Multi-walled carbon nanotube (MWCNT) has also been deployed during the machining of hard titanium alloy, like Ti-6Al-4V. Adding MWCNT results in the

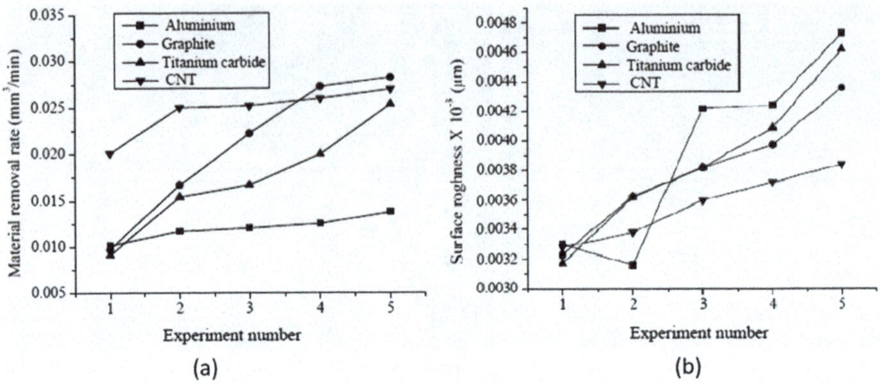

Fig. 5.17 Performance of CNT on **a** MRR and **b** surface roughness during electrical discharge machining of EN19 alloy steel [13]

improvement of machining stability. This happens due to the fact that the number of unnecessary sparks decreases in this system. The value of material removal rate is dependent on the pulse duration. For lower pulse duration case, the value of MRR is comparatively lower than conventional EDM process. But for increased pulse duration, addition of MWCNT results in better material removal rate. Figure 5.18a describes this relation of MRR with pulse duration for both conventional EDM and MWCNT-mixed EDM. The value of tool wear ratio (TWR) decreases after the addition of MWCNT in the dielectric. The difference of TWR between conventional EDM and MWCNT-mixed EDM is prominent for lower pulse duration cases. Figure 5.18b depicts this phenomenon. Adding MWCNT powder also decreases the length and size of surface microcracks on the machined surface [35], as seen in Fig. 5.19.

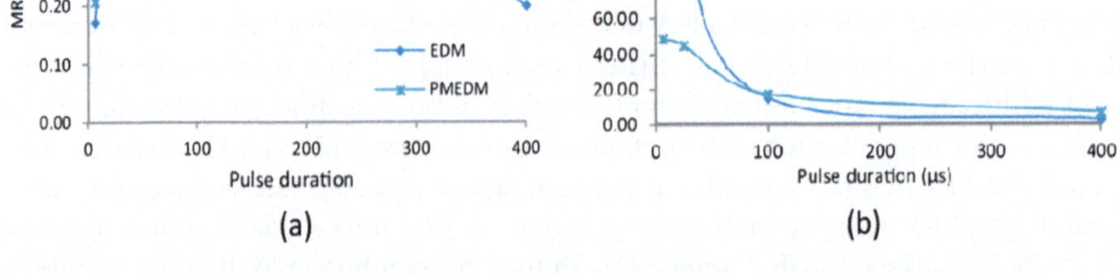

Fig. 5.18 Change of **a** MRR, and **b** TWR with pulse duration for conventional and MWCNT-mixed EDM of Ti-6Al-4V [35]

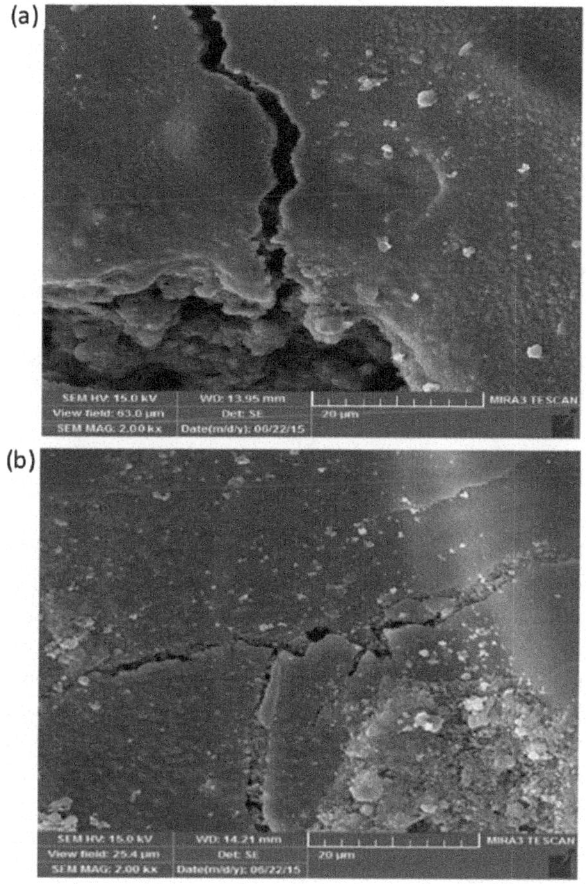

Fig. 5.19 Microcracks on the machined surface of Ti-6Al-4V alloy for **a** conventional EDM, **b** MWCNT-mixed EDM [35]

Researchers have incorporated MWCNT in the machining system of metal matrix composites (MMC). Addition of MWCNT has positive effects during the machining of Al-SiCp MMC. Adding MWCNT increases the material removal rate. The increment of MWCNT powder density in the dielectric also increases the material removal rate. Figure 5.20a shows the relation of material removal rate with MWCNT concentration in the dielectric. Addition of MWCNT to the system can also decrease the surface roughness [36]. The powder concentration has to be optimized to attain minimum surface roughness. Figure 5.20b depicts the relation between surface roughness and MWCNT concentration in the dielectric.

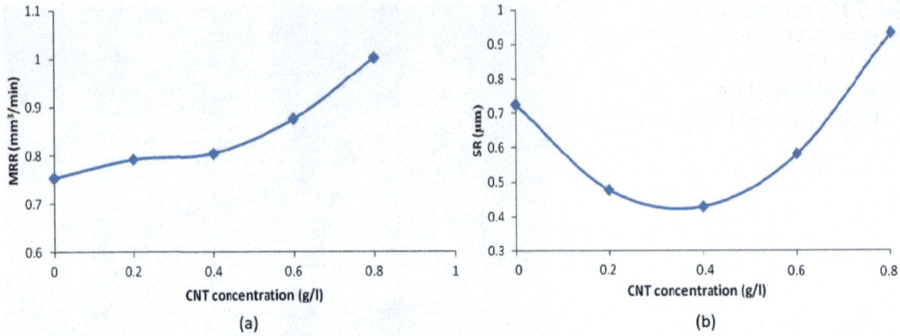

Fig. 5.20 Relation of **a** MRR, and **b** surface roughness with CNT concentration in dielectric for machining Al-SiCp alloy [36]

5.6.8 Molybdenum Disulfide

Molybdenum disulfide (MoS_2) is another material that can be used as a powder in powder mixed electrical discharge machining process. MoS_2 powder can lead to mirror-like finishing for steel alloys. Addition of MoS_2 powder reduces machining time and results in uniform dispersion of electrical sparks. So the machining becomes stable and efficient. This process can be applied on various tool steels, like SKH-54. Although the final machined surface is very smooth, it contains overlapping and well-formed small craters. The craters formed in this process are shallower and allow a reflective surface [12]. Figure 5.21a shows the machined surface of SKH-54 tool steel in conventional EDM process. Figure 5.21b shows the surface of machined SKH-54 steel with the help of MoS_2 powder mixed EDM process. The improvement in surface finish in MoS_2 powder mixed EDM is observed in Fig. 5.21b.

Researchers have also used MoS_2 powder in vibration-assisted electrical discharge machining process. Suspending micro MoS_2 powder in micro-EDM process seems

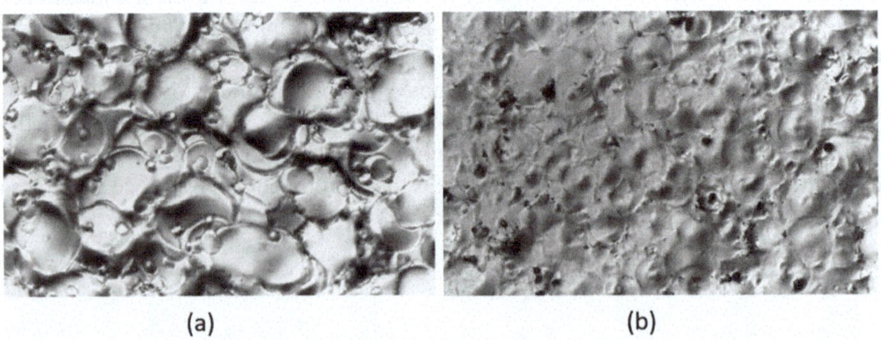

Fig. 5.21 Machined surface of SKH-54 by **a** conventional EDM, and **b** MoS_2 powder mixed EDM process [12]

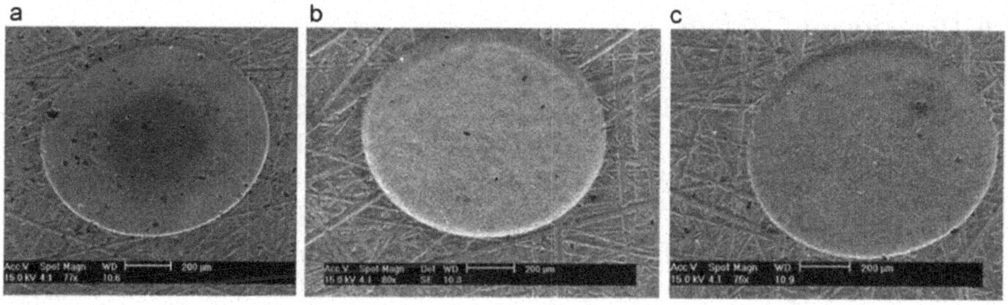

Fig. 5.22 Microstructure of Cu-W surface machined by micro-EDM process using **a** pure dielectric, **b** MoS_2 powder of concentration 2 g/l, **c** MoS_2 powder of concentration 5 g/l [37]

to bring positive effects. Vibration assistance improves the material removal rate compared to conventional EDM process. Adding MoS_2 powder to the dielectric further improves the material removal rate. The MRR also increases with the increment of MoS_2 powder density in the dielectric [37].

Other than improving the material removal rate, this modified vibration assisted MoS_2 powder mixed EDM process significantly improves the surface quality. Addition of MoS_2 powder creates a smooth path for discharge current over the surface, which gives rise to a flat surface. The surface is also free of black spots in the center [37]. Figure 5.22 compares the machined surface generated from conventional EDM and MoS_2 powder mixed EDM.

5.6.9 Copper

Copper is known for its good electrical and thermal conductivity. Addition of high conductivity copper powder in the dielectric improves the machinability of superalloys. Copper addition brings good results during the machining of Inconel 718 super alloy. Although this process does not improve the machined surface by drastic levels, the amount of surface irregularities seem to decrease. Mechanical properties of the machined surface are improved due to the formation of carbide, which is generated from disassociation of the dielectric oil. As addition of conductive copper powder allows for greater propagation of energy, the carbide formation on machined surface is enhanced. So the improvement of machined surface hardness is observed. Increasing peak current results in increases material removal rate in this case. Increase in pulse-on duration has quite the opposite effect, as the machining rate is reduced in this case. Increased pulse-on time also results in better surface quality [38].

5.6.10 Boron Carbide

Boron carbide abrasive powder can be used in abrasive-assisted EDM. Researchers have used boron carbide powder (B_4C) for machining hard titanium alloys, like Ti-6Al-4V. Boron carbide has been suspended in kerosene or deionized water and used as dielectric during the electrical discharge machining process. Adding B_4C powder to deionized powder increases the material removal rate. Figure 5.23a demonstrates the fact that using B_4C powder with deionized powder as dielectric improves material removal rate during EDM of Ti-6Al-4V. Addition of B_4C powder with kerosene dielectric results in decreased tool wear ratio [16]. Figure 5.23b shows that mixing B_4C powder with kerosene generates lower tool wear ratio.

This increment of material removal rate and decrement of tool wear ratio comes at a cost. Adding B_4C to the dielectric reduces the machining accuracy of micro-holes in Ti-6Al-4V. A non-uniformity arises in the machined holes in the form of circularity and taper as a result of adding B_4C to the dielectric [17]. Figure 5.24

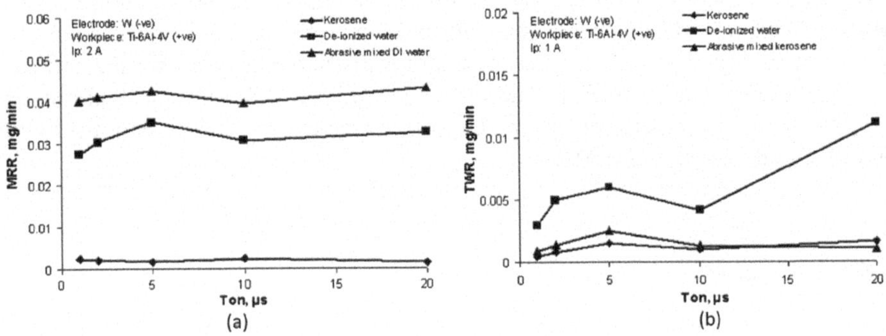

Fig. 5.23 **a** Material removal rate (MRR), and **b** tool wear ratio (TWR) with pulse-on time for various dielectric for Ti-6Al-4V alloy [16]

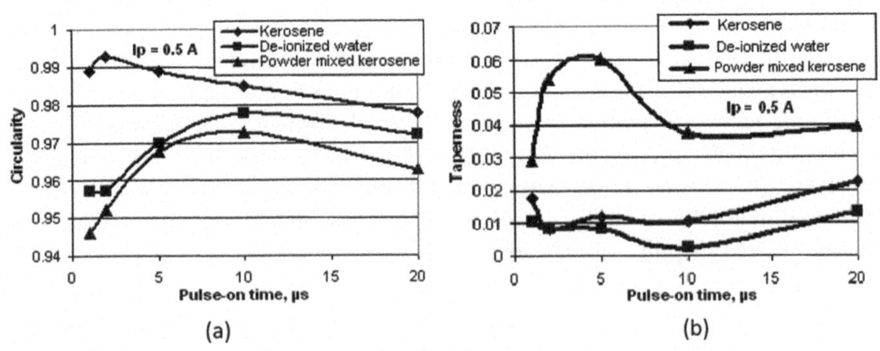

Fig. 5.24 **a** Circularity, and **b** taperness in machined micro-holes of Ti-6Al-4V alloy for different dielectric [17]

shows the decrement of circularity in micro-holes for B_4C powder mixed dielectric compared to pure kerosene or deionized water. Figure 5.24 shows the increment of taperness in micro-holes for B_4C powder mixed dielectric compared to pure kerosene or deionized water.

5.6.11 Tungsten

Tungsten powder is one of the hardest abrasive particles used in powder mixed electrical discharge machining process for the purpose of surface modification. Addition of tungsten powder can modify the surface finish and increase surface micro-hardness of the machined surface. Using a brass electrode and adding tungsten to the dielectric improves the surface finish. Using a copper-tungsten electrode and adding tungsten to the dielectric improves the micro-hardness of the machined surface. This phenomenon is observed because of the transfer of materials from electrode and powder to the machined surface during electrical discharge machining [39]. Significant amount of material transfer occurs from powder mixed dielectric to the machined surface. Suspended tungsten powder reacts with carbon in the dielectric and modifies the formation of plasma channel. These factors contribute to changing the surface properties. Peak current is the most important factor in controlling this phenomenon. Low discharge current and shorter pulse-on time attribute to the increment of surface micro-hardness [40]. Figure 5.25a shows the surface of un-machined D2 steel. Figure 5.25b shows the machined surface of D2 steel, machined by tungsten powder mixed EDM process.

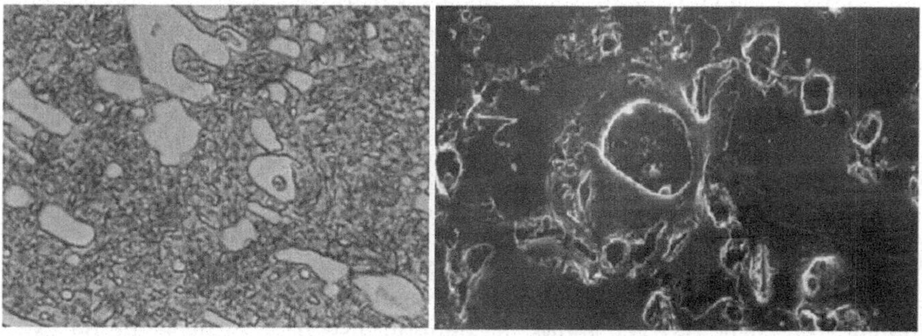

(a) (b)

Fig. 5.25 SEM micrograph of **a** un-machined, and **b** machined D2 die steel after tungsten powder mixed EDM [40]

5.7 Abrasive Assisted Machining of Hard Materials

5.7.1 Inconel Alloy

Inconel is a very useful alloy because of its resistance to oxidation and corrosion. This alloy can maintain its strength even at elevated temperatures. This alloy has been used for high-temperature applications in high-speed planes, supersonic jets, nuclear power plants, rockets. But these alloys are very difficult to machine because of their properties. Abrasive powder mixed EDM can be a viable solution to machine this material effectively.

Copper powder has been used to shape Inconel 718 super alloys. Copper powder does not enhance the surface quality. The surface has poke marks, cracks, and globules of debris. There is also a formation of white layer. Migration of carbon from the dielectric to the machined surface has been observed. So carbide is formed on the machined surface. Although this process does not allow phase transformation, layers of oxides can form. The material removal rate can be enhanced in this process by increasing the peak current. The pulse-on time has an inverse relation to material removal rate. The machining is efficient for low pulse-on time and high peak current, as it generates an optimized flushing condition [38].

Aluminum oxide powder can be used to enhance the machining performance of Inconel 825 alloy. Higher peak current and higher pulse-on time result in improvement of material removal rate. The increment of number of sparks leads to this phenomenon. The bridging effect comes into play here. The surface roughness increases with pulse-on time. Best surface finish is found when the spark gap is increased to create better flushing conditions. Surface roughness also decreases with increasing gap voltage. Peak current, pulse-on time, and gap voltage have to be optimized to maintain minimum surface roughness. For optimized case, the size of micro-cracks and micro-holes are smaller and the material deposition is less irregular [5].

5.7.2 Metal Matrix Composite

Multi-walled carbon nanotube (MWCNT) can be added to the dielectric to increase the machinability of Al-SiC_p metal matrix composite. Small specific gravity and straight pain shape of CNT allows the uniform dispersion of dielectric oil. As a result, the material removal rate is improved and the surface roughness is reduced. Increment of concentration of CNT powder in the dielectric leads to the increment of material removal rate of Al-SiC_p MMC. Higher thermal conductivity of CNT allows greater amount of effective current transfer to the workpiece. After a certain concentration value, the machining becomes unstable. Powder concentration also determines the surface quality. Surface roughness initially decreases with increasing CNT concentration in the dielectric. More uniform dispersion of energy allows decreased crater size and improved surface finish. Figure 5.26 shows the improve-

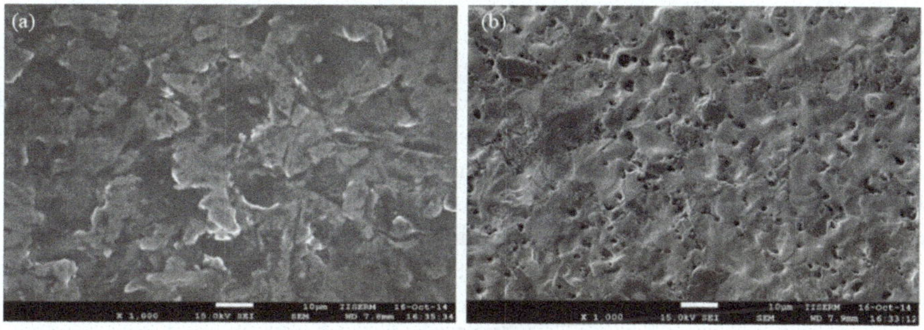

Fig. 5.26 SEM image of machined surface of Al-SiC$_p$ metal matrix composite for **a** conventional EDM, **b** MWCNT powder mixed EDM [36]

ment of surface roughness after addition of MWCNT to the dielectric. It can be seen that the crater sizes become smaller in CNT mixed EDM compared to that of conventional EDM. Just like the material removal rate, after a certain value of concentration, the surface roughness tends to increase [36].

Aluminum powder can also be used to enhance the machining of Al-SiC$_p$ metal matrix composite. The surface roughness can be reduced greatly. The corrosion resistance and wear resistance can also be improved in this process. Carbon transfer occurs from the dielectric to the machined surface. Aluminum powder allows a larger discharge channel which leads to the easier breakdown of the dielectric. This allowance of material transfer improves the surface hardness of Al-SiC$_p$ metal matrix composite. This surface modification also leads to higher corrosion resistance. The wear resistance also increases. So the Al-SiC$_p$ metal matrix composite becomes more suitable for application for rolling friction [24]. Figure 5.27 shows the decrease in wear mass for PMEDEM surface indicating the wear resistance after powder mixed EDM.

Fig. 5.27 Decrement of wear mass for Al-SiC$_p$ metal matrix composite machined aluminum powder added EDM process [24]

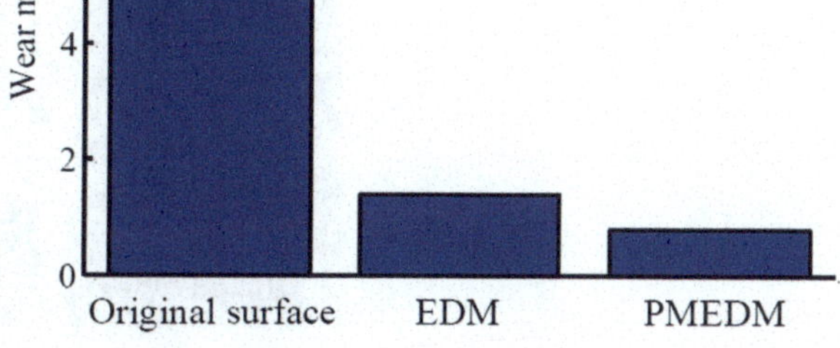

5.7.3 Titanium Alloy

Titanium alloys are difficult to machine because of their high strength and low thermal conductivity. Carbon nanotubes can be utilized as powder in the dielectric to improve the thermal and electrical conductivity of the system. This modified system is capable of reducing the surface roughness and distribute the spark energy. This PMEDM process allows longer pulse-duration without reducing the material removal rate. CNT particles reduce the breakdown voltage of dielectric. A larger dispersion of sparks can be observed by the addition of CNT. For longer pulse-on time, tool-wear ratio is lower for conventional EDM process than CNT powder mixed EDM process. Higher amount of heat generation leads to greater tool wear in this PMEDM process. The uniformity of spark generated by adding CNT particles also allows a better surface finish [35]. Figure 5.28 shows the improvement of surface finish by adding CNT to the dielectric. Achieved surface finish is better than conventional EDM process. The dimensions of micro-cracks reduce in this modified process. Addition of MWCNT increases the machining stability of Ti-6Al-4V alloy at lower settings with low pulse-on time. Addition of boron carbide powder in the machining

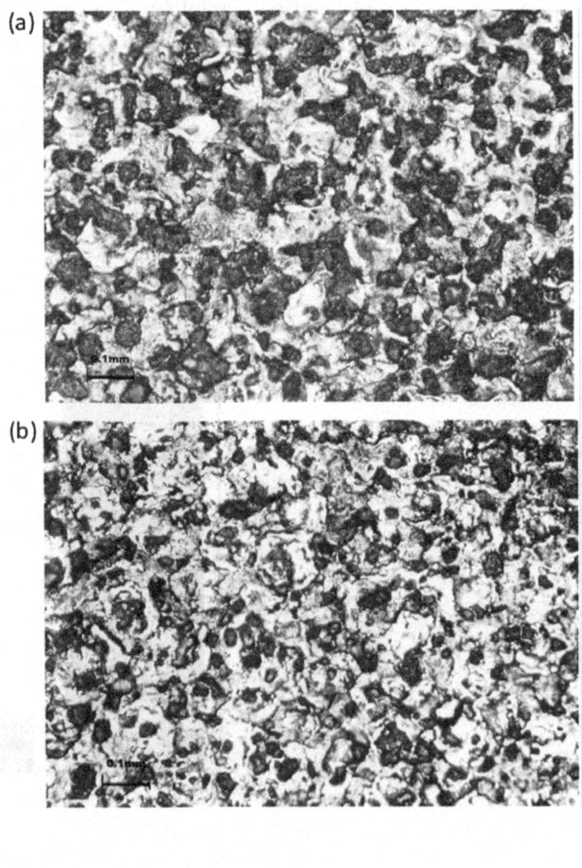

Fig. 5.28 Machined surface of Ti-6Al-4V alloy generated by **a** conventional EDM, **b** CNT powder mixed EDM [35]

of titanium alloy affects the machining accuracy [17]. Using deionized water instead of kerosene or hydrocarbon oil as the dielectric improves the material removal rate [16].

5.7.4 Carbides

Carbides are one of the most difficult-to-machine materials. Graphite nanopowder can be added to the dielectric to improve the machinability of cemented tungsten carbide (WC-Co). The addition of semi-conductive graphite powder improves the material removal rate and reduces the electrode wear ratio and surface roughness. The crater distribution also reaches a favorable condition. This occurs due to the increment of spark gap and uniform distribution of discharges. The material removal rate is improved due to the decrement of dielectric breakdown strength in powder-mixed EDM process. Powder concentration plays an important role in determining the outcome of this process. There exists an optimum powder concentration for which the best machining output is generated. Figure 5.29 shows that a lesser value of powder concentration can be more effective in terms of surface roughness while machining tungsten carbide by abrasive powder mixed EDM process [18].

Aluminum powder can also effectively lead to more uniform dispersion of energy in the machining process of tungsten-carbide by EDM. The material removal rate also increases as a result. The electrode wear ratio is also reduced in this process. An optimized powder concentration is present for which material removal rate is maximum or electrode wear ratio is minimum. The optimized powder concentration for MRR and EWR is not the same. So a compromise has to be reached to optimize the overall process [41]. Adding carbon nanofiber improves the machining of silicon carbide in terms of material removal rate and tool wear ratio [42].

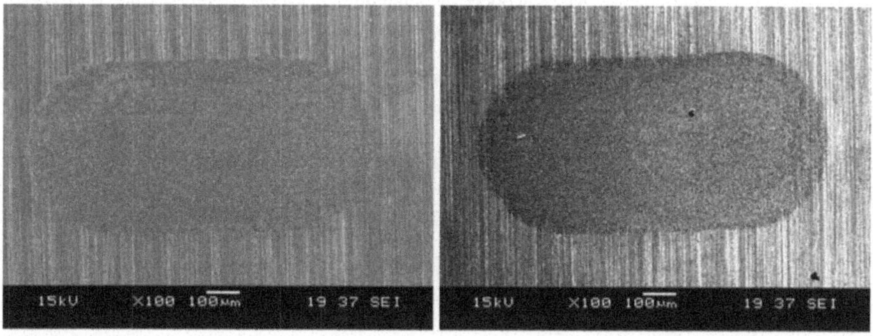

Fig. 5.29 Machined surface of tungsten carbide in powder mixed micro-EDM for **a** 0.2 g/l graphite concentration, **b** 2 g/l graphite concentration [18]

5.8 Summary

Abrasive assisted electrical discharge machining has opened up doors of possibilities for machining various difficult-to-cut metals and alloys efficiently. Performance parameters of machining can be positively influenced by electrical, thermal, and physical properties of the abrasive powder particles in this process. Apart from increased productivity, this process also allows for better surface finish. In some cases, surface modification can lead to impressive surface properties as well. But questions can be drawn about the feasibility of the process for newer applications. Measurements also need to be carried out to make this process economically and environmentally feasible. The powder-mixed process also affects the lifetime of the machine. Research investigations have been carried out to figure out the best procedure to add abrasives to the dielectric. Proper modifications are necessary to supply the abrasives in the machining zone in an effective manner. Maintaining a constant powder concentration throughout the system is a challenge that has limited solutions. Selection of appropriate powder is of major significance to the outcome of the overall process. Filtration process has to be modified be keeping in mind the environmental implications of this particular process.

References

1. Jahan MP, Md Ali Asad AB, Rahman M, Wong YS, Masaki T (2011) Micro-electro discharge machining (μEDM). In: Koç M, Özel T (eds) Micro-manufacturing. Wiley, Hoboken, NJ, USA, pp 301–346
2. Rajurkar KP et al (2006) Micro and nano machining by electro-physical and chemical processes. CIRP Ann 55(2):643–666
3. Kunieda M, Lauwers B, Rajurkar KP, Schumacher BM (2005) Advancing EDM through fundamental insight into the process. CIRP Ann 54(2):64–87
4. Selvarajan L, Manohar M, Amos Robert Jayachandran J, Mouri P, Selvakumar P (2018) A review on less tool wear rate and improving surface quality of conductive ceramic composites by spark EDM. Mater Today Proc 5(2):5774–5782
5. Kumar V, Kumar A, Kumar S, Singh NK (2018) Comparative study of powder mixed EDM and conventional EDM using response surface methodology. Mater Today Proc 5(9), Part 3:18089–18094
6. Kim Y-S, Chu C-N (2018) The effects of graphite powder on tool wear in micro electrical discharge machining. Procedia CIRP 68:553–558
7. Jeswani ML (1981) Effect of the addition of graphite powder to kerosene used as the dielectric fluid in electrical discharge machining. Wear 70(2):133–139
8. Marashi H, Sarhan AAD, Hamdi M (2015) Employing Ti nano-powder dielectric to enhance surface characteristics in electrical discharge machining of AISI D2 steel. Appl Surf Sci 357:892–907
9. Jyoti Saharia N, Sree Lakshmi T, Surekha B, Jena H (2018) Experimental investigations on the effect of hybrid aluminum and graphite powders mixed dielectric in EDM. Mater Today Proc 5(9), Part 3:20443–20448
10. Kumar S, Kumar A, Kumar V, Singh NK (2018) Study of machining of Inconel 825 super alloy using powder mixed EDM process. Mater Today Proc 5(9), Part 3:18129–18134

11. Wu KL, Yan BH, Huang FY, Chen SC (2005) Improvement of surface finish on SKD steel using electro-discharge machining with aluminum and surfactant added dielectric. Int J Mach Tools Manuf 45(10):1195–1201
12. Wong YS, Lim LC, Rahuman I, Tee WM (1998) Near-mirror-finish phenomenon in EDM using powder-mixed dielectric. J Mater Process Technol 79(1):30–40
13. Mondal G, Surekha B, Dev Choudhury S (2018) Investigation on the influence of different powder mixed dielectric in electric discharge machining. Mater Today Proc 5(9), Part 3:18281–18286
14. Chen S-L, Lin M-H, Huang G-X, Wang C-C (2014) Research of the recast layer on implant surface modified by micro-current electrical discharge machining using deionized water mixed with titanium powder as dielectric solvent. Appl Surf Sci 311:47–53
15. Yi S, Li G, Ding S, Mo J (2017) Performance and mechanisms of graphene oxide suspended cutting fluid in the drilling of titanium alloy Ti-6Al-4V. J Manuf Process 29:182–193
16. Kibria G, Sarkar BR, Pradhan BB, Bhattacharyya B (2010) Comparative study of different dielectrics for micro-EDM performance during microhole machining of Ti-6Al-4V alloy. Int J Adv Manuf Technol 48(5):557–570
17. Kibria G, Bhattacharyya B (2011) Analysis on geometrical accuracy of microhole during micro-EDM of Ti-6Al-4V using different dielectrics. AIP Conf Proc 1315(1):155–160
18. Jahan MP, Rahman M, Wong YS (2011) Study on the nano-powder-mixed sinking and milling micro-EDM of WC-Co. Int J Adv Manuf Technol 53(1):167–180
19. Kansal HK, Singh S, Kumar P (2005) Parametric optimization of powder mixed electrical discharge machining by response surface methodology. J Mater Process Technol 169(3):427–436
20. Kumar A, Kumar S, Mandal A, Rai Dixit A (2018) Investigation of powder mixed EDM process parameters for machining Inconel alloy using response surface methodology. Mater Today Proc 5(2), Part 1:6183–6188
21. Yih-fong T, Fu-chen C (2005) Investigation into some surface characteristics of electrical discharge machined SKD-11 using powder-suspension dielectric oil. J Mater Process Technol 170(1):385–391
22. Tzeng Y-F, Lee C-Y (2001) Effects of powder characteristics on electrodischarge machining efficiency. Int J Adv Manuf Technol 17(8):586–592
23. Peças P, Henriques E (2008) Effect of the powder concentration and dielectric flow in the surface morphology in electrical discharge machining with powder-mixed dielectric (PMD-EDM). Int J Adv Manuf Technol 37(11):1120–1132
24. Hu FQ et al (2013) Surface properties of SiCp/Al composite by powder-mixed EDM. Procedia CIRP 6:101–106
25. Mohanty G, Mondal G, Surekha B, Tripathy S (2018) Experimental investigations on graphite mixed electric discharge machining of EN-19 alloy steel. Mater Today Proc 5(9), Part 3:19418–19423
26. Surekha B, Gangadhara Rao P, Bijetha B, Srinivasa Sai V (2018) Surface characteristics of EN19 steel materials by EDM using graphite mixed dielectric medium. Mater Today Proc 5(9), Part 3:17895–17900
27. Singh AK, Kumar S, Singh VP (2015) Effect of the addition of conductive powder in dielectric on the surface properties of superalloy Super Co 605 by EDM process. Int J Adv Manuf Technol 77(1):99–106
28. Bai X, Zhang Q-H, Yang T-Y, Zhang J-H (2013) Research on material removal rate of powder mixed near dry electrical discharge machining. Int J Adv Manuf Technol 68(5–8):1757–1766
29. Kansal HK, Singh S, Kumar P (2007) Effect of silicon powder mixed EDM on machining rate of AISI D2 die steel. J Manuf Process 9(1):13–22
30. Peças P, Henriques E (2003) Influence of silicon powder-mixed dielectric on conventional electrical discharge machining. Int J Mach Tools Manuf 43(14):1465–1471
31. Ojha K, Garg RK, Singh KK (2011) Parametric optimization of PMEDM process using chromium powder mixed dielectric and triangular shape electrodes. J Miner Mater Charact Eng 10(11):1087–1102

32. Toshimitsu R, Okada A, Kitada R, Okamoto Y (2016) Improvement in surface characteristics by EDM with chromium powder mixed fluid. Procedia CIRP 42:231–235
33. Dubey V, Singh V (2018) Study of material removal rate in powder mixed EDM of AA7075/B_4C composite. Mater Today Proc 5(2), Part 2:7466–7475
34. Choudhury SD, Saharia NJ, Surekha B, Mondal G (2018) Study on the influence of hybridized powder mixed dielectric in electric discharge machining of alloy steels. Mater Today Proc 5(9), Part 3:18410–18415
35. Shabgard M, Khosrozadeh B (2017) Investigation of carbon nanotube added dielectric on the surface characteristics and machining performance of Ti-6Al-4V alloy in EDM process. J Manuf Process 25:212–219
36. Mohal S, Kumar H (2017) Study on the multiwalled carbon nano tube mixed EDM of Al-SiCp metal matrix composite. Mater Today Proc 4(2), Part A:3987–3993
37. Prihandana GS, Mahardika M, Hamdi M, Wong YS, Mitsui K (2009) Effect of micro-powder suspension and ultrasonic vibration of dielectric fluid in micro-EDM processes—Taguchi approach. Int J Mach Tools Manuf 49(12–13):1035–1041
38. Paul BK, Sahu SK, Jadam T, Datta S, Dhupal D, Mahapatra SS (2018) Effects of addition of copper powder in the dielectric media (EDM Oil) on electro-discharge machining performance of Inconel 718 super alloys. Mater Today Proc 5(9), Part 3:17618–17626
39. Bhattacharya A, Batish A, Kumar N (2013) Surface characterization and material migration during surface modification of die steels with silicon, graphite and tungsten powder in EDM process. J Mech Sci Technol 27(1):133–140
40. Kumar S, Batra U (2012) Surface modification of die steel materials by EDM method using tungsten powder-mixed dielectric. J Manuf Process 14(1):35–40
41. Kung K-Y, Horng J-T, Chiang K-T (2009) Material removal rate and electrode wear ratio study on the powder mixed electrical discharge machining of cobalt-bonded tungsten carbide. Int J Adv Manuf Technol 40(1–2):95–104
42. Liew PJ, Yan J, Kuriyagawa T (2013) Carbon nanofiber assisted micro electro discharge machining of reaction-bonded silicon carbide. J Mater Process Technol 213(7):1076–1087

Chapter 6
ELID Grinding for Final Finishing Operation

Tanveer Saleh and Rubina Bahar

Abstract Electrolytic In-Process Grinding (ELID) is an advanced grinding process that carries out in situ dressing of the grinding wheel using an electrochemical reaction, i.e. anodic dissolution. This process is targeted to achieve the final finished product by eliminating subsequent finishing processes such as polishing and lapping. The process mainly aims for hard and brittle materials to achieve ductile mode cutting with the best cutting parameters, and by doing so to achieve the nano-level surface finish. The typical grinding particle size of the ELID grinding wheel varies from 1 to 5 μm. A characteristic ELID wheel is metal bonded which is used as an anode. There is another electrode which usually used as the cathode, and a liquid is flown between the anode and the cathode, which functions both as grinding fluid and electrolyte. A DC pulsed power supply is generally used as the power source for the electrolytic in-process dressing of the grinding wheel. This chapter will shed light on the history, fundamental and the current status of the ELID grinding related researches.

Keywords ELID grinding · Electrolytic In-Process Dressing · Ductile mode · Surface roughness

6.1 Introduction

Grinding is a conventional machining method that is normally applied as a pre-finishing process. In grinding, coarse or fine abrasive particles are used which are harder than the workpiece material for the cutting. In this machining method, the

T. Saleh (✉)
Department of Mechatronics Engineering, International Islamic University Malaysia, Kuala Lumpur, Malaysia
e-mail: tanveers@iium.edu.my

R. Bahar (✉)
Department of Mechanical and Materials Engineering, Lee Kong Chian Faculty of Engineering and Science, University Tunku Abdul Rahman (UTAR), Sungai Long, Malaysia
e-mail: rubina@utar.edu.my

S. Das et al. (eds.), *Advances in Abrasive Based Machining and Finishing Processes*, Materials Forming, Machining and Tribology,
https://doi.org/10.1007/978-3-030-43312-3_6

surface speed of the grinding/abrasive wheel is considerably faster than that of other subtractive processes, for example, turning, and milling operations. Rapid development in electronic and optical devices (such as integrated chips, micro electromechanical system (MEMS), advanced sensors, ultra-smooth lenses) with the growing need of their performance, the necessity of ultra-smooth surface finish with greater form accuracy of the specialized glass, the silicon wafer is rising. However, typical grinding is unable to meet such requirement with the desired precision and accuracy. Henceforth, conventional grinding is still followed by final finishing operation such as polishing and lapping. But, these processes have their limitations, as described in [1, 2]. As such, ELID grinding with super abrasive wheel has shown enough credibility to eliminate lapping and polishing as the final finishing operation by achieving a nanometric surface finish with desirable form accuracy on hard and brittle materials.

6.1.1 Brief History

Back in 1985, Murata et al. [3] ELID for the first time to machine hard and brittle material like ceramics. Ceramics are more challenging to machine due to its excessive hardness and brittleness. Usually, when harder materials are mechanically machined, the cutting grits are worn out very fast and results in a poor grinding outcome. ELID gave the opportunity of continuous dressing. Hence, the problem of wheel loading was overcome. Later in the '90s, Ohmori [3] further investigated and improved ELID grinding by introducing super abrasive wheels with sub 5 μm grit (particle) size.

6.1.2 Basic Concept of ELID Grinding

The basic concept [4] of the ELID grinding is explained in Fig. 6.1. Usually, ELID grinding wheels contain extremely fine grits with a size of 1–5 μm which are firmly held by metal bond or metal-resin bond. As described in Fig. 6.1 a cathode electrode (usually Copper or Brass) is closely placed near the grinding wheel, and a grinding fluid cum electrolyte is forced to flow in the gap between the wheel and electrode. The other part of the wheel takes part in the actual grinding operation, as shown in Fig. 6.1. It is to note that a pulsed DC power supply is continuously applied between the grinding wheel and the cathode to carry out electrochemical dressing of the wheel (anodic dissolution). The details of the physics behind the ELID dressing is explained in the next paragraph.

Figure 6.2 describes how the wheel dressing takes place during the process of Electrolytic In-Process Dressing (ELID) [4]. In the very beginning, a new grinding wheel does not have any grit protuberance. All the grits are sintered inside the metal matrix, as shown in Fig. 6.2a. Therefore, the wheel needs to be pre-dressed without any grinding action for a period of 15–20 min. During the 1st phase of the dressing, the anodic dissolution of the metal bond takes place. The dressing current is high at

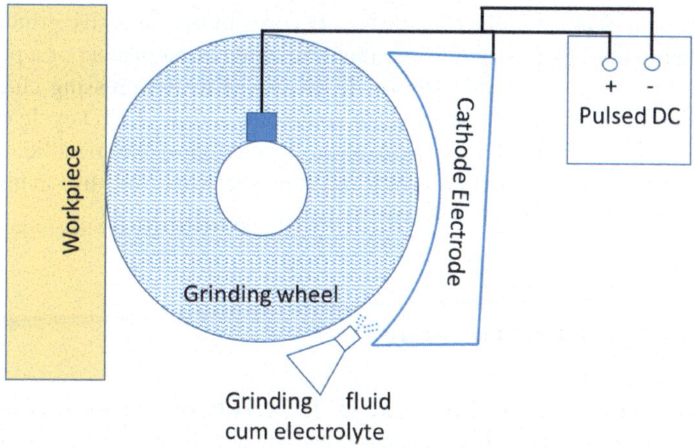

Fig. 6.1 Basic ELID grinding setup

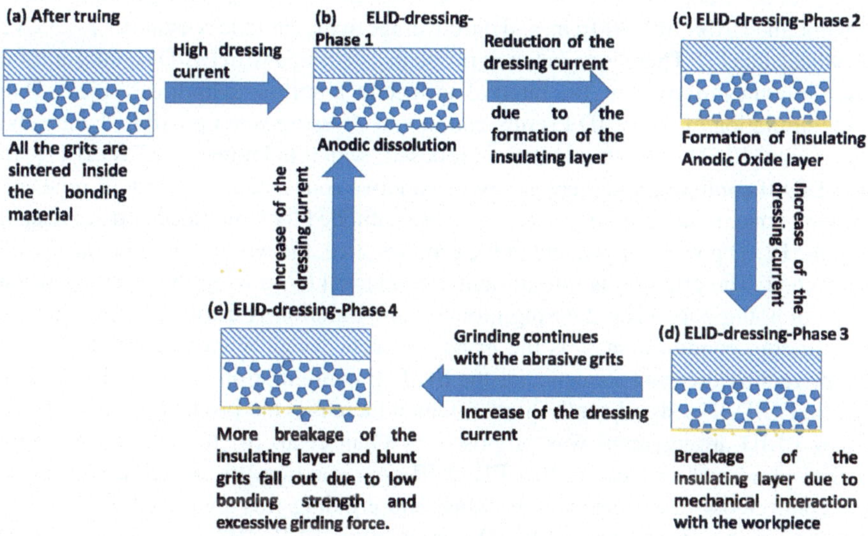

Fig. 6.2 Basic working principle of ELID dressing of the grinding wheel

this stage due to the absence of the metallic oxide. As time progresses, a metallic oxide layer is formed around the circumference of the grinding wheel, which retards excessive dressing current flow and saves the wheel from overdressing Fig. 6.2b. At this stage, the grits are protruded out and held partially by the soft and brittle metallic oxide. After the pre-dressing, the actual grinding starts as described in Fig. 6.1. At the same time dressing continuously on the other side of the wheel (Fig. 6.1). When the wheel comes in contact with the workpiece, the metal oxide layer partially breaks, and the sharp grits start to grind the workpiece (Fig. 6.2d). As the grinding progresses the grits become blunt and falls of the wheel because of the excessive grinding force and them being held by soft and brittle metallic oxide bond instead of a pure metal bond. This breakage of the metallic oxide gives rise to the dressing current, and anodic dissolution, hence new sharp grits come out, and the ELID cycle continues (Fig. 6.2e, b). It is to note that because of the formation of the metallic oxide, the ELID dressing has an internal control to prolong the wheel life by saving it from unneeded overdressing.

6.1.3 Types of ELID Grinding

There are several variants of ELID grinding. The very basic one is known as ELID I, which is explained in Fig. 6.1. In this case, the wheel is being dressed continuously in one side, and the grinding action is taken place on the other side of the wheel. In the case of ELID, I grinding can be carried out by the bottom [5] or the peripheral surface of the wheel [6]. The typical gap between the cathode and the grinding wheel is maintained to be 100–300 μm, and the coverage of the cathode is usually 17–25% of the wheel area [4]. There is an inherent problem associated with ELID I configuration when it comes to grinding the internal surface of a hollow cylindrical object. Due to the limitation of the ELID I setup, continuous dressing is not possible; hence, the intermittent ELID configuration was proposed, which is known as ELID II [7]. In the ELID II configuration, there are two distinctive zones, namely grinding zone and dressing zone. In the dressing zone, there is a cathode block mounted and connected with the ELID power supply, and in the grinding area, the workpiece to be machined is attached. The dressing is intermittently conducted by moving the grinding when to the dressing zone. The dressing moment and duration is decided by monitoring the grinding condition such as the grinding force. As an obvious consequence of the alternating dressing sequence of the ELID II configuration, the overall grinding time in ELID II is significantly higher, which increases the production time. Thus, a new ELID arrangement was proposed by Qian et al. [8] for internal grinding of the cylindrical surface. In this ELID III configuration, the metallic workpiece was connected as a cathode and dressing, and grinding was continued without any interruption in a similar way of ELID I preparation. ELID IIIA is a subset of ELID III grinding setup. In ELID IIIA, the power supply that was used was AC instead of DC. During the ELID III operation sometimes there will be cases where some unwanted spark may occur due to the breakdown of the dielectric barrier of the electrolytes which may cause localized damage on the workpiece and grinding wheel. By introducing the AC power source, the workpiece becomes anode in the half-cycle, and an anodic oxide layer is formed on both the workpiece and the grinding wheel. This layer act as a protective coating so that the no unwanted discharge may happen between the workpiece and the grinding wheel. In 2007, Ohmori et al. [9] another electrode less arrangement of ELID grinding, which is mostly used for fine grinding operation. In this case, the nozzle was attached with two plates charged positively and negatively, respectively, as shown in Fig. 6.3. As a result of this, the liquid that

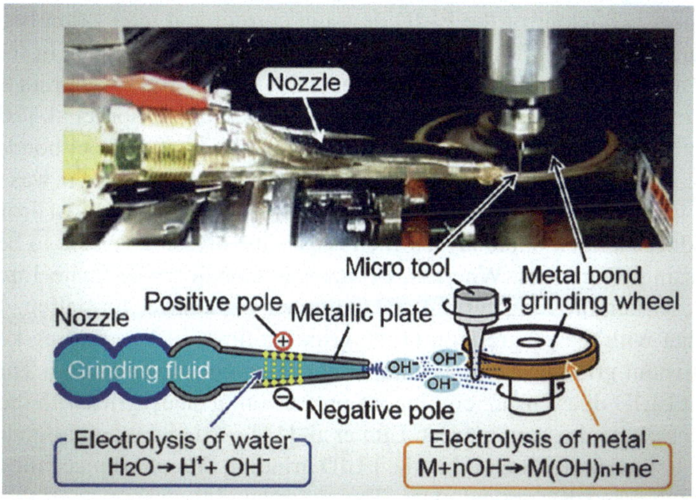

Fig. 6.3 Ion shot ELID grinding (ELID IV) [9]

was forced out of the nozzle was already charged creates an oxide layer on the ELID wheel. And the ELID cycle continues as described in Fig. 6.2.

In summary, we can state that ELID I, ELID II are ELID configuration with an extra cathode electrode, whereas others are electrodeless ELID arrangements. Al types of ELID setups use pulsed DC power source except for ELID IIIA. ELID II is the only variant of ELID grinding that uses intermittent dressing. Finally, except ELID IV, all other types of ELID use non-ionized electrolyte/grinding fluid.

6.2 Research on ELID Grinding

As mentioned earlier, ELID grinding was first introduced by Murata et al. [3] back in '85. However, it was later extensively researched by the chief scientist of Riken Japan Dr. Hitoshi Ohmori [5] and other researchers around the world. The research related to ELID grinding can primarily be divided into three categories, they are, research related to fundamental physics of ELID grinding, the study of ELID grinding on different types of materials and investigation on various instrumentation and control aspects of ELID grinding.

6.2.1 Fundamental Investigation on ELID Grinding

ELID grinding was thoroughly investigated by Ohmori et al. [5] to understand the physics behind the ELID mechanism. They were the first [5] to report and explain the non-linear behaviour of the ELID dressing current and interaction between the grinding and breakage of the oxide layer, as described in Fig. 6.2. In their work, Ohmori et al. [5] used a super abrasive wheel with sub-micron sized grain to achieve the average surface of the Si wafer as low as 8 nm. Later Ohmori et al. [10] studied ELID grinding more comprehensively to understand how the wheel bond material, grinding fluid, and power supply affect the dressing performance. It was observed that bond material had a profound role in the dressing current. Cast Iron bonded wheel produces a thick oxide layer which causes the dressing current to be reduced as the dressing progresses. Whereas, Bronze and Cobalt bonded wheel maintained relatively constant dressing current due to thinner ELID layer formation. They also reported that with the progression of the grinding time, ELID grinding keeps relatively consistent grinding force as compared to conventional grinding. This finding proves that ELID does ensure continuous protrusion of sharp grits with the application of electrolytic dressing current. Lim et al. [11] carried out an in-depth study of a various phenomenon observed in the ELID grinding. The workpiece they used for their research was BK7 optical glass. They correlated the variation of the grinding force with the initiation and end of the dressing (i.e. the breakage of the insulating layer). They also observed that higher dressing pulse frequency causes more dressing hence reduces the tool life. Conventional grinding without ELID dressing caused the grinding force to be increased because of the wheel loading occurrence. During ELID grinding higher feed speed was observed to be detrimental because of the slower dressing rate, which eventually caused a burning mark on the workpiece. Fathima et al. [12] also studied the wear mechanism of the grinding wheel during the ELID process. They defined a factor, which is a function of dressing current, voltage and duty ratio. This factor determines the mode of machining, whether it is the ductile mode or brittle mode. They also suggested for coarse grinding operation shorter Ton time is preferred as opposed to longer Ton time during the fine grinding. Fathima et al. [13] further continued the research in pioneering to model the exerted grinding force for ELID grinding. They showed [13] when the grinding force was simulated using cast iron as the bond material, the experimental finding was significantly lower than the simulated grinding force. Which demonstrates convincingly that the ELID layer plays a vital role in reducing the cutting force. However, experimental and simulated results were quite similar when they used micro/nanomechanical properties of ELID layer as the bond material which is the actual case. Chen [14, 15] carried out modelling for anodic dissolution during the ELID dressing process for both the two and three dimensional cases. Their observation suggests that the dressing rate is higher at the boundary of diamond and the metal bond. They also concluded from their study that for the similar wheels and same targeted dressing rate the dressing current necessary for a denser diamond concentration tool is a lesser amount of than that needed for a lower diamond concentration tool. Fathima et al. [16] further investigated the wear mechanism of the ELID wheel during the grinding of the Silicon wafer. They found that wheel wear is influenced by the depth of cut, wafer surface condition and the dressing condition. Zhuet al. [17] analyzed the electrolytic flow in the gap between the ELID wheel and the cathode electrode. They compared the flow analysis for rigid electrode and flexible foil electrode and concluded that foil electrode performs better in the case of low electrolytic flow rate as compared to the conventional rigid electrode.

6.2.2 Study on the Application of ELID Grinding

Use of ELID grinding is something that many researchers around the globe have investigated extensively. Most of the researchers tried ELID grinding on hard and brittle material to achieve ductile mode surface finish. The notable materials that were investigated with the application of ELID grinding are ceramics, Silicon (Si) wafer, metal, optical glasses etc.

Ohmori et al. [18] first tried ELID grinding to carry out finishing operation of structural ceramics. They managed to achieve sub 10 nm average surface roughness (Ra) using the super abrasive diamond wheel with grit size 4000–8000. Bandyopadhyay et al. [19] studied the application of ELID grinding on SiN ceramics to show how ELID dressing reduces the grinding force significantly. However, it was found in the study [19] that even in the ELID, the initial grinding force was higher before it settles to a lower stable value. Bandyopadhyay et al. [19] proposed a two-stage dressing method to address this problem. In his solution, authors [19] carried out first pre-dressing and removed the oxide layer mechanically. Then they carried out second pre-dressing and performed the ELID grinding as usual as explained in Fig. 6.1. As a result of this improvised technique, the grinding force was observed to below from the beginning of the grinding process. Katahira et al. [20] carried out a fascinating study of ELID operation on Aluminum Nitride (AlN) ceramics to show that finer wheels with a mesh size of #4000 and above could generate a ductile surface with a surface roughness of as low as 8 nm and without any brittle fracture. They also concluded that because of the diffusion of Oxygen, the end surface after the ELID operation is superior over the conventionally polished surface in terms of surface hardness, frictional coefficient etc.

Ohmori et al. [21] first tested ELID lap grinding using the metal resin-bonded wheel to polish Si wafer. They successfully achieved a very smooth finish with a surface roughness of as low as ~3 nm. Later Islam et al. [22, 23] used ELID grinding to conduct the thinning of the Si wafer. The thinning ratio was more than ten times with a fine average surface roughness of 6 nm. They also developed a new type of injection nozzle for better coverage of the electrolytes between the wheel and electrode gap. Injection electrode outperformed the conventional ELID system by reducing the grinding force as it enhances the dressing and constant protrusion of the sharper grits.

Saito et al. [24] investigated ELID grinding for lens mould materials (cemented carbide) using different grinding wheels with varying bond materials namely Copper (Cu), Chromium (Cr) and hybrid Resinoid bond. It was found that Cr series wheel showed the best result in terms of ground surface roughness. The resinoid performance was the worst among the three. Ohmori et al. [25] used ELID grinding to achieve a mirror finish on Stainless Steel. The study confirms that ELID grinding helps to diffuse grinding particles into the workpiece and produce a stable oxide layer on the ground surface. This provided better surface, tribological and corrosion resistance property as compared to the conventionally polished surface. In another study, Raffleset al. [26] use ELID grinding with metal/resin bonded wheel to achieve the super finish on the spherical bearings.

Direct finishing of glass using ELID grinding for various optical applications has been investigated by numerous researchers. Kumar et al. [27] investigated ELID performance on peripheral grinding of BK7 optical glass. They [27] observed that the current duty ratio plays a vital role in minimizing surface crack on the ground sample. They also reported that higher current duty ratio usually reduces the grinding force because of more dressing and renewal of the sharp grits. Stephenson et al. [28] experimented ELID grinding of BK7 and Zerodur glass along with Acoustic emission (AE) signal. Stephenson et al. [28] concluded that the AE signal is well correlated with wheel loading and aggressive grinding, which can be used as an effective method of monitoring technique for ELID grinding. Yin et al. [29] also studied the feasibility of grinding Zerodur glass using ELID grinding. They successfully achieved ductile mode grinding with an average surface roughness of ~10 nm. Saleh et al. [30] also experimented ELID grinding for vertical channel grinding and spherical lens fabrication. At lower feed speed, they [30] managed to achieve an average surface roughness of 15 nm with #4000 series wheel.

6.2.3 Study on the Feedback and Intelligent Control of ELID Grinding

To improve the performance of the ELID grinding, few researchers have worked on incorporating various sensory and intelligent control to the conventional ELID system. Boland et al. [31] introduced a control algorithm to make the dressingcurrentuniform. It is a well-understood phenomenon that dressing current is high in the absence of ELID layer and reduces down as the oxide layer forms. Boland et al. [31] wanted to minimize the variation of the dressing current by observing it in frequency domain thus controlling the feed speed of the motion. Lee et al. [32, 33] proposed another interesting control algorithm to keep the thickness of the ELID layer constant throughout the grinding process. This confirms the constant grinding mechanics throughout the operation. In their study, Lee et al. [32, 33] showed that layer thickness is a linear function of the actual dressing current. Hence, by continuous monitoring of the actual dressing current and controlling of the peak dressing current the thickness of the ELID layer was maintained to be constant during the entire grinding operation. In ELID grinding tool wear is significant and would affect the form accuracy during the grinding, of surface if not compensated. To address this problem Rahman et al. [34] developed an on-machine measurement (OMM) system for sequential compensation of the ELID wheel wear during the grinding of the curved lens. To increase the wheel life, Saleh et al. [6] proposed an efficient dressing method by monitoring the grinding force ratio K (normal grinding force/tangential grinding force). The dressing voltage was varied as a linear function of K. As the K increased the dressing voltage was also increased as this indicated worn-out condition of, the wheel. However, if the K value is low and the wheel is still sharp enough, the dressing voltage was lowered down to avoid excessive dressing. This method helped to minimize the dressing current variation and to increase the wheel life without compromising the grinding output (i.e. the surface roughness). Later, Saleh et al. [35] also proposed an in-process wheel truing method by applying the pulse width control (PWC) of the ELID power. At first they [35] monitored the wheel profile with the help of an inductive and optical sensor. Inductive sensor was used to measure the distance between the sensor head and the wheel's metallic bond (hence the wheel profile) whereas the optical sensor was acted as the trigger to start the measurement of the inductive sensor. Then the inductive sensor data was arranged along the circumference of the wheel to understand its profile. By knowing the wheel profile the dressing power duty ratio was varied by implementing PWC control. This provides a programmed non uniform dressing and ELID layer formation along the circumference of the ELID wheel. Eventually as the wheel came into contact with the workpiece the excess ELID layer was broken off and the wheel became trued. This method helped to eliminate a need for a separate truing mechanism of the ELID wheels. Fathima et al. [36] proposed a knowledge based feedback control system for the ELID grinding of curvilinear surface. In their proposed system the optimized ELID parameters were chosen based on the knowledge base and then feedback control was applied to maintain ELID parameter (namely the dressing current) constant. This reduced the number of correction passes due to the wheel wear. The feedback control mechanism also improved the surface roughness by a margin of 2 × to 3 × times.

6.3 Recent Trend on the Research of ELID Grinding

This section will discuss how the recent research (for the last five years or so) is being evolved for ELID grinding. Kersschot et al. [37] developed the ELID layer growth model using the Helmholtz model for electrolytic passive layer formation. It was observed in their study [37] that growth rate of the passivation layer is positively influenced by aggressive dressing power setting, larger grit size and slower wheel speed. Hence, authors suggested to reduce the wheel speed during the time of pre-dressing to form a thicker layer in a shorter time. The authors also concluded that proper electrolyte with the right oxidizing substance is very important for the growth of the ELID layer which may not be achievable with the commercial electrolytes. Wei et al. on the other hand, [38] implemented fuzzy logic control mechanism in order to maintain uniform ELID oxide layer.

Kuai [39] investigated the composition of the formed ELID layer and observed the presence of α-Fe_2O_3 in the passive ELID layer. He also described the method of generation and transformation of α-Fe_2O_3 in the ELID layer. Kuai et al. [40] further investigated the oxide layer that is formed during the ELID process in a more detail manner using scanning electron microscopy, tunneling electron microscopy, infrared ray spectrum analysis. Their results revealed the presence of α-Fe_2O_3 in the passive ELID layer, which provides the polishing effect alongside actual grinding. Moreover, it was also observed that the α-Fe_2O_3 particle in oxide film is nearly spherical with a diameter of 5–50 nm which further explains why the polishing effect could generate ultrafine smooth surface during the ELID grinding operation.

Wu et al. [41, 42], in their study, pointed out the non-uniform contact between the wheel and workpiece during ELID groove grinding. They studied ELID grinding for bearing raceway under several dissimilar grinding conditions. They were convinced by their study that non-uniform contact does persist during the ELID groove grinding process which affects certain surface roughness properties of the ground surface. Figure 6.4 explains four different stages of non-uniform contact for ELID groove grinding. In order to explain the non-uniform contact the authors imaginarily divided the grinding wheel into three regions; left, middle and the right region. During the pre-dressing time, the formation of the ELID layer is uniform throughout the region (stage 1 in Fig. 6.4). As the grinding progresses the grits in the middle region mostly takes part in the machining action (stage 2 in Fig. 6.4) not the two side regions. As time progresses grits and layer in the middle region wears out and electrolytic process restarts, however, the two regions on the sides now take place in the grinding action (stage 3 in Fig. 6.4). In the final stage, the middle region has more oxide layer and the side regions have worn out grits and oxide layer. These stages are recursive for the whole grinding process. Authors found that this non-uniform contact during the ELID grinding causes adverse effects on some of the surface roughness properties of the ground workpiece. This non-uniform contact could be eliminated by the active control of the ELID layer formation as pointed out by the authors in their subsequent study [41].

Zhao et al. [43] described that wheel truing is very important in order to achieve precision grinding with ductile mode machining. They also emphasized that the protrusion of the cutting grits should be less than the critical depth of cut for ductile mode machining. To achieve the conditioning of the grinding wheel, they [43] have developed a setup as described in Fig. 6.5. In the work spindle, a conditioning cup wheel was mounted which was dressed using conventional ELID technique. Then the grinding wheel was mechanically trued using the cup wheel. The grinding wheel was

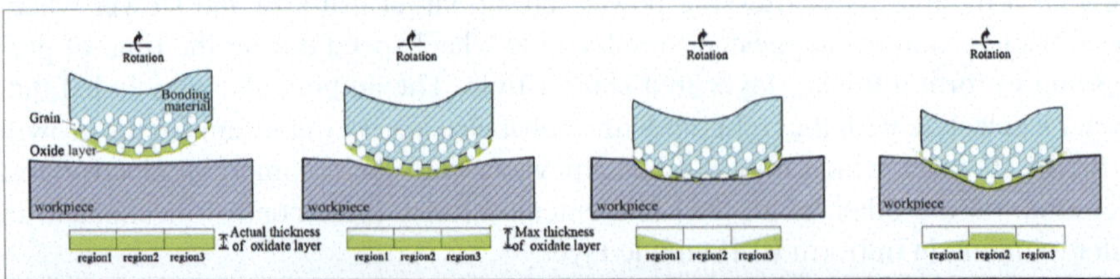

Fig. 6.4 Stages of non-uniform contact between the grinding wheel and bearing raceway [41]

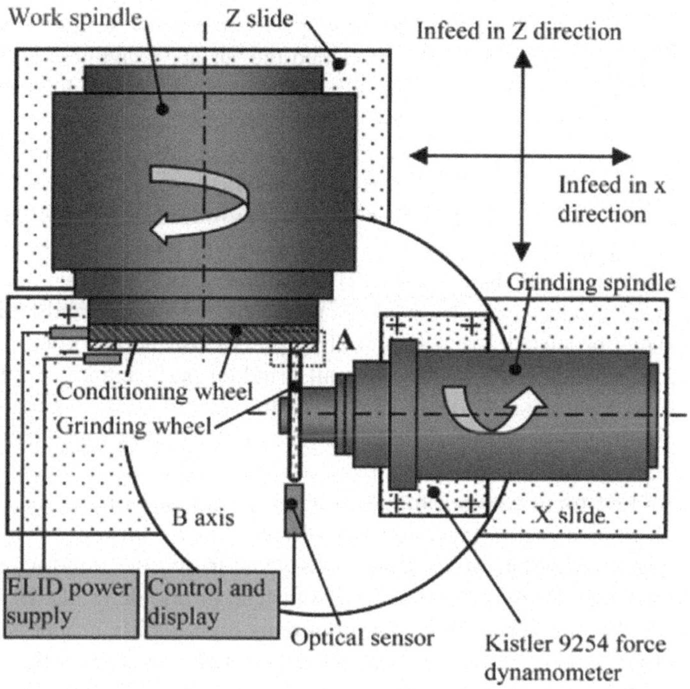

Fig. 6.5 Grinding wheel conditioning setup [43]

constantly monitored until desired run-out tolerance was achieved. By applying this wheel conditioning technique researchers managed to improve the wheel run-out by a margin of ~18 x. In their second stage of the research Zhao et al. [44] tested the trued wheel for surface and profile grinding. It was observed in the study that even coarse grinding wheel could achieve stable and smooth grinding condition provided the wheels are conditioned as explained in [43]. The material removal rate was observed to be substantially high. However, the surface roughness was very low (~10 nm) which is counterintuitive. Moreover, the wheel did not require continuous dressing as opposed to conventional ELID system, hence the wheel life was prolonged. The subsurface damage was observed to be in the micrometer range. The author concluded that the newly developed wheel conditioning system is feasible for precision grinding of optical glasses.

Tang et al. [45] investigated and modelled accumulated depth of cut (DoC) error for ELID grinding and compared it with the conventional grinding. It was found in their study that after a certain number of grinding passes the accumulated error in the DoC gets saturated and the saturation is faster if the DoC for the single grinding pass is higher. Authors also reported that ELID grinding provides better stability and higher material removal rate as compared to conventional grinding specially during large volume material removal. Moreover, faster grinding wheel rotational speed was found to be beneficial for the grinding process.

Zhao et al. [46] studied ELID grinding performance for grinding SiC bearing rollers. They also experimentally modelled the surface roughness and material removal rate (MRR) as a function of grinding wheel speed, applied load and eccentricity of the workpiece. The authors concluded that among the three factors the applied load demonstrated the most significant effect in influencing surface roughness and material removal rate. Both increases as the load are increased. Wheel speed was found to have no or little effect on the surface roughness and MRR. Finally, eccentricity was found to have a strong correlation with surface roughness but no significant effect on MRR. Kuai et al. [47] studied the oxide film that is developed during the ELID process and observed the presence of absorbed and lattice water in the film. They proposed a two-step cooling process of the absorbed and lattice water in the film which improved the overall grinding performance. Yu et al. [48] conducted an experimental study of ELID grinding on SiCp/Al composites. They studied the grinding force variation under ELID condition. They found that grinding force increase with the increase of the depth of cut and feed speed which is natural. However, grinding force was not found to be proportional with the feed speed as the increase rate of the grinding force was observed to be slowed down at higher feed speed. Surface roughness was found to have an increasing trend with the feed speed as it facilitates brittle mode machining instead of ductile mode. ELID dressing was found of being very effective in producing high-quality ground surface for SiCp/Al workpiece. Gao et al. [49] conducted thorough research on the formation mechanism of the passive oxide layer during the process of ELID grinding. They also investigated the parameters that affect the layer formation. They successfully achieved mirror surface finish on silicon nitride ceramics by controlling the oxide layer thickness in real-time.

Zhao et al. [50] introduced a new way of conducting ELID grinding using the assistance of ultrasonic vibration. Authors studied the process experimentally and also modelled the process to predict the ground surface quality. Figure 6.6 [50] shows the predicted and actual surface roughness variation for different depth of

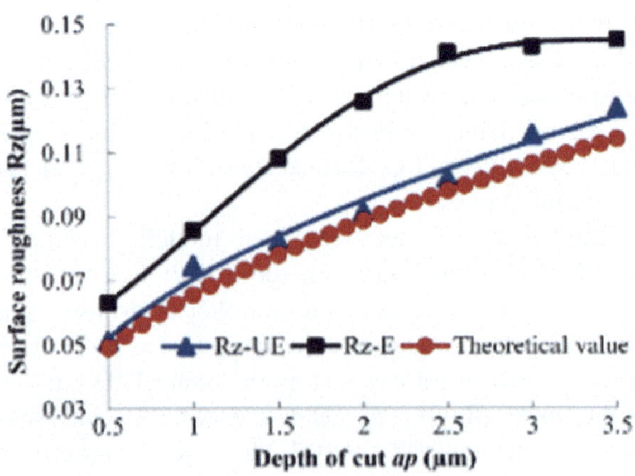

Fig. 6.6 Surface roughness for different grinding depth of cut [50]

cut of grinding. As can be observed from Fig. 6.6, the fit with the experimental result is better for ultrasonically assisted ELID grinding. However, for other grinding parameters such as grinding wheel and workpiece linear speed and axial feed speed the variation between the theoretical prediction and experimental finding for the surface roughness was quite significant.

Recently Wang et al. [51] carried out a comparative study on the ELID layer for tool electrode and workpiece electrode case. Authors conducted experiments to investigate the different properties of ELID layer (such as thickness and surface morphology) for the case of workpiece-cathode and tool-cathode with various dressing conditions. They also studied the grinding forces for the said conditions. The authors [51] observed that as the dressing current increased the oxide layer became discontinuous for the case of workpiece-cathode whereas layer became smoother for the tool-cathode condition. Interestingly it was observed that grinding force was smaller for workpiece-cathode as compared to tool cathode as the ELID layer was thinner for the later. Kuai et al. [52] continued their investigation on the ELID layer by studying its mechanical properties. They observed that the hardness and stiffness of the oxide film were higher than the wheel's bond matrix. Huang et al. [53] continued their study of ELID grinding on SiC/Al composite to investigate the variation of the grinding force for various grinding condition. They found that for ELID and non ELID, dry and wet grinding; in all the cases grinding force increased with the table speed and depth of cut.

Wu et al. [54] highlighted that ELID wheels have two stages of wear. Primarily it has mechanical grit wear and at the same time, it has anodic dissolution due to the electrolytic dressing. Hence, the wear prediction of the grinding wheel becomes more uncertain. As such, Wu et al. [54] proposed a new method of wear characterization quantitatively to understand the wear life of the ELID wheels more effectively. Shao et al. [55] studied the material removal mechanism for ultrasonically assisted ELID grinding by proposing a mathematical model for this. Experimentally they [55] observed that MRR increased with higher DoC and feed speed. They [55] also found that ultrasonically assisted ELID improved the MRR as compared to conventional ELID grinding. Kuai et al. [56] investigated an ELID grinding performance on a new material (CoCrMo Alloy) which is widely used in various biomedical applications. They found that ELID grinding is quite effective on the said material moreover ELID grinding caused to form a corrosion-resistant oxide layer on the alloy. Very recently Bader et al. [57] proposed a regression model to predict the oxide layer thickness during the ELID process. The polynomial model takes voltage, gap dimension and electrolyte flow rate as the input to predict the layer thickness. The influences of all the three parameters were equally found to have significant influence for the formation of the ELID layer.

6.4 Summary

ELID grinding has been in the research interest for almost thirty years. It has been proven to be useful for obtaining sub-ten-nanometer surface finish on the hard and brittle material such as glass, ceramics, tough metallic alloys etc. Several reported researches have established the fact that ELID grinding can achieve ductile mode grinding on the hard and brittle material as it ensures constant protrusion of the sharp grits by conducting in situ dressing. The main objective of ELID grinding was to remove the subsequent final finishing process like polishing and lapping. However, the fact is ELID grinding has not been accepted yet by the industry players even after thirty years of research. There are some challenges that need to be addressed properly before ELID grinding could be accepted widely by the industry. There is no ELID grinding machine available commercially. Most of the researchers use conventional ELID grinding and retrofit it to conduct ELID dressing of the grinding wheels. However, this is not good for the machine as the flow of electrolytes with power supply can initiate corrosion of the machine parts which eventually will reduce the grinding machine's life. This can be one reason for the industry to be not interested in introducing ELID grinding in their operation. Therefore, it is a necessity to design an ELID grinding machine by introducing proper flushing and draining system for the electrolytes so that the machine's structural body can be saved. Another challenge is a proper model (machine-learning based) is required to predict the optimum input condition for the ELID grinding for various workpiece and grinding wheel combination. If this is available then the industry would be more interested in using ELID grinding as their final finishing operation.

References

1. Pei ZJ, Fisher GR, Bhagavat M, Kassir S (2005) A grinding-based manufacturing method for silicon wafers: An experimental investigation. Int J Mach Tools Manuf 45:1140–1151. https://doi.org/10.1016/j.ijmachtools.2004.12.006
2. Rahman M, Lim HS, Neo KS, Senthil Kumar A, Wong YS, Li XP (2007) Tool-based nanofinishing and micromachining. J Mater Process Technol 185:2–16. https://doi.org/10.1016/j.jmatprotec.2006.03.121
3. Rahman M, Kumar AS, Lim HS, Fatima K (2003) Nano finish grinding of brittle materials using electrolytic in-process dressing (ELID) technique. Sadhana 28:957–974. https://doi.org/10.1007/BF02703325
4. Saleh T, Rahman M (2014) Electrolytic in-process dressing (ELID) grinding for nano-surface generation. In: Comprehensive materials processing. Elsevier Ltd, pp 483–522
5. Ohmori H, Nakagawa T (1990) Mirror surface grinding of silicon wafers with electrolytic in-process dressing. CIRP Ann Manuf Technol 39:329–332. https://doi.org/10.1016/S0007-8506(07)61065-8
6. Saleh T, Bishwas I, Rahman M (2010) Efficient dressing of the wheel in ELID grinding by controllable voltage with force feed back. Int J Adv Manuf Technol 46:123–130. https://doi.org/10.1007/s00170-009-2076-x

7. Zhang C, Ohmori H, Li W (2000) Small-hole machining of ceramic material with electrolytic interval-dressing (ELID-II) grinding. J Mater Process Technol 105:284–293. https://doi.org/10.1016/S0924-0136(00)00648-8
8. Qian J, Ohmori H, Lin W (2001) Internal mirror grinding with a metal/metal-resin bonded abrasive wheel. Int J Mach Tools Manuf 41:193–208. https://doi.org/10.1016/S0890-6955(00)00067-5
9. Ohmori H, Katahira K, Naruse T, Uehara Y, Nakao A, Mizutani M (2007) Microscopic grinding effects on fabrication of ultra-fine micro tools. CIRP Ann Manuf Technol 56:569–572. https://doi.org/10.1016/j.cirp.2007.05.136
10. Ohmori H, Nakagawa T (2007) Utilization of nonlinear conditions in precision grinding with ELID (electrolytic in-process dressing) for fabrication of hard material components. CIRP Ann 46:261–264. https://doi.org/10.1016/s0007-8506(07)60821-x
11. Seok Han L, Fathima K, Kumar AS, Rahman M (2002) A fundamental study on the mechanism of electrolytic in-process dressing (ELID) grinding. Int J Mach Tools Manuf 42:935–943. https://doi.org/10.1016/S0890-6955(02)00023-8
12. Fathima K, Senthil Kumar A, Rahman M, Lim HS (2003) A study on wear mechanism and wear reduction strategies in grinding wheels used for ELID grinding. Wear 254:1247–1255. https://doi.org/10.1016/S0043-1648(03)00078-4
13. Fathima K, Rahman M, Senthil Kumar A, Lim HS (2007) Modeling of ultra-precision ELID grinding. J Manuf Sci Eng 129:296. https://doi.org/10.1115/1.2515382
14. Chen H, Li JCM (2000) Anodic metal matrix removal rate in electrolytic in-process dressing I: two-dimensional modeling. J Appl Phys 87:3151–3158. https://doi.org/10.1063/1.372313
15. Chen H, Li JCM (2002) Anodic metal matrix removal rate in electrolytic in-process dressing II: protrusion effect and three-dimensional modeling. J Appl Phys 87:3159–3164. https://doi.org/10.1063/1.372314
16. Fathima K, Schinhaerl M, Geiss A, Rascher R, Sperber P (2010) Wear analysis of electrolytically dressed wheels for finishing substrate materials. Tribol Int 43:245–251. https://doi.org/10.1016/j.triboint.2009.05.029
17. Zhu Z, Wang X, Thangam S (2004) Simulation and analysis of rigid/foil electrolytic in-process dressing (ELID) systems for grinding. Trans ASME J Manuf Sci Eng 126:565–570. https://doi.org/10.1115/1.1765152
18. Forks G (1996) Ultra-precision grinding of structural ceramics by electrolytic in-process dressing (ELID) grinding. J Mater Process Technol 57:272–277
19. Bandyopadhyay BP, Ohmori H, Takahashi I (1997) Efficient and stable grinding of ceramics by electrolytic in-process dressing (ELID). J Mater Process Technol 66:18–24. https://doi.org/10.1016/S0924-0136(96)02454-5
20. Katahira K, Ohmori H, Uehara Y, Azuma M (2005) ELID grinding characteristics and surface modifying effects of aluminum nitride (AlN) ceramics. Int J Mach Tools Manuf 45:891–896. https://doi.org/10.1016/j.ijmachtools.2004.10.017
21. Itoh N, Ohmori H, Moriyasu S, Kasai T, Karaki-Doy T, Bandyopadhyay BP (1998) Finishing characteristics of brittle materials by ELID-lap grinding using metal-resin bonded wheels. Int J Mach Tools Manuf 38:747–762
22. Islam MM, Senthil Kumar A, Balakumar S, Lim HS, Rahman M (2006) Performance evaluation of a newly developed electrolytic system for stable thinning of silicon wafers. Thin Solid Films 504:15–19. https://doi.org/10.1016/j.tsf.2005.09.023
23. Islam MM, Kumar AS, Balakumar S, Lim HS, Rahman M (2008) Characterization of ELID grinding process for machining silicon wafers. J Mater Process Technol 198:281–290. https://doi.org/10.1016/j.jmatprotec.2007.06.077
24. Saito T, Katahira K, Ohmori H, Komotori J, Mizutani M, Nemoto A (2006) Fabrication of high-quality surfaces on precise lens mold materials by a new ELID grinding wheel. Towards synthesis of micro-/nano-systems. Springer, London, pp 315–318
25. Ohmori H, Katahira K, Komotori J, Mizutani M (2008) Functionalization of stainless steel surface through mirror-quality finish grinding. CIRP Ann Manuf Technol 57:545–549. https://doi.org/10.1016/j.cirp.2008.03.131

26. Raffles MH, Stephenson DJ, Shore P, Jin T (2011) Electrolytic in-process dressing superfinishing of spherical bearings using metal-resin bond ultra-fine CBN wheels. Proc IMechE Part B J Eng Manuf 225:112–122. https://doi.org/10.1243/09544054JEM2003
27. Kumar AS, Lim HS, Rahman M, Fathima K (2002) A study on the grinding of glass using electrolytic in-process dressing. J Electron Mater 31:1039–1046. https://doi.org/10.1007/s11664-002-0040-8
28. Stephenson DJ, Sun X, Zervos C (2006) A study on ELID ultra precision grinding of optical glass with acoustic emission. Int J Mach Tools Manuf 46:1053–1063. https://doi.org/10.1016/j.ijmachtools.2005.08.013
29. Yin S, Ohmori H, Dai Y, Uehara Y, Chen F, Tang H (2009) ELID grinding characteristics of glass-ceramic materials. Int J Mach Tools Manuf 49:333–338. https://doi.org/10.1016/j.ijmachtools.2008.10.013
30. Saleh T, Rahman MS, Lim HS, Rahman M (2007) Development and performance evaluation of an ultra precision ELID grinding machine. J Mater Process Technol 192–193:287–291. https://doi.org/10.1016/j.jmatprotec.2007.04.034
31. Boland R, Rd LB (1999) Computer control and process monitoring of electrolytic in-process dressing of metal bond fine diamond wheels for NIF optics: Fgp, A. 3782:61–69
32. Lee ES (2000) Study on the mirror-like grinding of die steel with optimum in-process electrolytic dressing. J Mater Process Technol 100:200–208. https://doi.org/10.1016/S0924-0136(99)00462-8
33. Leer E, Kimt J (1997) A study on the analysis of grinding mechanism and development of dressing system by using optimum in-process electrolytic dressing. Int J Mach Tools Manuf 37:1673–1689
34. Sazedur Rahman M, Saleh T, Lim HS, Son SM, Rahman M (2008) dressin. Int J Mach Tools Manuf 48:887–895
35. Saleh T, Rahman M (2011) In-process truing for ELID (electrolytic in-process dressing) grinding by pulsewidth control. IEEE Trans Autom Sci Eng 8:338–346
36. Fathima K, Schinhaerl M, Geiss A, Rascher R, Sperber P (2010) A knowledge based feed-back control system for precision ELID grinding. Precis Eng 34:124–132. https://doi.org/10.1016/j.precisioneng.2009.05.004
37. Kersschot B, Qian J, Reynaerts D (2013) On the dressing behavior in ELID-grinding. Procedia CIRP 6:632–637. https://doi.org/10.1016/j.procir.2013.03.089
38. Wei L, Fangyong Z, Minjie X (2015) Fuzzy control theory and technique of ELID grinding. J Zhejiang Univ Technol 2015
39. Kuai JC (2014) Discovery of α-Fe2O3 in the oxide film on ELID grinding wheel. Adv Mater Res 1061–1062:446–449. https://doi.org/10.4028/www.scientific.net/amr.1061-1062.446
40. Kuai J, Ardashev DV, Zhang H (2017) Study of α-Fe 2 O 3 formation and its measurement in oxide films of wheel surface during ELID grinding process. Mod Phys Lett B 31:1750025. https://doi.org/10.1142/s0217984917500257
41. Wu ML, Zhang K, Ren C (2015) Study on the non-uniform contact during ELID groove grinding. Precis Eng 39:116–124. https://doi.org/10.1016/j.precisioneng.2014.07.011
42. Wu ML, Ren C, Zhang K (2015) ELID groove grinding of ball-bearing raceway and the accuracy durability of the grinding wheel. Int J Adv Manuf Technol 79:1721–1731. https://doi.org/10.1007/s00170-015-6853-4
43. Zhao Q, Guo B (2015) Ultra-precision grinding of optical glasses using mono-layer nickel electroplated coarse-grained diamond wheels. Part 1: ELID assisted precision conditioning of grinding wheels. Precis Eng 39:56–66. https://doi.org/10.1016/j.precisioneng.2014.07.006
44. Zhao Q, Guo B (2015) Ultra-precision grinding of optical glasses using mono-layer nickel electroplated coarse-grained diamond wheels. Part 2: investigation of profile and surface grinding. Precis Eng 39:67–78. https://doi.org/10.1016/j.precisioneng.2014.07.007
45. Tang H, Deng ZH, Guo YS, Qian J, Reynaerts D (2015) Depth-of-cut errors in ELID surface grinding of zirconia-based ceramics. Int J Mach Tools Manuf 88:34–41. https://doi.org/10.1016/j.ijmachtools.2014.08.003

46. Zhao L, Bafakeeh O, Liu H, Marinescu ID (2016) ELID fine grinding of SiC bearing rollers. Key Eng Mater 686:204–211. https://doi.org/10.4028/www.scientific.net/kem.686.204
47. Kuai J, Wang J, Jiang C (2016) Experimental research on two-step cooling mechanism of the oxide film of ELID grinding wheel. MATEC Web Conf 67:4004. https://doi.org/10.1051/matecconf/20166704004
48. Yu X, Huang S, Xu L (2016) ELID grinding characteristics of SiCp/Al composites. Int J Adv Manuf Technol 86:1165–1171. https://doi.org/10.1007/s00170-015-8235-3
49. Gao DX, Liu Q, Zhang B, Xiao CX, Ren CZ (2016) Research on formation mechanism of grinding wheel oxide film in ELID ultra precision mirror grinding. Mater Sci Forum 861:90–96. https://doi.org/10.4028/www.scientific.net/msf.861.90
50. Zhao B, Chen F, Jia XF, Zhao CY, Wang XB (2017) Surface quality prediction model of nano-composite ceramics in ultrasonic vibration-assisted ELID mirror grinding. J Mech Sci Technol 31:1877–1884. https://doi.org/10.1007/s12206-017-0335-6
51. Wang Z, Ren C, Chen G, Zhang L, Deng X (2018) A comparative study on state of oxide layer in ELID grinding with tool-cathode and workpiece-cathode. Int J Adv Manuf Technol 94:1299–1307. https://doi.org/10.1007/s00170-017-0931-8
52. Kuai JC, Wang JW, Jiang CR, Zhang HL, Yang ZB (2018) Mechanical properties of oxide films on electrolytic in-process dressing (ELID) copper-based grinding wheel. In: IOP Conference series: materials science and engineering. IOP Publishing, p 012013
53. Huang S, Yu X (2018) A study of grinding forces of SiCp/Al composites. Int J Adv Manuf Technol 94:3633–3639. https://doi.org/10.1007/s00170-017-1115-2
54. Wu ML, Ren CZ, Zhang KF (2018) Wear life characterization of the grinding wheel for electrolytic in-process dressing (ELID) grinding of ball bearing raceways: a new perspective based on a moving normal distribution curve of the grit state variation. Int J Adv Manuf Technol 96:1919–1928. https://doi.org/10.1007/s00170-018-1694-6
55. Shao SJ, Zhao B, Bian PY (2018) Removal mechanism of ultrasonic vibration and ELID composite plane grinding based on ultrasonic vibration of nano-ceramic workpiece. J Vibroeng 20:417–426. https://doi.org/10.21595/jve.2017.18968
56. Kuai JC, Jiang CR, Wang JW, Ardashev DV (2018) The ELID grinding of CoCrMo alloy bioprosthesis and the formation mechanism of its corrosion-resistant oxide layer. Mater Sci Forum 934:140–144. https://doi.org/10.4028/www.scientific.net/msf.934.140
57. Alqahtani B, Zhang M, Marinescu I, Bafakeeh OT, Al Sofyani S (2019) Microscopic characterization and modeling of oxide layer for electrolytic in-process dressing (ELID) grinding with focus on voltage, electrode-wheel gap, and coolant flow. Int J Adv Manuf Technol 1–10. https://doi.org/10.1007/s00170-019-03435-5

Chapter 7
Advances in Abrasive Flow Finishing

Vivek Rana, Anand C. Petare, and Neelesh Kumar Jain

Abstract Abrasive flow finishing (AFF) is a non-traditional advanced fine finishing process using an abrasive-laden viscoelastic polymer for finishing, polishing, deburring, radiusing of the component having a complex geometrical shape with inaccessible areas and difficult to machine materials. AFF is a consistent, accurate and repeatable process compared to another finishing process. It was initially developed in 1960 by Extrude Hone Corporation, USA for finishing valves of radar-guided vehicles. It uses a flexible abrasive laden viscoelastic finishing medium, which makes it most popular with nanofinishing process for simple as well as complex geometrical components. Continual research in many areas of AFF is in underway to explore its applications for different components, finishing medium developments and applications, process modeling and online control, automation with robotic interface for mass finishing operation, development of new mechanisms and types for fourth-generation AFF. This article reports on advances in process modeling and optimization, alternative finishing medium and rheological characterization, hybrid, derived and hybrid-derived variants and applications of AFF for finishing advanced materials and complicated geometry. This article will be beneficial for researchers, academicians, and industrialists who are working in the field of precision finishing of complex geometrical mechanical components and willing to implement AFF.

Keywords Abrasive finishing · Material removal · Nano-finishing · Putty · Process modeling · Rheological · Surface roughness

V. Rana · A. C. Petare · N. K. Jain (✉)
Discipline of Mechanical Engineering, Indian Institute of Technology Indore, Simrol, MP 453552, India
e-mail: nkjain@iiti.ac.in

V. Rana
e-mail: mtphd1806103001@iiti.ac.in

A. C. Petare
e-mail: phd1401103001@iiti.ac.in

S. Das et al. (eds.), *Advances in Abrasive Based Machining and Finishing Processes*, Materials Forming, Machining and Tribology,
https://doi.org/10.1007/978-3-030-43312-3_7

7.1 Introduction

Advancement in technology required highly finished products in specific areas such as space technology, automobile components, production of surgical equipment, aerospace components, marine and defense components, power plants and turbine equipment, dies and mold, additive manufacturing, in micro/nano electro-mechanical systems (MEMS/NEMS) and micro-systems, etc. Precision finishing of complex geometrical components and difficult to machine materials are costly labor intensive and uncontrollable. Improving surface finishing manufactured components below 1 μm accounts for 15% of overall production cost [1]. To finish components various conventional finishing processes such as polishing, grinding, lapping, honing, burnishing, peening and shot blasting used. These conventional finishing processes suffer some inherent limitations while finishing complicated geometry and difficult to finish materials. Polishing is applicable for finishing external cylindrical and planer surfaces by using emery paper, belt containing abrasives of alumina, silicon carbide or diamond. The finishing result of polishing depends upon manual pressure and type of handling workpiece. It is unable to finish internal cylinder or complex geometries. Grinding wheel geometry not permit it to go inside the deep holes, intricate sharp corner the complicated geometry and it also produces burn marks on finished surface. Lapping is slow process to improve minor surface irregularities and not applicable to finish complex geometrical components. Honing can improve surface finish cylindrical as well as planer surface but it having very short life of honing sticks. Shot blasting removes burrs effectively, but it causes deformation in soft materials and embedding of abrasive particles on finished surface required cleaning through hand file or blowing. Burnishing causes localized hardening over finished surface and not applicable to finish thin-walled components. The deficiencies of conventional finishing processes have enforced development of advanced finishing solutions for finishing complicated geometries, difficult to machine materials, curved holes and sharp corners, etc.

Abrasive flow finishing (AFF) comes into existence to overcome the limitations of conventional finishing processes by using high pressurized flexible finishing medium called putty. The putty can easily go inside intricate spacing, confined passages of complex geometrical components and imparted nano finishing up to 0.05 μm. Figure 7.1 illustrates the schematic of the working principle of AFF. It consists of hydraulic or pneumatic cylinders mounted at top and bottom inside the structure of bars and platen. Top and bottom hydraulic or pneumatic cylinders coupled with medium cylinders. A finishing medium of abrasive particle, mixing oil and viscoelastic polymer prepared and filled in medium cylinder. The workpiece is kept sandwiched in between both medium cylinders inside the fixture and finishing medium passes over the workpiece. The back and forth motion of finishing medium over workpiece surface causes abrasion and impart fine finishing to the workpiece. A hydraulic or pneumatic power unit used to provide motion to the hydraulic or pneumatic cylinder.

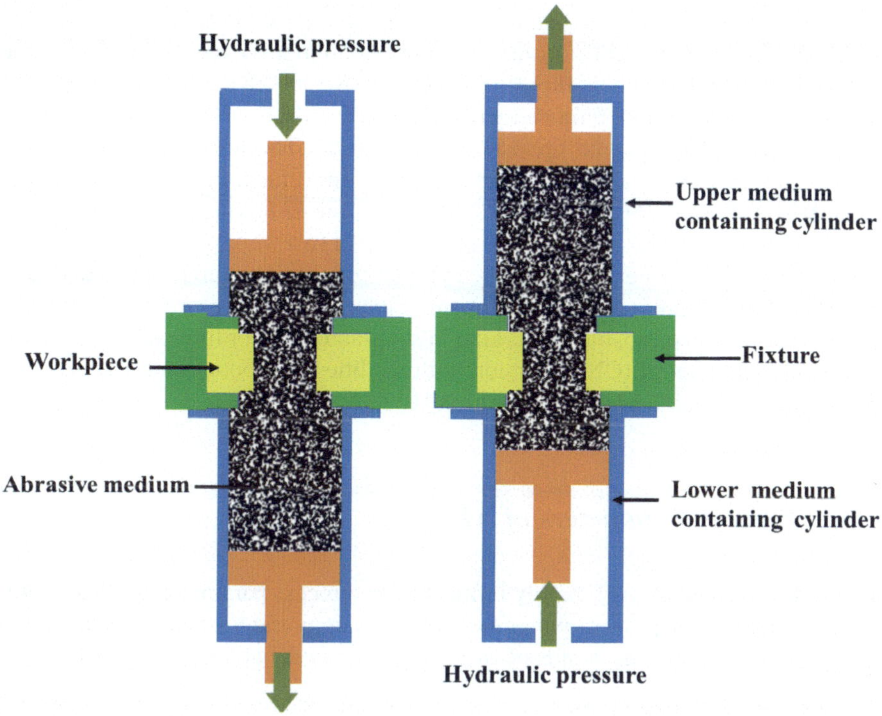

Fig. 7.1 Schematic of the two-way abrasive flow finishing (AFF) process and its working

The major component of the AFF system divided into three parts: (a) AFF structure; (b) Finishing medium; and (c) Tooling and fixture. *AFF structure* consists of hydraulic or pneumatic cylinders, power supply unit, medium cylinder, supports plates, bars, stroke counter, limit switches to control stroke length of the medium cylinder and pressure regulators. *Finishing medium* is mixture of viscoelastic polymer, abrasive particle and mixing oil. Abrasive particle size, concentration and types selected based on finishing requirement. Aluminum oxide, silicon carbide (SiC), cubic born nitride (CBN) and diamond abrasive particles are generally used. *Tooling and fixture* hold workpiece to withstand in high extrusion pressure and guide finishing to reach the desired surface of workpiece where finishing is desired. It also serves purpose to block the space selectively where finishing does not require. Nylon, Teflon, metlon and stainless steel generally used for fabrication of fixtures. AFF frequently used for finishing variety of industrial components in aerospace, automobile, pump manufacturing, turbine, additive manufacturing, recast layer removal, pipe manufacturing, casting, medical industry, food processing and industry producing complex structure with accessible areas. The AFF is applied for removal of materials from soft aluminum to tough nickel alloys, ceramics and carbides effectively [1]. AFF system is of three types (a) *One way AFF*: In this type system finishing medium is introduced at entrance of workpiece by medium cylinder and it collected at the exit of workpiece. Finishing medium again collected and reinserted for next cycle. (b) *Two way AFF*: This type of AFF finishing medium continuously reciprocates between two medium cylinders and workpiece placed in between both medium cylinders. (c) *Orbital AFF*: In orbital AFF finishing medium back and forth between two medium cylinders. A displacer placed opposite to work piece. The finishing medium flow in between workpiece and displacer and vibratory movement provided to displacer which deflects flow of finishing medium. (d) *Multi-flow AFF*: This system is extended version of two ways AFF in which one more setup of two way AFF placed near to it. The direction of movement of hydraulic cylinders of second setup is opposite to first one. This system employed for mass finishing jobs and intricate castings. Presently, AFF systems are also capable of working in semi-automatic or fully automatic computer numerical control (CNC) version with facilities to robotic interface for mass finishing as well as job finishing operation [1, 2].

7.1.1 Process Parameters of AFF

The result produced by AFF highly influenced by process parameter applied, therefore the selection of process parameters are very important aspect to achieve the desired improvement. The details of various AFF process parameters are as follows:

- **Extrusion pressure**: Extrusion pressure is a force exerted by hydraulic or pneumatic power supply unit over per unit area of finishing medium to extrude it through restriction, created between workpiece to fixture. Extrusion pressure ranges from 0.7 to 22 MPa [3] and MRR increases with an increase in percentage concentration of abrasives in finishing medium and extrusion pressure, but with mesh size of abrasive grains, its value decreases [4]. Higher extrusion pressure causes decrease in average surface roughness (R_a) and increases microhardness due increase of axial forces over abrasive particle. Some parts of the total applied extrusion pressure lost to overcome internal resistance caused by finishing medium [3].
- **Number of cycles or finishing time**: Cycle is the distance travelled by the medium piston during its movement from initial position to return to again initial position and it varies from one to several according to finishing requirement. A review of literature suggests that rate of abrasion is very high during initial cycle due to the large peak to valley ratio. Decrement in surface roughness increases with increases in finishing cycle and MRR decreases due to reduction in peak height [5, 6].
- **Flow volume of medium**: The volume of the finishing medium contained in the medium cylinder is called the flow volume of the medium. MRR related to the slug length of the flow of medium and it connected with flow volume. Slug length of flow of medium is ratio of flow of medium to the cross-sectional area of the confined section. The higher flow volume of medium results in higher MRR [6].

- **Flow rate of medium**: Flow rate of the medium is the ratio of flow volume of medium to finishing time, and it depends on extrusion pressure and the viscosity of finishing medium. Slow flow rate causes uniform removal of materials and produces good surface finish whereas high flow rate causes the formation of edges. The flow rate generally exceeded 380 L per minute for extrusion pressure ranges between 0.7 and 22 MPa [1].
- **Medium temperature**: Medium temperature profoundly influences the physical properties of the finishing medium. As number of cycles increases, temperature increases hence viscosity of the medium decreases [7]. It must be in the range 30–70 °C increasing beyond 100 °C causes a permanent change in medium properties [8–10].
- **Abrasive type and size**: Generally used abrasive particles are silicon carbide (SiC), aluminum oxide (Al_2O_3), diamond, cubic boron nitride (c-BN), boron carbide (B_4C) used for preparing finishing medium. Its size ranges from 0.005 to 1.5 mm [1] and it selected according to finishing and material removal requirement. Coarse abrasive particles are recommended for higher MRR and fine abrasive particles for fine surface finish [3].
- **Abrasive concentration**: It is a volumetric ratio of the volume of abrasive particles in finishing medium to total volume of finishing medium. The abrasive concentration varies from 5 to 50% in a finishing medium [10]. An increase in the volume of abrasive concentration in finishing medium increases surface finish and MRR due to availability of more active grain but high percentage of abrasive concentration causes increase in stiffness of medium and chances of choking of the finishing medium [11].
- **Mixing oil type and concentration**: Mixing oil is used to form a bond between viscoelastic medium and abrasive particles. Concentration highly influenced the finishing results produced by AFF. An increase in the concentration of mixing oil decreases surface finish and MRR whereas decrease in the concentration of mixing oil causes chocking of finishing medium [12].
- **Viscosity of medium**: Viscosity of medium is a highly influential parameter, it depends on type and concentration of viscoelastic polymer, mixing oil, abrasive particle, and other additives. The higher viscosity of medium gives a higher reduction in surface roughness and MRR but the more chances of chocking of the medium. For deburring and radiusing low viscosity medium is applicable [2]. Higher viscosity medium is preferred when the channel length is beneath the two times of channel width or diameter; otherwise, low viscosity medium is preferred [13]. The viscosity of medium is increased by increasing the concentration of abrasive particles, and it results in higher surface finish and MRR [11].
- **Surface condition**: Surface quality of workpiece before AFF significantly affect finishing results because, in initial AFF cycles, abrasive particles try to flattening of roughness peaks having high height. After some AFF cycle when surface peaks come to average size than abrasive particles shear off all peaks altogether. Loveless et al. [2] examined the finishing mechanism of AFF with different types of surface produced by wire electro-discharge machining, lathe, grinding and milling. They found that surface produced by WEDM was best for AFF because of less height

of peaks, micro-cracks, and formation of loose connections of white layer with the workpiece surface.
- **Hardness**: Hardness of material significantly affects the finishing results generated by AFF. Past studies show that while contacting with the soft material depth of depression of abrasive particle is more as compared to hard materials, therefore more MRR and improvement in surface finish [11].

7.2 Advantages

- AFF can finish concave areas, convex areas and objects having a complex geometrical shape which are unable to finish by using other finishing processes.
- AFF can finish edges and radius very quickly compared to another finishing process.
- Drilled holes of size from 0.22 to 1000 mm can be finished/deburred easily by AFF. Also, multiple holes of diameter 3 mm and depth 10 mm can be finished simultaneously [14].
- AFF can produce a surface finish up to 0.05 μm with a close dimensional tolerance of ±5 μm within a short duration of time [14].
- Finishing results produced by AFF are consistent, repeatable, and predictable.
- AFF offers higher flexibility compared to another finishing process in terms of abrasive particle type, size, concentration, finishing medium type, fixture type, workpiece (external and internal both, conductive and nonconductive both).
- AFF constitutes a generalized finishing solution that gives highly accurate and precise results.
- CNC controlled AFF setup has the capability to process thousands of parts per day which can save employment costs for manual finishing up to a great extent.

7.3 Limitations

- Unable to correct taper deep surface irregularities, large counter profile, out-of-roundness and surface cracks etc.
- The material removal rate is very slow, therefore not applicable for mass material removal application.
- Controlling the viscosity of finishing medium is very less due to the mixing of scarf continuously which affects finishing result.

7.4 Advances in AFF

Available research initiatives reported following advances in the areas of modeling and optimization, development of finishing medium and rheological characterization, AFF based hybrid and derived processes development and applications.

7.4.1 Advances in Modeling and Optimization

Prediction of material removal and finishing result in AFF is the complex and stochastic due to following reason:

- Variation of physical conditions properties and viscosity of finishing medium due to mixing of removed materials and increase of temperature of the medium.
- Unable to detect movement and contact pattern of the abrasive particle with workpiece due to close contact of the fixture.
- Variability material removal pattern due to change of extrusion pressure, forces, abrasive particle size, type, and concentration.
- Variation of the viscosity of medium is because of change in the pattern due to change of abrasive particle size, type and concentration, mixing oil type and concentration and additives.
- Variation in material properties and geometry of workpiece to be finished.

The above variation attracts the researcher to attempt for modeling and optimization of AFF [10]. The advances reported in modeling and optimizations are as follows:

Williams and Rajurkar [5] were first attempted for modeling of AFF using data-dependent Systems (DDS) to find the ratio of surface roughness peak height (R_Z) to the average surface roughness value (R_a). They reported that the ratio of R_z/R_a in between 1.4 and 2.2 mm. They found surface profile having large wavelength (associated with path of abrasive particle) and small wavelength (associated with cutting edge). A coefficient (C_d) ratio abrasive grain wear and the number of active grains proposed. The increasing value of C_d means process is stable whereas decrease in value of C_d means cutting edges blunted and need to replace medium. Loveless et al. [15] applied DDS to model surface roughness profile of milled, turned, WEDM surface finished by AFF and reported similarities exist in finishing mechanism of grinding and AFF.

Rajeshwar et al. [16] used constitutive equations of the Maxwell model to find out the properties of the medium flow during AFF. The authors found a simple relationship between the layer thickness of material removed and shear stress acting on the surface. Petri et al. [17] established a predictive neural network model (NNM) using search algorithm for polishing and surface removal applications. They reported NNM as a very useful tool to examine the effect of variation of AFF process parameters. To predict abrasive wear in AFF Haan and Stief [18] developed a mathematical model and found abrasive particles have wear efficiency in the range of 10^5–10^3. Williams [18] compared mechanism of surface generation in AFF with grinding process by using acoustic emission signals. They reported that acoustic emission generated during AFF is highly correlated to the restriction created between workpiece and fixture, flow volume of the finishing medium and material removal rate.

Jain et al. [19] proposed a FEM strategy to predict forces and stresses established on the cylindrical workpiece during the flow of abrasive media. They also developed a model for MRR and average surface roughness in terms of abrasive particle size, number of abrasive particles contacting with workpiece surface, size and hardness of material and extrusion pressure. They reported (a) Increase in reduction ratio normal stresses increases on workpiece surface; (b) Increase in abrasive concentration and extrusion pressure results in increase in MRR; (c) Decrease in abrasive size decreases MRR; and (d) Roughness of machined surface reduces with increase in concentration of abrasive particle, extrusion pressure and abrasive particle size.

For selecting the optimum value of process parameters in AFF to predict the material removal and surface roughness, Jain and Jain [20] developed NNM and genetic algorithm. They reported that (a) The results produced by NNM and GA were very close agreement with each other; (b) possibility of use NNM to predict surface roughness and MR confirmed. The same author [21] developed hypothetical simulation model to decide active grain density in AFF medium. They reported (a) increase in concentration and size of abrasive results in increase in grain density of active abrasive particle; and (b) the developed model can simulate grain density active abrasive particle for any size and concentration.

Jain and Jain [22] compared experimental results of MRR and surface roughness with NNM and reported (a) good agreement between experimental result and result produced by NNM; and (b) NNM as a useful tool to observe the effect of variation of AFF parameters before experimenting. To predict specific energy and tangential forces in AFF the same author [23] proposed a model by changing number of cycles, extrusion pressure, concentration and size of abrasive and workpiece hardness. They also developed a heat transfer model to predict workpiece temperature. They concluded that (a) specific energy for grinding and AFF is overlapping; (b) specific energy remains constant respective to increase of abrasive particle size, but it remains higher for higher hardness materials; and (c) developed heat transfer model was capable to predict workpiece temperature. Jain and Jain [24] compared the result of material removal and surface finish produced by simulation and response surface analysis (RSA). They reported that (a) Increase in abrasive concentration causes increase in active number of particles hence enhance the material removal and surface finish; and (b) increasing the, number of cycles, extrusion pressure, abrasive concentration and reduction ratio reduces the surface roughness and increases material removal.

A multivariable regression analysis (MVRA) model to estimate material removal (MR) and surface roughness developed by Jain and Adsul [11]. They stated (a) MR is based on the initial surface of material and softer material gives more MR and surface finish compared harder material; (b) Increase in abrasive concentration increases surface finish and MR; and (c) Increase in abrasive particle size decreases MR and surface roughness.

Goranaet al. [25] modeled forces on a single abrasive particle in AFF. They measured radial, axial forces, active grain density and indentation depth experimentally and compared it with the result produced by the model. They reported that (a) radial, axial forces, and active grain density have a significant impact on deformation of materials; (b) rubbing and plough mode of deformation observed. The same author [26] developed an analytical model that simulate surface roughness by changing abrasive concentration, abrasive size, and extrusion pressure. They reported that increasing extrusion pressure and concentration of abrasive particle increases active grain density which decreases surface roughness.

Jain et al. [10] optimize process parameters of AFF using genetic algorithm for cylindrical surface and developed an optimization model for surface roughness. They found clŏse agreement between predicted and experimental results and validated developed surface roughness model by the experimental result of Jain and Adsul [11].

Wan et al. [27] developed a FEM simulation using a zero-order semi-mechanistic approach to evaluate surface roughness in two-way AFF. They reported that (a) zero-order methodology is advantageous to predict surface roughness; and (b) increasing extrusion pressure and finishing time increases surface finish.

Singh et al. [28] studied the theoretical model of AFF for change in surface roughness with respect variation in process parameters of AFF and developed the FEM model to predict finishing forces. Simulation on finishing forces used to predict radial forces. They concluded that (a) increasing extrusion pressure causes increase in finishing forces which increases surface finish; (b) 10% error among predicted and experimental result of surface finish found at extrusion pressure of 5 MPa; (c) increasing number of cycle decreases surface roughness; (d) both predicted and minimum experimental value of surface roughness was 0.238 μm at 800 number of cycle; and (e) increasing weight percentage of mixing oil decreases change in SR.

Wei et al. [29] proposed a MR model in terms of change in profile height and mass based on combined single abrasive and statistical active abrasive particle. They concluded that (a) MR model is related to the finish medium and material of workpiece; and (b) difference between the experimental results and theoretical results was 6.9% and 6.4% respectively.

Lv et al. [30] developed a model using the finite element method (FEM) and smoothed particle hydrodynamics (SPH) to investigate the erosion mechanism of aluminum nitride material subjected to abrasion. They observed that (a) radial and lateral cracks were the damage mechanisms in material and it is due to oblique impact of abrasive particle; (b) depth of indentation increases with increase of velocity of impact and angle of incident; (c) contact time of abrasive with workpiece increases with impact velocity and decreases with impact angle; and (d) increase in impact angle increases restitution coefficient of particle velocity.

Shao and Cheng [31] integrated micro-cutting mechanics modeling (MCMM) with the Monte Carlo (MC) simulation using computational fluid dynamics (CFD) approach to predict surface profile, roughness and MR of the integrally bladed rotor

(IBR) of turbine. They reported that (a) surface profile, roughness and MR influenced by volume and type of medium. Volume can be easily changed compared to changing medium, therefore mass production companies required prohibitively higher batch size before changing medium; and (b) CFD simulation is an effective tool to predict profile and surface roughness of workpiece using abrasion model.

7.4.2 Advances in Development of Finishing Medium and Rheological Characterization

AFF medium consists of mixture viscoelastic polymer, abrasive particle, mixing oil and other additives. AFF medium is a key element to achieve desired surface finish and material removal because of its capability to accurately erode the chosen areas ahead its flow length. AFF process performance depends on viscoelastic polymer used, abrasive size, concentration, mixing oil content, viscosity etc. The viscoelastic polymer utilized in AFF for finishing operation is a mixture of silicone oil and boric acid chemically it named as polyboroxilane (PBS). The AFF medium must able self-deformable on application of external force, flexible to follow complex passage and transfer force to abrasive particle for abrasion of work surface which is to be finished. Hull et al. [32] attempted for rheological characterization of PBS and they reported (a) PBS as rheotropic and complex time dependent polymer having shear thinning characteristics at high value of shear rate whereas shear thickening characteristics at low value of shear rate; and (b) finishing medium with higher shear rate produces more finishing result compared to medium having low shear rate.

Davies and Fletcher et al. [33] evaluated rheological characteristics of abrasive medium for low, medium and high viscosity. They reported that (a) increase of viscosity form low to high result an increased pressure reduction within the fixture and workpiece as well as decrease in rate temperature rise; (b) pressure drop and average viscosity decreases as the abrasive concentration increases; and (c) increase in temperature result an increase in volumetric flow rate and reduction in the viscosity which decreases surface finish and MR.

Jain et al. [34] investigated the rheological characterization of commercial grade putty and found a significant effect of medium viscosity on surface finish and MR. They reported (a) medium viscosity decreases with increased shear rate and wall shear stress; (b) medium viscosity increases with increase in abrasive concentration, but when temperature increases medium viscosity decreases and (c) material removal increases and the surface roughness decreases with increase in viscosity.

Wang and Weng et al. [35] attempted to develop a new finishing medium using vinyl-silicone polymer, SiC abrasive particle and additive. They reported (a) developed medium was capable to remove recast layer produced by WEDM; and (b) higher viscosity yield higher improvement surface finish. Wang et al. [36] compared silicon rubber (P-silicon) with silicon rubber with additives (A-silicon) based on rheological properties and finishing behavior. They reported that A-type silicone rubber having higher viscosity compared to P-type, therefore it finished complex hole easily compared to P-type silicone rubber having low viscosity.

Kar et al. [37] developed a finishing medium using SIC abrasive particle and styrene butadiene rubber (SBR) for AFF. They performed compression, hysteresis losses, tensile and tear testing of developed finishing medium. They stated developed medium is mechanically and thermally stable and give higher reduction in surface roughness. The same author [38] also compared performance of butyl rubber and natural rubber and reported butyl rubber gives higher surface finish compared to natural rubber.

Rajeshha et al. [39] developed medium which consists SiC as abrasive particles, natural polymer, and naphthenic oil and reported (a) Increasing concentration of abrasive and extrusion pressure results in better surface finish and higher MR; (b) No change in MR with increase of medium flow rate whereas surface finish increases with medium flow rate; and (c) Increase in medium viscosity increases surface finish and MR.

Sankar et al. [40] developed a finishing medium using styrene butadiene polymer, SiC abrasives particle and processing oil and used it for finishing Aluminum alloy and. They reported (a) developed finishing medium shows shear thinning phenomenon; (b) change in surface roughness increases with increase in viscosity of medium; (c) increasing in processing oil concentration results in better surface finish; and (d) change in surface roughness is higher in case of Al alloy/SiC_p (10%) compared to Al alloy and Al alloy/SiC_p (15%).

Sankar et al. [41] evaluated the styrene-butadiene based medium which hasSiC abrasives particle and processing oil. They finished Al alloy, Al alloy/SiC (10%) and Al alloy/SiC (15%) using developed medium and reported (a) increase in storage modulus and shear stress, resulting in higher MR; (b) SR increases with increase in shear stress and attain maximum value than start decreasing due to high radial forces; and (c) SR increases with decrease in amount of stress relaxation.

Bremerstein et al. [42] examined used and fresh medium consist of PBS, hydrocarbon oil, metal soap and SiC abrasive. They reported (a) medium viscosity and elastic percentage changes due addition swarf and fine abrasive particle; (b) no chemical alteration in medium properties seen during finishing; (c) continues finishing action causes blunting of sharp cutting edges and tips of abrasive particle; (d) used finishing medium result 20% poorer surface finish and 30% lesser MR; and (e) process efficiency reduced due to mixing of swarf and blunting of cutting edges of abrasive particle which result ploughing mechanism.

7.4.3 Advances in the Development of Hybrid and Derived Processes of AFF

AFF is very well established process for finishing variety of components but it having certain limitation such as (a) prolonged finishing time and less material removal rate; (b) non-interchangeability of fixture for finishing mass components; (c) unpredictability of wear of abrasive particle which finishing components; and (d) unable to visualize component and finishing mechanism while finishing. These inherent limitations overcome by hybridization AFF with other processes, modification of AFF in form of derived process and both to enhance the performance and productivity. The brief description of each process presented in following paragraphs.

7.4.3.1 Derived Processes of AFF

In the derived process of AFF some amendments in the existing AFF to improving its performance by providing rotational motion to workpiece, inserting drill bit in medium flow path, providing helical passage, movable rotatable mandrel, centrifugal force to rod and rotating spiral-shaped tool in finishing zone of AFF.

Rotational Abrasive Flow Finishing (R-AFF)

In R-AFF process, a rotational movement is provided to the workpiece contained fixture and to and fro motion to finishing medium. Combined rotational motion of workpiece and reciprocation of finishing medium improves surface finish and gives more material removal. In R-AFF due to rotary motion of work sample, tangential force (F_t) also act with axial force (F_a) and radial force (F_r) over abrasive particles contained in finishing medium (Fig. 7.2). These cause more abrasion which gives higher material removal with less finishing time [43].

Drill Bit Assisted Abrasive Flow Finishing (DBAFF)

Sankar et al. [44] used a helical drill bit in a medium flow path that provides irregular movement to the abrasives and causes reshuffling of the abrasive medium which results in higher surface finish and more MR. Figure 7.3 represent working principle of DBAFF. They used twin slot fixtures which bifurcated the abrasive media while entering to finishing the workpiece, and at exit, combine again and also assist the drill bit alignment along the workpiece axis. In this process intermixing of abrasive depends self-deformability of medium as well as on the extrusion pressure exerted on drill bit. Combined effect of different flows of abrasive medium, the workpiece-abrasive contact length increases and, becomes curved from the straight. Therefore, the number of sheared peaks increased and it results higher material removal.

Sankar et al. [44] reported that (a) as the drill bit diameter increases the surface finish and the material removal increased; and (b) surface roughness decreases as abrasive size, drill diameter, and number of cycles increase. Butola et al. [45] reported that using three start drill bit results in higher metal removal rates compared to two start drill bit.

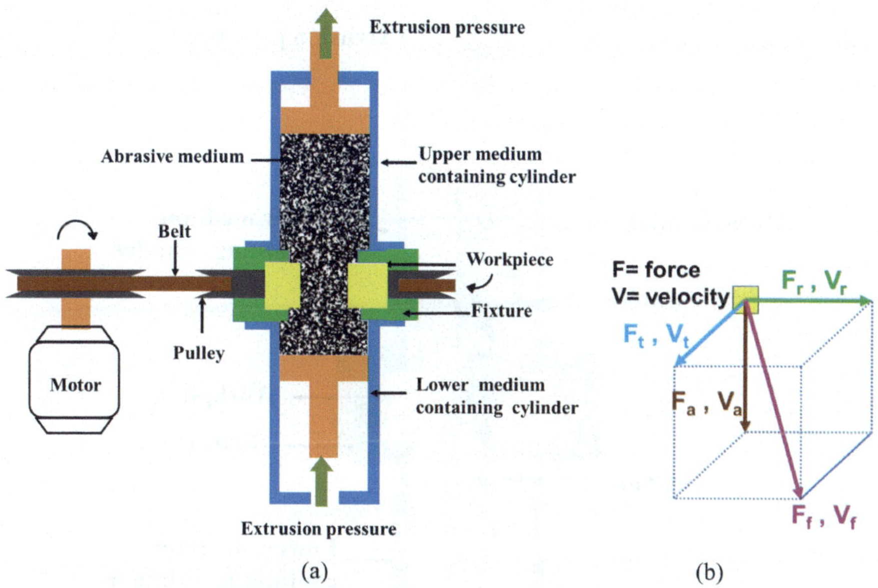

Fig. 7.2 Schematics of **a** arrangement for rotating workpiece contained fixture in R-AFF process, **b** components of the resultant force

Helical Passageway in Abrasive Flow Finishing (HAFF)

Wang et al. [46] placed the helical rod in the finishing zone of AFF to create multiple flow paths for AFF medium and to deflect the movement of abrasive for creating tangential force. The addition of tangential force with axial force and radial force causes improvement in surface roughness and material removal. Figure 7.4 depicts schematic of helical passageway in abrasive flow finishing. They reported that space between helical edge and workpiece surface was the significant factor to determine the level of surface roughness in AFF and five-helix passageway augment the polishing effectiveness more with regard to surface roughness and MR.

Spiral-Rotating Abrasive Flow Finishing (SAFF)

Yuan et al. [47] placed a grinding rod with rotating helical flute by an external motor in finishing zone to generate a centrifugal force that causes abrasive particles to impact over work surface frequently and randomly. This spiral-rotating motion of the abrasive medium with centrifugal force result higher surface finish due to homogeneous distribution of abrasive particles in the abrasive medium. Figure 7.5 depicts a schematic representation of spiral-rotating abrasive flow finishing. SAFF is applicable for finishing internal surfaces only. They reported that (a) increasing the rotation speed increases the tangential velocity of the abrasive particles which

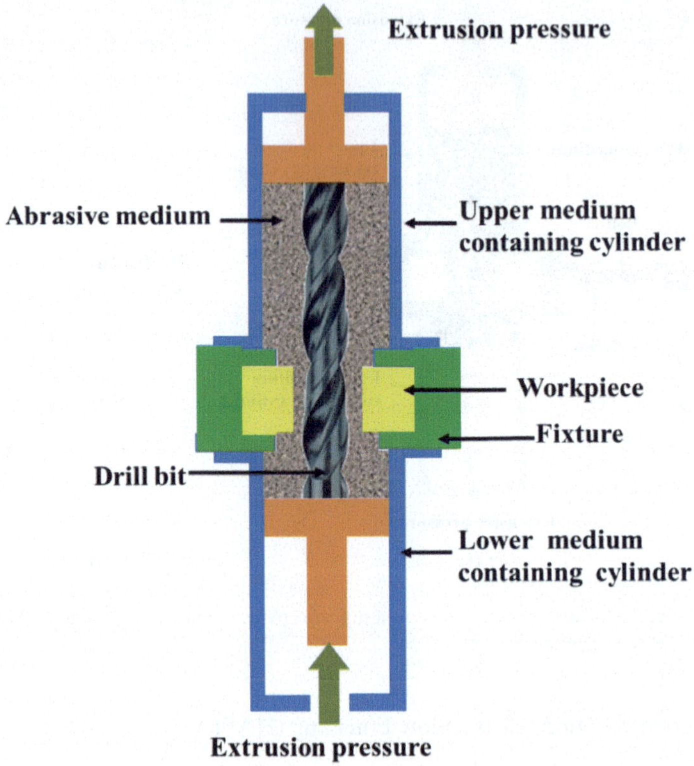

Fig. 7.3 Schematic of the drill bit assisted abrasive flow finishing

increases active grain density of abrasive particles on workpiece surface which result higher surface finish; (b) material removal rate higher compared to AFF due to higher kinetic energy in SAFF; and (c) extrusion pressure acting on abrasive having no effect on increase of the rotation speed of the helical flute.

Abrasive Flow Finishing with Movable Rotatable Mandrel (AFF-MRM)

Kenda et al. [48] used a movable/rotatable mandrel in the finishing zone of AFF to increase abrasive medium speed which results in the increase of active grain density towards work surface and result high surface roughness and efficiency of the process. Figure 7.6 depicts schematics of abrasive flow finishing with movable rotatable mandrel (AFF-MRM). They reported that AFF-MRM capable of controlling the microgeometry of the product with uniform polished surface and material removal in less time compared to AFF.

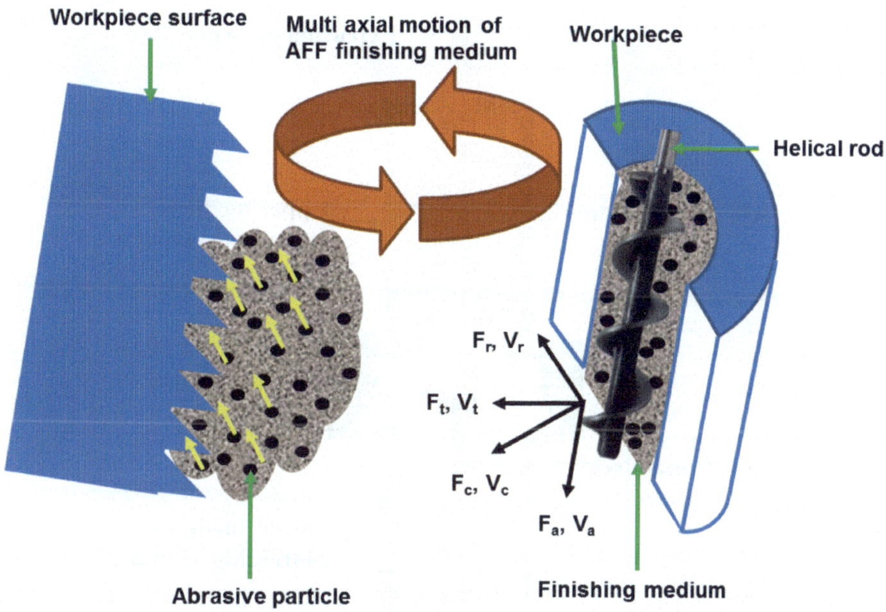

Fig. 7.4 Schematic of the working principle of helical passageway in abrasive flow finishing

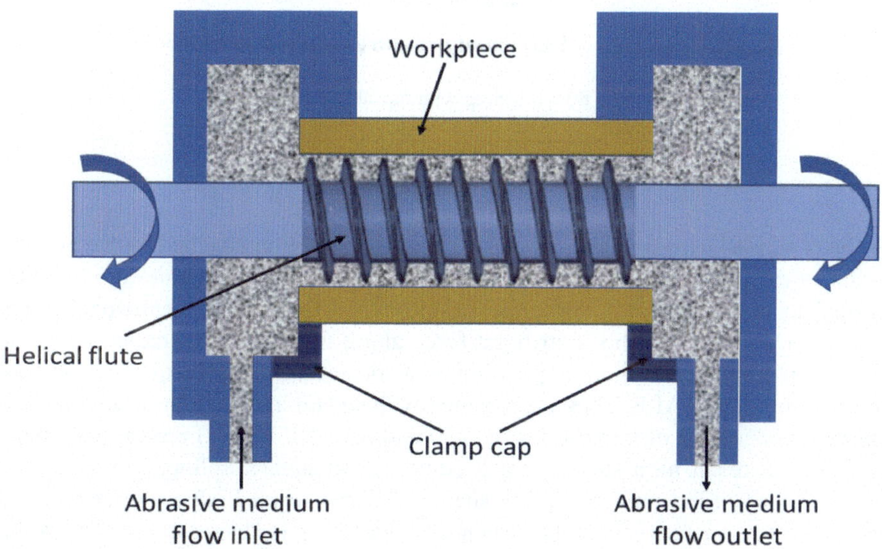

Fig. 7.5 Diagrammatic depiction of spiral-rotating abrasive flow finishing

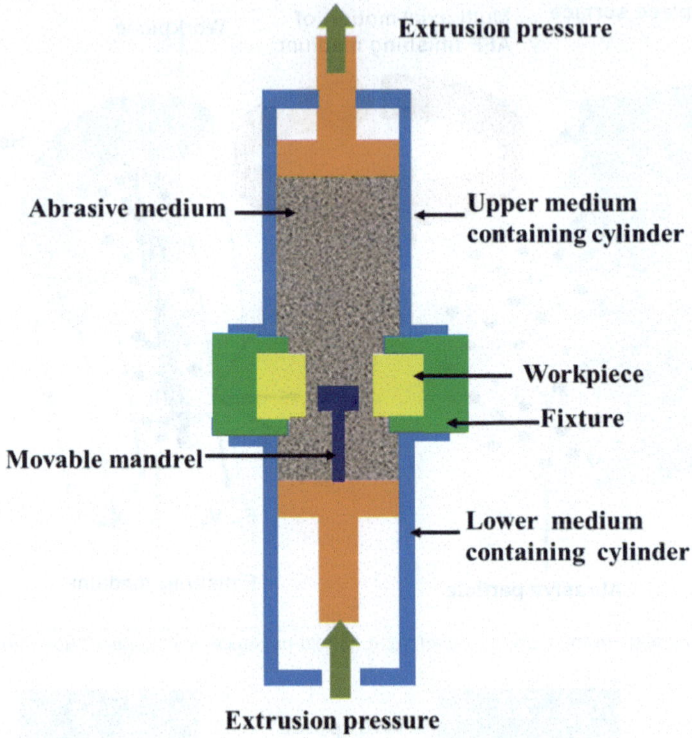

Fig. 7.6 Schematics of abrasive flow finishing with movable rotatable mandrel (AFF-MRM)

Centrifugal Assisted Abrasive Flow Finishing (CFA-AFF)

Walia et al. [49] used a rotating rod in AFF finishing region attached with a gearing system to apply centrifugal force on the abrasive particles towards the direction normal to the axis of the workpiece. They named this rod as Centrifugal Force Generator (CFG) rod. The centrifugal force along with axial and radial acting on abrasive particle causes more impression on work surface which results high surface finish compared to AFF. They investigated with square, rectangular, triangular and spline cross-section of CFG rod (Fig. 7.7) and revealed that (a) rectangular shape of CFG rod result high surface finish compared to square, spline, and triangular in decreasing order; and (b) Spline shape CFG rod result high material removal compared to square, rectangular, triangular decreases in decreasing order. Walia et al. [50] found that CFA-AFF process does not affect the surface micro-layer with the help of X-ray diffraction (XRD) analysis and optical micrographs.

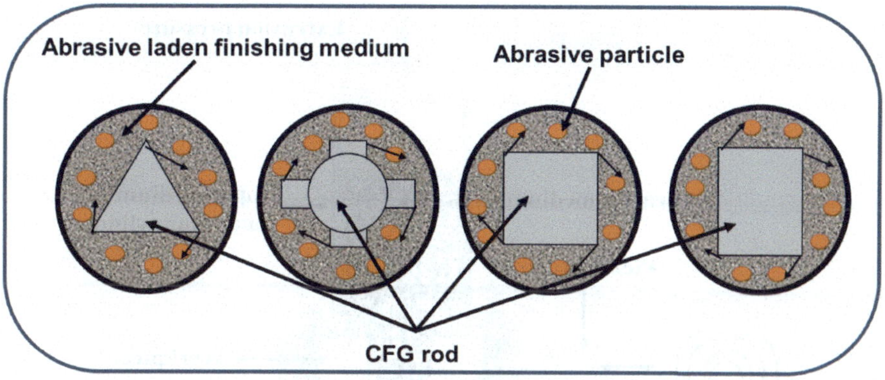

Fig. 7.7 Different type of CFG rod and their effect on abrasive particles

7.4.3.2 Hybrid Processes of AFF

Hybrid processes are the combination of two processes in order to produce highly efficient and more productive components by taking advantage of the worthiness of the constituent processes or overcome the limitation of constituent processes [51]. In AFF hybrid processes are developed by combining ultrasonic machining (USM), magnetic abrasive finishing (MAF), electrochemical machining (ECM) and magnetorheological finishing (MRF). The detailed descriptions of hybrid processes of AFF are reported in succeeding paragraphs.

Ultrasonic-Assisted Abrasive Flow Finishing (UA-AFF)

Jones et al. [52] combined ultrasonic machining with AFF for finishing blind dies and named this process as ultrasonic flow polishing (UFP). They reported that UFP could finish the surface of the cavity with a very high-quality finish while maintaining the workpiece dimensional accuracy. In UA-AFF workpiece is subjected to ultrasonic vibration with high-frequency piezo actuator in the perpendicular to the AFF medium flow direction which causes the abrasive particle in the abrasive medium to strikes the workpiece surface at an angle with high velocity [53]. Figure 7.8 shows the schematic of UAAFF process. Ultrasonic vibration added a radial force (F_r) perpendicular to the direction of abrasive medium flow which assists the abrasive particles to impinge into the asperities of the workpiece surface. Which results in good surface finish, high surface finishing and material removal rate in respect of AFF process. The aggressiveness of abrasive particles depends on the vibration frequency, extrusion pressure, medium viscosity, number of cycles, and vibration amplitude. They also report that increases in the applied vibration frequency and extrusion pressure increases material removal.

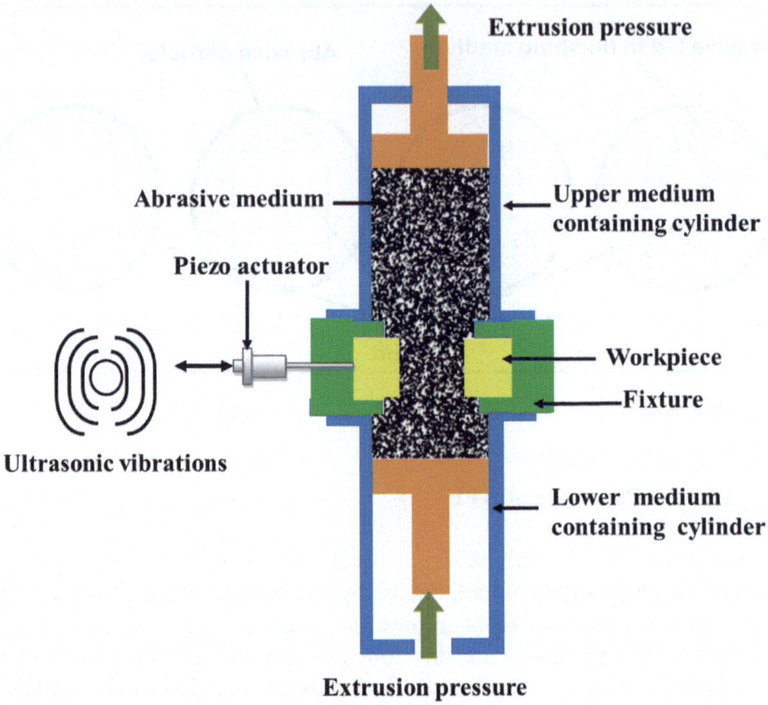

Fig. 7.8 Schematic representation of UAAFF process

Electrochemical-Aided Abrasive Flow Finishing (ECA^2FF)

Brar et al. [54] established ECA^2FF process using abrasive laden polymeric electrolytic finishing medium in AFF. They prepared a finishing medium with NaI salt (molar concentration), silicon based polymer, hydrocarbon gel with Al_2O_3 abrasive particle. A unique fixture containing copper rod (cathode) placed in inside cylindrical workpiece of brass (anode). Both workpiece and a copper rod placed inside nylon fixture and connectivity made to them for DC power supply of 0–30. When prepared finishing medium extruded between workpiece and copper rod it causes combined machining action due to abrasion by abrasive particle as well as electrolytic action result high surface finish and increased material removal. Figure 7.9 depicts (a) Schematic of electrochemical aided abrasive flow finishing (ECA^2FF) process. (b) Cathode and anode position in ECA^2FF process. They reported that (a) increase in voltage results in increase in MR; and (b) Higher operating voltage result rough surface.

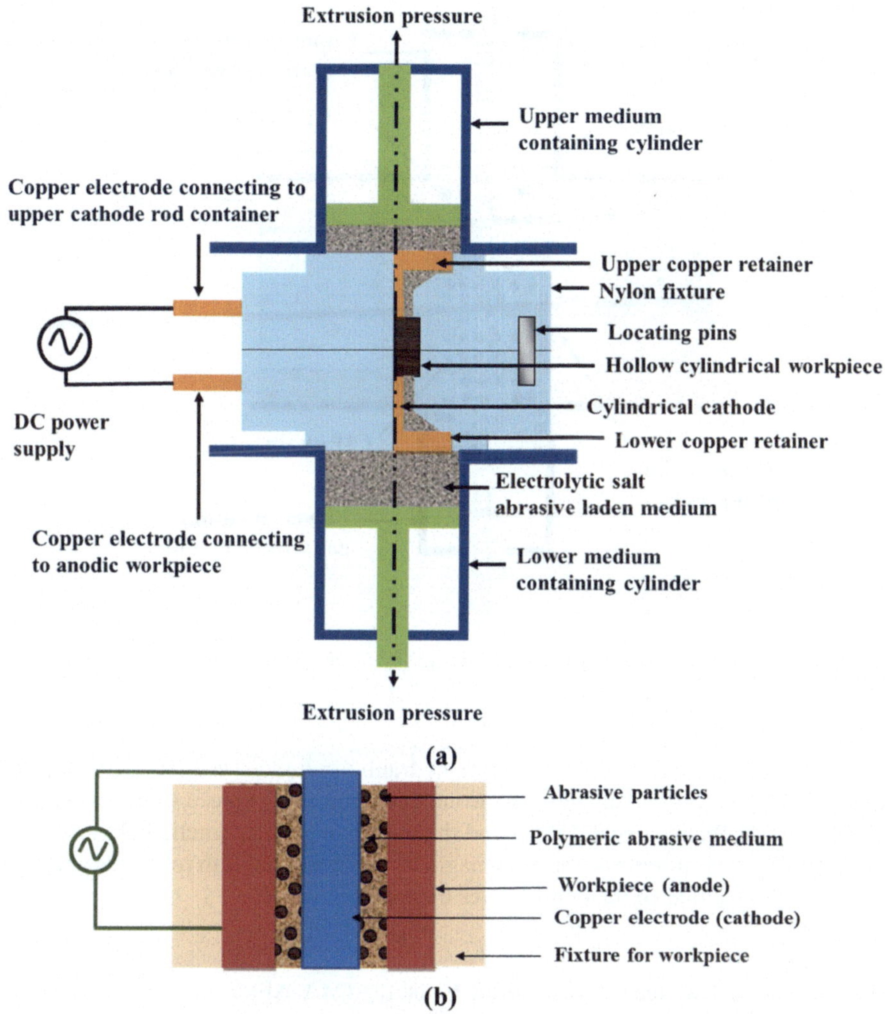

Fig. 7.9 Schematic of **a** ECA^{2}FF process, **b** Cathode and anode position in ECA^{2}FF process (*Source* [3] Reprinted with permission from Springer @ 2019)

Magneto Rheological Abrasive Flow Finishing (MRAFF)

This process uses the magnetic force to control the metal removal effectively using ferromagnetic particles and abrasive particles. This process has better control over the rheological properties of the abrasive magnetorheological media. Rheological properties of the abrasive magnetorheological media changes in the presence of a magnetic field. The viscosity of the abrasive magnetorheological media is strongly dependent on the strength of the applied magnetic field and it can be varied according to the application and workpiece material to obtain surface finish as desired. The

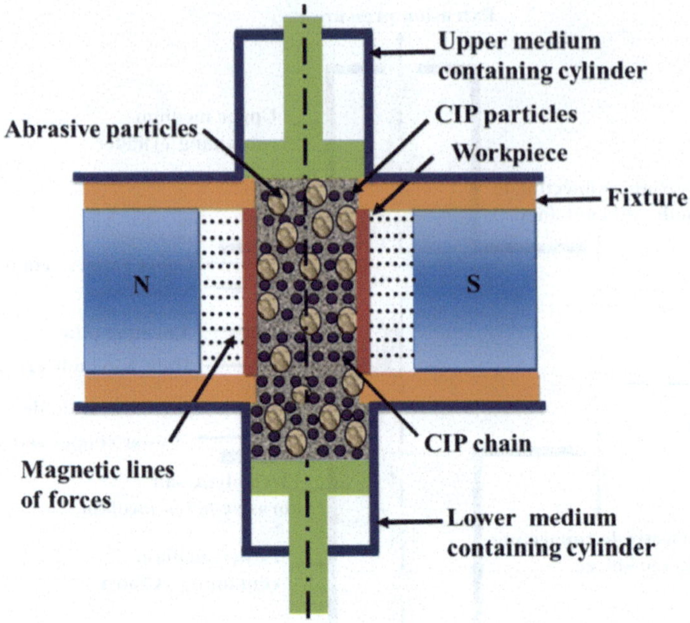

Fig. 7.10 Schematic of MRAFF process (*Source* [3] Reprinted with permission from Springer © 2019)

MRAFF experimental setup consists of hydraulic cylinders, magneto rheological abrasive (MRA) fluid containers, electromagnet, and a workpiece fixture as shown in Fig. 7.10. Kathiresan et al. [55] found that increasing the magnetic field increases the capability and strength of the ferromagnetic particles chains to retain the abrasive particles which enhance the control over the metal removal.

Magnetic Field Assisted Abrasive Flow Finishing (MA-AFF)

In this process, a fixture is used, which can also hold the electromagnets. The abrasive media composed of magnetic abrasive particles (containing 40% ferromagnetic ingredient, 40% Al_2O_3, and 15% Si_2O_3), a silicone-based polymer, and hydrocarbon gel. Magnetic field lines pull the abrasive particles towards the workpiece surface. The magnetic field puts a magnetic pressure on the abrasive media along with extrusion pressure. Schematic of magnetic field assisted abrasive flow finishing (MAAFF) process shown in Fig. 7.11. This process took less time to finishing of workpiece because it enhances surface finish and MRR, as compared to AFF. Singh et al. [56] reported the combined effect of low flow rates and high magnetic flux density produces higher MR in MAAFF than AFF.

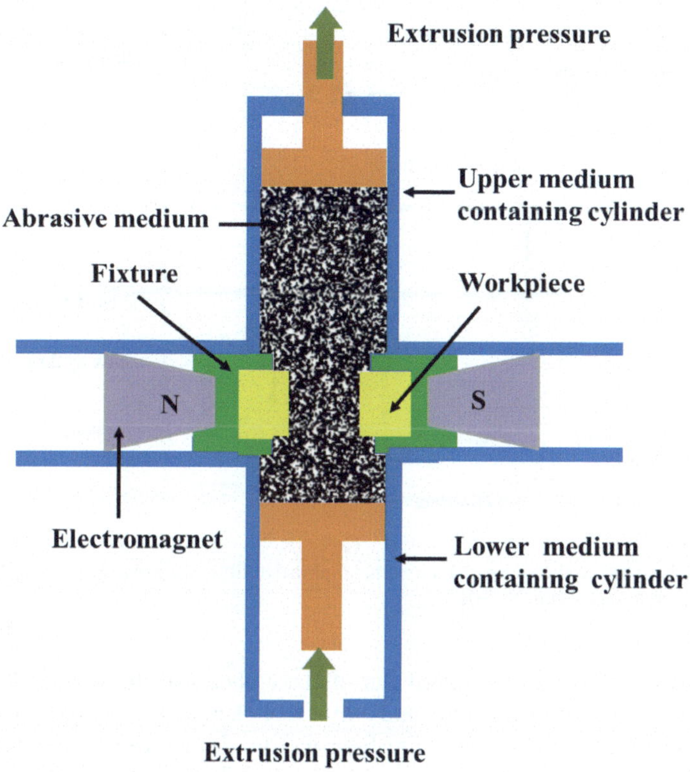

Fig. 7.11 Schematic depiction of MAAFF process

Chemical Abrasive Flow Finishing (CAFF)

In this process, fluid consist of acid and abrasive particles pumped to the inside of a hollow cylindrical workpiece. The schematic of CAFF process depicted in Fig. 7.12. The centrifugal pump tank mixer (1) is used for the pumping and mixing of chemicals and abrasives simultaneously. The valve (3) is used manually to change the fluid flow direction from the pump towards either the specimen tank (4) or depletion tank (5). The flow rate is measured by the ultrasonic flow sensor (6). The cylindrical workpiece (7) is placed inside the specimen tank (4). A corrosion-resistant thermocouple (8) measure the temperature of the fluid inside the feeding tank (2). After completion of the process, the fluid flows into the depletion tank (5).

Mohammadian et al. [57] developed this process to finish the inside surface of cylindrical IN625 components which were used in the aerospace industry. They performed static chemical polishing for 8 h with different chemical solutions and found 50% HF + 50% HNO_3 produce the better surface finish. They used above chemical solution diluted with water (40 volume% HF, 40 volume% HNO_3 and 20 volume% of distilled water) and alumina Al_2O_3 as abrasive particles size of 420 μm.

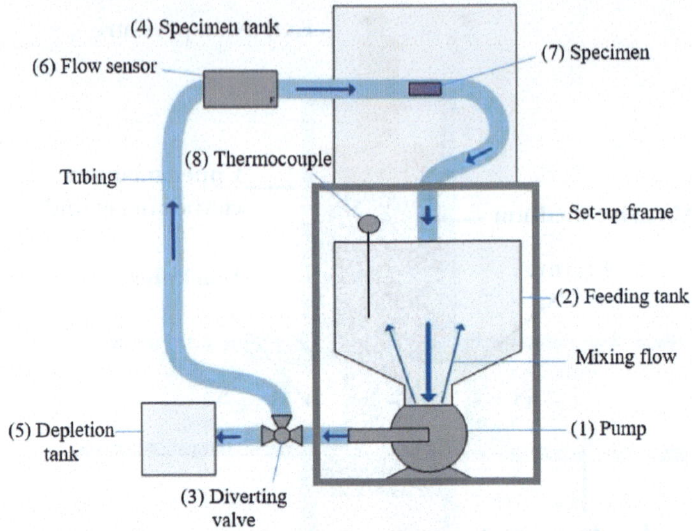

Fig. 7.12 Schematic of the setup of the chemical-abrasive flow finishing (*Source* [57] Reprinted with permission from Elsevier © 2018)

They found the workpiece required 8 h of static chemical finishing, 3 h of either chemical or abrasive flow finishing and only 1 h of chemical-abrasive flow finishing. When the chemical abrasive flow finishing was used, then time was reduced by a factor of 3 as a replacement of chemical or abrasive flow finishing processes.

The chemical finishing and abrasive flow finishing processes together enhance the outcomes of the process, such as material removal rate and surface finish. In the chemical abrasive flow finishing process, a passive layer formation has taken place at the workpiece surface because of chemical reactions which regularly removed by abrasive particles. This effect of the chemical solution and abrasive particles together makes the finishing process more efficient.

7.4.3.3 AFF Based Hybrid and Derived Processes

To enhance the performance of AFF more some hybrid and derived processes are also combined. The developed hybrid and derived processes of AFF are magnetorheological abrasive flow finishing with swirling-assisted inlet flow (MRAFF-SIF), rotational-magnetorheological abrasive flow finishing (R-MRAFF), magneto assisted spiral abrasive flow finishing (MS-AFF), electrochemical aided centrifugal force assisted abrasive flow finishing (EC^2A^2FF), ultrasonic-assisted double-disk magnetic abrasive flow finishing (UDDMAFF), centrifugal assisted abrasive flow finishing (CFA-AFF). The detailed working principle are described in below.

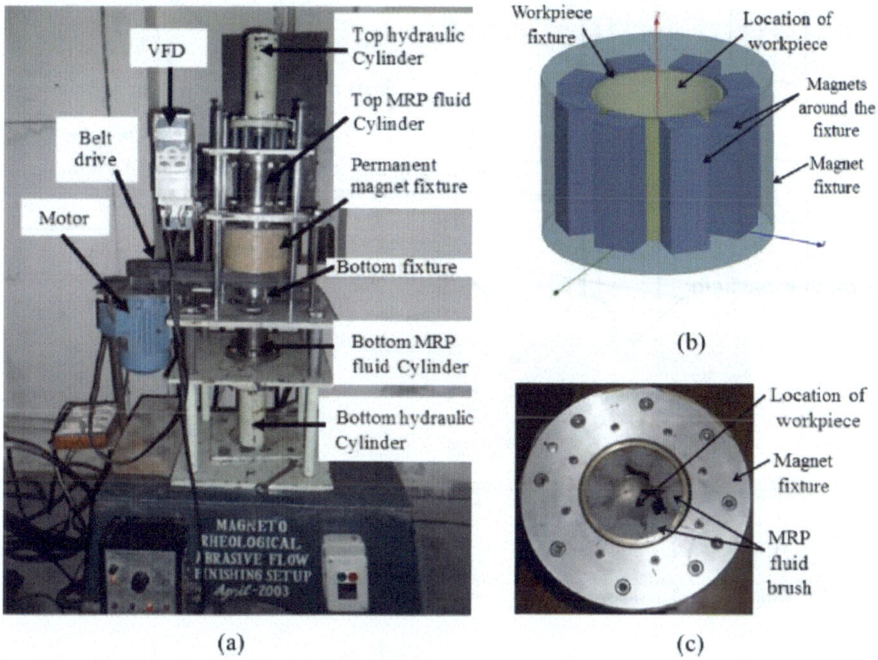

Fig. 7.13 Picture of **a** R-MRAFF experimental setup; **b** arrangement of workpiece and magnet fixture; and **c** location of the workpiece and MRP fluid brush in magnet fixture (*Source* [58] reprinted with permission from Elsevier © 2015)

Rotational-Magnetorheological Abrasive Flow Finishing (R-MRAFF)

Das et al. [58] combined rotary abrasive finishing (R-AFF) and magnetorheological Abrasive Flow Finishing (MRAFF) by providing rotary motion to MRP fluid contained inside of workpiece surrounded by the permanent magnet. Schematics of developed rotational-magnetorheological abrasive flow finishing (R-MRAFF) shown in Fig. 7.13. Permanent magnets are placed surrounding to workpiece and MRP fluid reciprocates inside the workpiece contained fixture. A rotary motion provided to permanent magnet which causes superimposition of rotary and reciprocating motion to MRP fluid. The superimposition of both motion causes centrifugal force with axial and redial force over MRP fluid contained abrasive particle. Therefore, contact of abrasive particles increases with workpiece surface which results in better surface finish than MRAFF.

Nagdeve et al. [59] reported non-uniform surface roughness and material removal in R-MRAFF due to non-uniform magnetic field. They introduced inverse replica of femoral surface of workpiece in fixture of R-MRAFF to obtain the uniform surface finish. This keeps velocity of MRP fluid constant in finishing region and gives uniform surface with constant material removal.

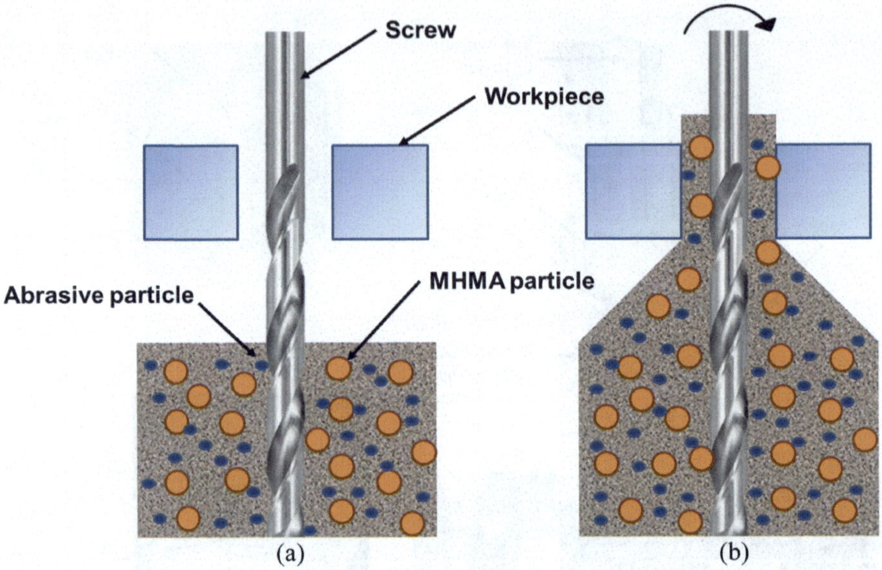

Fig. 7.14 Magneto assisted spiral abrasive flow finishing **a** screw stop, **b** screw rotate

Magnetorheological Abrasive Flow Finishing with Swirling-Assisted Inlet Flow (MRAFF-SIF)

Kheradmand et al. [60] introduced swirling vanes in finishing region to increase centrifugal force on the abrasive particle which enhances surface finish and material removal. They reported MRAFF-SIFas energy efficient and economical method to improve the productivity of MRAFF compared to rotating workpiece as in the case of R-MRAFF.

Magneto Assisted Spiral Abrasive Flow Finishing (MS-AFF)

In this process the abrasive medium was driven upward along the groove of the screw with the help of rotation of screw as shown in Fig. 7.14. When the abrasive medium quantity increased at the top side then it starts flow downward to the bottom of the mold because of the gravity. Chen et al. [61] invented a new type of magnetic hot melt adhesive particles (MHMA) which were transparent and colorless, these particles are magnetic and coated with steel grit. Steel grit is coated when the particle is hot melted (Fig. 7.15). A very strong magnet is placed around the workpiece to create magnetic field which attracts the magnetic hot melt adhesive (MHMA) particle with a great force and these particles force the abrasive particle towards the workpiece surface. When abrasive particles come into the contact of the workpiece, they start remove the asperities of the surface and result in the smooth surface of the workpiece. According

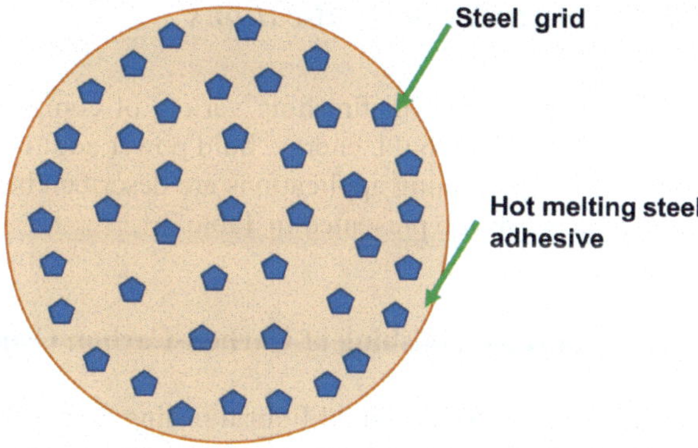

Fig. 7.15 Magnetic hot melt adhesive particle

to Chen et al. [61] magnetic flux density, MHMA particles concentration, and silicone oil viscosity are important variables.

Centrifugal Magnetic Force Assisted Abrasive Flow Finishing (CMFA-AFF)

Singh et al. [62] developed centrifugal magnetic force assisted abrasive flow finishing (CMFA-AFF) using centrifugal force generator (CFG) rod in finishing area of workpiece surrounded by magnets. They used three-part fixture which contained CFG rod, workpiece and magnets. The CFG rod inserted inside the workpiece and finishing medium is passed surrounding to it and special fixture contained magnets used to apply magnetic The CFG road having mechanism to provide rotary motion without hindering flow of finishing medium. On application of magnetic field abrasive partial moves towards workpiece and CFG rod apply extra centrifugal force over it which causes higher surface finish and material removal.

Electrochemical Aided Centrifugal Force Assisted Abrasive Flow Finishing (EC^2A^2FF)

Vaishya et al. [63] used electrolytic finishing medium in centrifugal assisted abrasive flow finishing. They developed a nylon fixture which holds a rotating CFG rod of copper retainer is attached to it. CFG rod placed inside a hollow cylindrical workpiece with a rotating mechanism. A DC power supply is given to rotating CFG rod (cathode) and workpiece (anode). A finishing medium consists of electrolytic salts and abrasives (Al_2O_3), viscoelastic medium of KI salt (a weak alkali) passed in between the gap of rotating CFG rod and workpiece. Electrolytic medium causes electrolytic dissolution of materials and centrifugal force act over abrasive particle. The electrolytic dissolution causes softening of work surface and centrifugal force contained abrasive removes more materials from workpiece surface. EC^2A^2FF is suitable for finishing of prismatic surfaces.

7.4.4 *Advances in Applications*

AFF frequently used for finishing variety of components in areas of automotive, aerospace, die and mould, energy, fluid power additive manufacturing, medical etc. Some worth mentioning applications are described below and detailed summary of AFF applications are presented in Table 7.1.

7.4.4.1 Surface Finishing of Carbon-Carbon Composite

Ravikumar et al. [64] used AFF for finishing carbon-carbon composites (C/Cs) and found that 32% improvement in surface finish at 150 cycles, an extrusion pressure of 6 MPa, and media having 70 wt% SiC abrasive particles (mess size of 220) and 12 wt% processing oil.

7.4.4.2 Microscopic Geometry Changes of a Direct-Injection Diesel Injector Nozzle

Jung et al. [65] used AFF to finish direct-injection diesel injector nozzle and AFF as efficient way to finish difficult to reach areas of injector nozzle which will enhance engine performance and emission.

7.4.4.3 Blade Surface Uniformity of Blisk

Fu et al. [66] used AFF for blisk surface aero engine and reported reduction surface roughness from 0.436 to 0.235 μm of blade surface without guide block. The surface finish of blade surface decreases from 0.513 to 0.141 μm with guide block and tool marks removed completely.

7.4.4.4 Finishing of a Hip Joint

Subramanian et al. [67] used AFF for finishing of a hip joint made of ASTM grade Co-Cr alloy. They reported 39 nm surface finish by using AFF technique with ability of finishing symmetrical object.

7.4.4.5 Nanofinishing of Knee Joint Implant

Kumar et al. [68] evaluate the effect of extrusion pressure and mesh size of abrasive particles for finishing knee joint implant by R-MRAFF process. The authors measured the surface roughness before and after finishing. They obtained nanometer

Table 7.1 Application summary of AFF and AFF based derived, hybrid, hybrid and derived processes

Applications	Process	Workpiece shape	Material	Remarks
Nano finishing of Surgical stainless-steeltube [71]	AFF	Cylindrical	Stainless steel (SS-316L)	• The average surface roughness (R_a) of 48 nm has been achieved
Polishing of rectangular microgroove [72]	AFF	Rectangular	Copper and stainless steel (SUS-304)	• Achieved average surface roughness (R_a) 4.8 nm and 12.7 nm respectively
Polishing of diamond-coated complicated tools [73]	AFF		Diamond-coated WC–Co milling cutter	• AFF efficiently finished and resharpen cutting edges of the nano-sized diamond coated milling cutters without affecting graphitization and purity diamond purity
Residual stress profiles induced [74]	AFF		Stainless steel (15-5PH)	• Compressive stress induced by AFF
Finishing of T-tube [75]	AFF	T shape	Not available	• Reported AFF as an effective method to improve surface finish of T pipe
Enhancing the microgeometry of laser textured spur gears [76]	AFF	Gear	Alloy steel (20MnCr5)	• Achieved percentage reduction of 28.49; 40.20; 24.85%; 4.80;69.12; 68.9 respectively in profile error, lead error, pitch error, radial runout, average surface roughness (R_a) and maximum surface roughness (R_{max}) by AFF
Wettability of machined biomaterials [77]	AFF		Stainless steel (SS-316L) and Titanium alloy (Ti-6Al-4V)	• AFF improved surface finish and wetting characteristics which reduced chances of bacterial infection in implants
Bacterial adhesion study on stainless steel [78]	MRAFF	Plate	Stainless steel (SS-316L)	• Achieved average surface roughness (R_a) in range of 10.52–37.4 nm
Finishing of the mold for manufacturing convex gear tooth [48]	AFFMM	Mold for convex gear tooth	Heat-treated steel grade (AISI H11)	• Surface roughness significantly reduces with desired concavity

(continued)

Table 7.1 (continued)

Applications	Process	Workpiece shape	Material	Remarks
Enhancing the surface finish, wear characteristics and microgeometry of straight bevel gears [79]	AFF	Gear	Alloy steel (20MnCr5)	• Reported finishing time and viscosity of medium having huge impact on simultaneous improvement in wear characteristics, surface finish and microgeometry of bevel gears
Surface finishing of convergent-divergent nozzle [80]	AFF	Nozzle (nonlinear tube)	Aluminum and mild steel	• Reported 56.14 and 40.55% of reduction in average surface roughness value for aluminum and mild steel on the convergent side respectively and 48.07 and 36.63% on divergent side
Finishing of gear mold used for manufacture plastic gears [81]	AFF	Gear	Heat-treated tool steel (AISI H11)	• Reported 89% reduction in average surface roughness (R_a)
Deburring of micro burrs in spring collets [82]	AFF	Spring collet	Chrome-molybdenum	• AFF completely removed micro burrs inside spring collets with improving surface finish • SiC abrasive particle gives higher improvement compared to Al_2O_3 media in terms of the removal of the burrs

scale surface roughness (35–78 nm range) at different locations of finished implants. Figure 7.16 depicts photographs of finished knee joint implants.

7.4.4.6 Finishing of Micro-Hole Fabricated by EDM

Lin et al. [69] manufactured microholes on SUS 304 and Ti-6Al-4V plates by using electrical discharge machining (EDM). They reported AFF process could enhance the dimensional accuracy and roughness of the machined microholes. The recast layers on the wall of the microhole generated were eliminated by the AFF process, so AFF is an effectual finishing method for microholes manufactured by EDM, increase the shape accuracy and reduction of surface roughness.

(a) (b)

Fig. 7.16 Photograph of the knee joint **a** before finishing, **b** after finishing (*Source* [68] Reprinted with permission from Elsevier © 2015)

7.4.4.7 Finishing of Microchannel Through AFF

Tzeng et al. [70] used self-modulating abrasive media which can change viscosity and fluidity during the process for finishing microchannels. They reveal that the average roughness of the microchannel was enhanced from 2.42 to 0.99 μm at extrusion pressure 6.7 MPa. The edge of the microchannel is clearly defined, which has been subjected to AFF.

They also reported that the viscosity of the abrasive medium decreases with finishing time. They also revealed that fluidity property of the medium is automatically adjusted while finishing the microchannels.

7.5 Future Research Work

- Additively manufactured components, shape memory alloy, composites, silicon carbide, titanium alloys, stellite, superalloy, Kevlar, dyneema and ceramics can be attempted to finish by AFF.
- Accurate prediction method for optimum AFF process parameters selection for respective geometry, materials prior to experimental investigation still not established besides many research attempts. More research initiatives for development of accurate modelling and simulation method for optimum selection of process parameters in AFF required.
- Disposal of used finishing medium and management of waste finishing medium is still an unexplored area of AFF.
- Research initiatives are required to develop low cost, readily available, non-toxic and easy to prepare finishing medium.

- Development of flexible and multi-utility fixture to finish variety of component are still not explored. A transparent fixture can be developed to visualize movement and finishing action abrasive particle which will help to simulate material removal mechanism.
- Finishing external surfaces of the workpiece such as spur gear, helical gear, bevel gears, worm and worm wheel by AFF based derived processes can be attempted.
- In the future, precision components produced by advanced precision manufacturing processes will need to be finished by newly developed abrasive flow finishing processes to maintain their dimensional accuracy and surface integrity.
- There is a need to develop a new abrasive flow finishing technology to finish very complex components made using additive manufacturing.

7.6 Summary

The following are the summary can be drawn from the advancement in the abrasive flow finishing process presented in this paper:

- Finishing results produced by AFF is highly influenced by selection abrasive particle size, concentration, mixing oil, type of polymer, viscosity, fixture design and selection of proper process parameters. Fixtures design changes according geometry, shape, size and complexity of workpiece. The finishing result strongly influenced by selective restrictions provided in the fixture.
- Many research initiatives attempted to optimize process parameters (i.e., abrasive concentration and size, extrusion pressure, number of cycles, medium viscosity, etc.) for finishing cylindrical, square, prismatic and rectangular shape and reported dominant variable in AFF to enhance material removal and surface finish.
- To improve of productivity AFF, many hybrid AFF processes were developed and investigated. AFF based hybrid processes having some limitations such as (i) design of fixture to provide ultrasonic vibrating to workpiece at vertical direction of medium flow in UA-AFF is very critical task; (ii) design of fixture to power supply workpiece and copper rod for electrochemical action, preparation of electrolytic finishing medium in ECA^{2}FF is very sophisticated and these process having limitation to finish only hollow electrical conductive workpiece; (iii) Non-uniform and less finishing rate in MRAFF; (iv) MA-AFF yield higher finishing result for non-magnetic workpiece compared magnetic workpiece; (v) CAFF required masking to prevent surfaces where finishing not desired. Derived process of AFF such as R-AFF, DBAFF, HAFF,SAFF,AFF-MRM and CFA-AFF required complicated design of fixture for various arrangements and these processes are limited finish internal surface of hollow workpiece. Hybrid and derived process of AFF such as R-MRAFF, MRAFF-SIF, MS-AFF, CMFA-AFF, EC^{2}A^{2}FF required very complicated arrangement in fixture to adopt workpiece, external mechanism and to pass finishing medium through workpiece without leakage.

References

1. Rhoades LR (1991) Abrasive flow machining: a case study. J Mater Process Technol 28(1–2):107–116. https://doi.org/10.1016/0924-0136(91)90210-6
2. Vector abrasive flow machining product data sheet (2019), Kennametal Extrude Hone, Irwin, PA, USA. https://extrudehone.com/wp-content/uploads/2015/12/B-11-02634_Vector_Product-Data-Sheet_US.pdf
3. Petare AC, Jain NK (2018) A critical review of past research and advances in abrasive flow finishing process. Int J Adv Manuf Technol 97:741–782. https://doi.org/10.1007/s00170-018-1928-7
4. Dabrowski L, Marciniak M, Szewczyk T (2006) Analysis of abrasive flow machining with an electrochemical process aid. Proc Inst Mech Eng Part B J Eng Manuf 220(3):397–403. https://doi.org/10.1243/095440506X77571
5. Williams RE, Rajurkar KP (1992) Stochastic modeling and analysis of abrasive flow machining. Trans ASME J Eng Ind 114(1):74–81. https://doi.org/10.1115/1.2899761
6. Hull JB, Jones AR, Heppel ARW, Fletcher AJ, Trengove SA (1992) The effect of temperature rise on the rheology of carrier media used in abrasive flow machining. Surf Eng Eng Appl II:235–244
7. Fang L, Zhao J, Sun K, Zheng D, Ma D (2009) Temperature as sensitive monitor for efficiency of work in abrasive flow machining. Wear 266:678–687. https://doi.org/10.1016/j.wear.2008.08.014
8. Przyklenk K (1986) Abrasive flow machining: a process for surface finishing and deburring of workpiece with a complicated shape by means of an abrasive laden medium. ASME, PED 22:101–110
9. Jain RK, Jain VK, Kalra PK (1999) Modeling of abrasive flow machining process: a neural network approach. Wear 231(2):242–248. https://doi.org/10.1016/s0043-1648(99)00129-5
10. Jain NK, Jain VK, Jha S (2007) Parametric optimization of advanced fine-finishing processes. Int J Adv Manuf Technol 34:1191–1213. https://doi.org/10.1007/s00170-006-0682-4
11. Jain VK, Adsul SG (2000) Experimental investigations into abrasive flow machining (AFM). Int J Mach Tools Manuf 40(7):1003–1021. https://doi.org/10.1016/S0890-6955(99)00114-5
12. Petare AC, Jain NK (2018) Improving spur gear microgeometry and surface finish by AFF process. Mater Manuf Process 33:923–934. https://doi.org/10.1080/10426914.2017.1376074
13. Kohut T (1988) Surface finishing with abrasive flow machining. In: SME Technical Paper, Proceedings of the 4th international aluminum extrusion technology seminar, Washington DC, New York, USA, pp 35–42
14. Jain VK (2008) Advanced (non-traditional) machining processes. In: Machining: fundamentals and recent advances. Springer London, pp 299–327. https://doi.org/10.1007/978-1-84800-213-5_11
15. Loveless TR, Williams RE, Rajurkar KP (1994) A study of the effects of abrasive-flow finishing on various machined surfaces. J Mater Process Technol 47:133–151. https://doi.org/10.1016/0924-0136(94)90091-4
16. Rajeshwar G, Kozak J, Rajurkar KP (1994) Modeling and computer simulation of media flow in abrasive flow machining process. In: American society of mechanical engineers, production engineering division (publication) PED, vol 68–2 New York, NY, United States: ASME, pp 965–971
17. Haan JJ, Steif PS (1998) Abrasive wear due to the slow flow of a concentrated suspension. Wear 219(2):177–183. https://doi.org/10.1016/S0043-1648(98)00191-4
18. Williams RE (1998) Acoustic emission characteristics of abrasive flow machining. Trans ASME J Manuf Sci Eng 120(2):264–271. https://doi.org/10.1115/1.2830123
19. Jain RK, Jain VK, Dixit PM (1999) Modeling of material removal and surface roughness in abrasive flow machining process. Int J Mach Tools Manuf 39(12):1903–1923. https://doi.org/10.1016/S0890-6955(99)00038-3
20. Jain RK, Jain VK, Kalra PK (1999) Modelling of abrasive flow machining process: a neural network approach. Wear 231(2):242–248. https://doi.org/10.1016/S0043-1648(99)00129-5

21. Jain RK, Jain VK (1999) Simulation of surface generated in abrasive flow machining process. Robot Comput Integrated Manuf 15(5):403–412. https://doi.org/10.1016/S0736-5845(99)00046-0
22. Jain RK, Jain VK (2000) Optimum selection of machining conditions in abrasive flow machining using neural network. J Mater Process Technol 108:62–67. https://doi.org/10.1016/S0924-0136(00)00621-X
23. Jain RK, Jain VK (2001) Specific energy and temperature determination in abrasive flow machining process. Int J Mach Tools Manuf 41(12):1689–1704. https://doi.org/10.1016/S0890-6955(01)00043-8
24. Jain RK, Jain VK (2004) Stochastic simulation of active grain density in abrasive flow machining. J Mater Process Technol 152:17–22. https://doi.org/10.1016/j.jmatprotec.2003.11.024
25. Gorana VK, Jain VK, Lal GK (2006) Forces prediction during material deformation in abrasive flow machining. Wear 260:128–139. https://doi.org/10.1016/j.wear.2004.12.038
26. Gorana VK, Jain VK, Lal GK (2006) Prediction of surface roughness during abrasive flow machining. Int J Adv Manuf Technol 31:258–267. https://doi.org/10.1007/s00170-005-0197-4
27. Wan S, Ang YJ, Sato T, Lim GC (2014) Process modeling and CFD simulation of two-way abrasive flow machining. Int J Adv Manuf Technol 71:1077–1086. https://doi.org/10.1007/s00170-013-5550-4
28. Singh S, Kumar D, Ravi Sankar M (2017) Experimental, theoretical, and simulation comparative study of nano surface roughness generated during abrasive flow finishing process. J Manuf Sci Eng 139:061012–061014. https://doi.org/10.1115/1.4035417
29. Wei H, Peng C, Gao H, Wang X, Wang X (2019) On establishment and validation of a new predictive model for material removal in abrasive flow machining. Int J Mach Tools Manuf 138:66–79. https://doi.org/10.1016/j.ijmachtools.2018.12.003
30. Lv Z, Hou R, Huang C, Zhu H, Qi H (2019) Meshfree analysis on dynamic behavior of hard brittle material in abrasive flow machining. Int J Adv Manuf Technol 100(5):2021–2030. https://doi.org/10.1007/s00170-018-2849-1
31. Shao Y, Cheng K (2019) Integrated modelling and analysis of micro-cutting mechanics with the precision surface generation in abrasive flow machining. Int J Adv Manuf Technol. https://doi.org/10.1007/s00170-019-03595-4
32. Hull J, Jones A, Heppel A, Fletcher A, Trengove S (1993) The effects of temperature rise on the rheology of carrier media used in abrasive flow machining. Special Publ Royal Soc Chem 127:235–235
33. Davies PJ, Fletcher AJ (1995) The assessment of the rheological characteristics of various polyborosiloxane/grit mixtures as utilized in the abrasive flow machining process. Proc Inst Mech Eng Part C J Mech Eng Sci 209(6):409–418. https://doi.org/10.1243/PIME_PROC_1995_209_171_02
34. Jain VK, Ranganatha C, Muralidhar K (2001) Evaluation of rheological properties of medium for AFM process. Mach Sci Technol 5(2):151–170. https://doi.org/10.1081/MST-100107841
35. Wang AC, Weng SH (2007) Developing the polymer abrasive gels in AFM process. J Mater Process Technol 192:486–490. https://doi.org/10.1016/j.jmatprotec.2007.04.082
36. Wang AC, Liu CH, Liang KZ, Pai SH (2007) Study of the rheological properties and the finishing behavior of abrasive gels in abrasive flow machining. J Mech Sci Technol 21(10):1593–1598. https://doi.org/10.1007/BF03177380
37. Kar KK, Ravikumar NL, Tailor PB, Ramkumar J, Sathiyamoorthy D (2009) Preferential media for abrasive flow machining. Trans ASME J Manuf Sci Eng 131(1):011009–011011. https://doi.org/10.1115/1.3046135
38. Kar KK, Ravikumar NL, Tailor PB, Ramkumar J, Sathiyamoorthy D (2009) Performance evaluation and rheological characterization of newly developed butyl rubber based media for abrasive flow machining process. J Mater Process Technol 209(4):2212–2221. https://doi.org/10.1016/j.jmatprotec.2008.05.012
39. RajeshaS Venkatesh G, Sharma AK (2010) Performance study of a natural polymer based media for abrasive flow machining. Indian J Eng Mater Sci 17:407–413

40. Sankar MR, Jain VK, Ramkumar J, Kar KK (2010) Rheological characterization and performance evaluation of a new medium developed for abrasive flow finishing. Int J Precision Technol 1(3–4):302–313
41. Sankar MR, Jain VK, Ramkumar J, Joshi YM (2011) Rheological characterization of styrene-butadiene based medium and its finishing performance using rotational abrasive flow finishing process. Int J Mach Tools Manuf 51(12):947–957. https://doi.org/10.1016/j.ijmachtools.2011.08.012
42. Bremerstein T, Potthoff A, Michaelis A, Schmiedel C, Uhlmann E, Blug B, Amann T (2015) Wear of abrasive media and its effect on abrasive flow machining results. Wear 342:44–51. https://doi.org/10.1016/j.wear.2015.08.013
43. Sankar MR, Jain VK, Ramkumar J (2016) Nano-finishing of cylindrical hard steel tubes using rotational abrasive flow finishing (R-AFF) process. Int J Adv Manuf Technol 85(9):2179–2187. https://doi.org/10.1007/s00170-015-8189-5
44. Sankar MR, Mondal S, Ramkumar J, Jain VK (2009) Experimental investigations and modeling of drill bit-guided abrasive flow finishing (DBG-AFF) process. Int J Adv Manuf Technol 42(7):678–688. https://doi.org/10.1007/s00170-008-1642-y
45. Butola R, Murtaza Q, Walia RS, Kumar P (2017) Two start and three start helical abrasive flow machining for brittle materials. Mater Today Proc 4(2):3685–3693. https://doi.org/10.1016/j.matpr.2017.02.263
46. Wang AC, Cheng K-C, Chen K-Y, Chien C-C (2015) Elucidating the optimal parameters of a helical passageway in abrasive flow machining. Int J Surf Sci Eng 9(2/3):145–158. https://doi.org/10.1504/IJSURFSE.2015.068239
47. Yuan Q, Qi H, Wen D (2016) Numerical and experimental study on the spiral-rotating abrasive flow in polishing of the internal surface of 6061 aluminium alloy cylinder. Powder Technol 302:153–159. https://doi.org/10.1016/j.powtec.2016.08.047
48. Kenda J, Pušavec F, Kopac J (2014) Modeling and energy efficiency of abrasive flow machining on tooling industry case study. Procedia CIRP 13:13–18. https://doi.org/10.1016/j.procir.2014.04.003
49. Walia RS, Shan HS, Kumar P (2008) Determining dynamically active abrasive particles in the media used in centrifugal force assisted abrasive flow machining process. Int J Adv Manuf Technol 38(11):1157–1164. https://doi.org/10.1007/s00170-007-1184-8
50. Walia RS, Shan HS, Kumar P (2008) Morphology and integrity of surfaces finished by centrifugal force assisted abrasive flow machining. Int J Adv Manuf Technol 39(11):1171–1179. https://doi.org/10.1007/s00170-007-1301-8
51. Gupta K, Jain NK, Laubscher, RF (2016) Hybrid machining processes: perspectives on machining and finishing. Springer International Publishing AG, Switzerland. https://doi.org/10.1007/978-3-319-25922-2 (eBook ISBN: 978-3-319-25922-2)
52. Jones A, Hull J (1998) Ultrasonic flow polishing. Ultrasonics 36(1–5):97–101. https://doi.org/10.1016/S0041-624X(97)00147-9
53. Sharma AK, Venkatesh G, Rajesha S, Kumar P (2015) Experimental investigations into ultrasonic-assisted abrasive flow machining (UAAFM) process. Int J Adv Manuf Technol 80(1–4):477–493. https://doi.org/10.1007/s00170-015-7009-2
54. Brar BS, Walia RS, Singh VP (2015) Electrochemical-aided abrasive flow machining (ECA^2FM) process: a hybrid machining process. Int J Adv Manuf Technol 79(1):329–342. https://doi.org/10.1007/s00170-015-6806-y
55. Kathiresan S, Mohan B (2018) Experimental analysis of magneto rheological abrasive flow finishing process on AISI stainless steel 316L. Mater Manuf Process 33(4):422–432. https://doi.org/10.1080/10426914.2017.1279317
56. Singh S, Shan HS (2002) Development of magneto abrasive flow machining process. Int J Mach Tools Manuf 42(8):953–959. https://doi.org/10.1016/S0890-6955(02)00021-4
57. Mohammadian N, Turenne S, Brailovski V (2018) Surface finish control of additively-manufactured Inconel 625 components using combined chemical-abrasive flow polishing. J Mater Process Technol 252:728–738. https://doi.org/10.1016/j.jmatprotec.2017.10.020

58. Das M, Jain VK, Ghoshdastidar PS (2012) Nanofinishing of flat workpieces using rotational–magnetorheological abrasive flow finishing (R-MRAFF) process. Int J Adv Manuf Technol 62(1):405–420. https://doi.org/10.1007/s00170-011-3808-2
59. Nagdeve L, Jain VK, Ramkumar J (2019) Preliminary investigations into nano-finishing of freeform surface (femoral) using inverse replica fixture. Int J Adv Manuf Technol 100(5–8):1081–1092. https://doi.org/10.1007/s00170-017-1459-7
60. Kheradmand S, Esmailian M, Fatahy A (2016) A novel approach of magnetorheological abrasive fluid finishing with swirling-assisted inlet flow. Result Phys 6:568–580. https://doi.org/10.1016/j.rinp.2016.08.014
61. Chen W-C, Wu K, Yan B-H, Tsao M-C (2013) A study on the magneto-assisted spiral polishing on the inner wall of the bore with magnetic hot melt adhesive particles (MHMA particles. Int J Adv Manuf Technol 69(5–8):1791–1801. https://doi.org/10.1007/s00170-013-5139-y
62. Singh R, Walia RS, Suri NM (2015) Parametric optimization of centrifugal-magnetic force assisted abrasive flow machining process using utility concept. Int J Res Eng Technol 4(8):382–388. https://doi.org/10.15623/ijret.2015.0408065
63. Vaishya R, Walia RS, Kalra P (2015) Design and development of hybrid electrochemical and centrifugal force assisted abrasive flow machining. Mater Today Proc 2(4):3327–3341. https://doi.org/10.1016/j.matpr.2015.07.158
64. Ravikumar NL, Kar KK, Sathiyamoorthy D, Kumar A, Devi R (2012) Surface finishing of carbon-carbon composites using abrasive flow machining. Fullerenes, Nanotubes, Carbon Nanostruct 20(2):170–182. https://doi.org/10.1080/1536383X.2010.533595
65. Jung D, Wang WL, Hu SJ (2008) Microscopic geometry changes of a direct-injection diesel injector nozzle due to abrasive flow machining and a numerical investigation of its effects on engine performance and emissions. Proc Inst Mech Eng Part A J Power Energy 222(2):241–252. https://doi.org/10.1243/09576509JPE421
66. Fu Y, Wang X, Gao H, Wei H, Li S (2016) Blade surface uniformity of blisk finished by abrasive flow machining. Int J Adv Manuf Technol 84:1725–1735. https://doi.org/10.1007/s00170-015-8270-0
67. Subramanian KT, Balashanmugam N, Shashi Kumar PV (2016) Nanometric finishing on biomedical implants by abrasive flow finishing. J Inst Eng C 97(1):55–61. https://doi.org/10.1007/s40032-015-0190-0
68. Kumar S, Jain VK, Sidpara A (2015) Nanofinishing of freeform surfaces (knee joint implant) by rotational-magnetorheological abrasive flow finishing (R-MRAFF) process. Precis Eng 42:165–178. https://doi.org/10.1016/j.precisioneng.2015.04.014
69. Lin YC, Chow HM, Yan BH, Tzeng HJ (2007) Effects of finishing in abrasive fluid machining on microholes fabricated by EDM. Int J Adv Manuf Technol 33:489–497. https://doi.org/10.1007/s00170-006-0485-7
70. Tzeng HJ, Yan BH, Hsu RT, Lin YC (2007) Self-modulating abrasive medium and its application to abrasive flow machining for finishing micro channel surfaces. Int J Adv Manuf Technol 32:1163–1169. https://doi.org/10.1007/s00170-006-0423-8
71. Singh S, Sankar MR, Jain VK (2018) Simulation and experimental investigations into abrasive flow nanofinishing of surgical stainless steel tubes. Mach Sci Technol 22(3):454–475. https://doi.org/10.1080/10910344.2017.1365897
72. Chen F, Hao S, Miao X, Yin S, Huang S (2018) Numerical and experimental study on low-pressure abrasive flow polishing of rectangular microgroove. Powder Technol 327:215–222. https://doi.org/10.1016/j.powtec.2017.12.062
73. Wang X-C, Wang C-C, Wang C-Y, Sun F-H (2018) Approach for polishing diamond coated complicated cutting tool: abrasive flow machining (AFM). Chin J Mech Eng 31(1):97. https://doi.org/10.1186/s10033-018-0296-4
74. Han S, Ferdinando S, Joël R (2019) Residual stress profiles induced by abrasive flow machining (AFM) in 15-5PH stainless steel internal channel surfaces. J Mater Process Technol 267:348–358. https://doi.org/10.1016/j.jmatprotec.2018.12.024
75. Li J, Zhou Z, Wei L, Zhang X, Xu Y (2017) Quality influence and process parameter optimization of T-pipe in abrasive flow finishing. Adv Mech Eng 9(8):1687814017718980. https://doi.org/10.1177/1687814017718980

76. Petare AC, Mishra A, Palani IA, Jain NK (2019) Study of laser texturing assisted abrasive flow finishing for enhancing surface quality and microgeometry of spur gears. Int J Adv Manuf Technol 101(1–4):785–799. https://doi.org/10.1007/s00170-018-2944-3
77. Hiremath SS (2019) Effect of surface roughness and surface topography on wettability of machined biomaterials using flexible viscoelastic polymer abrasive media. Surf Topogr Metrol Prop 7(1):015004. https://doi.org/10.1088/2051-672X/aaf6f6
78. Kathiresan S, Mohan B (2017) In-vitro bacterial adhesion study on stainless steel 316L subjected to magneto rheological abrasive flow finishing. Biomed Res (0970-938X) 28(7)
79. Petare AC, Jain NK (2018) On simultaneous improvement of wear characteristics, surface finish and microgeometry of straight bevel gears by abrasive flow finishing process. Wear 404–405:38–49. https://doi.org/10.1016/j.wear.2018.03.002
80. Guru G, Kumar S, Hiremath SS (2016) Investigation of abrasive flow finishing while machining convergent-divergent nozzle of different engineering materials. J Res Sci Technol Eng Manag 2(4):94–100
81. Kenda J, Duhovnik J, Tavčar J, Kopač J (2014) Abrasive flow machining applied to plastic gear matrix polishing. Int J Adv Manuf Technol 71:141–151. https://doi.org/10.1007/s00170-013-5461-4
82. Kim JD, Kim KD (2004) Deburring of burrs in spring collets by abrasive flow machining. Int J Adv Manuf Technol 24:469–473. https://doi.org/10.1007/s00170-002-1536-3

Chapter 8
Magnetic Abrasive Finishing Process

Sachin Singh, Vishal Gupta, and M. R. Sankar

Abstract The demand for high finish with accuracy combined with difficulty to finish components by conventional methods lead to the growth of advanced finishing processes. Magnetic abrasive finishing process is an advanced finishing process capable of finishing, deburring and precision chamfering of components. Magnetic abrasive finishing process can economically achieve nanometre surface finish on flat and cylindrical workpieces both internal and as well as externally. During magnetic abrasive finishing process, very low magnitude of finishing forces is applied on the magnetic abrasive particles to obtain nanometre surface finish. Therefore, it is particularly useful in finishing brittle materials without developing micro cracks and other defects. Various researchers modified the process and developed its hybrid variants. Their prime objective is to make the process more efficient and also capable of finishing micro features. To understand the physics of the process a considerable amount of work has been carried out in the form of experimental study with few attempts in the modeling of the magnetic abrasive finishing process.

Keywords Magnetic abrasive finishing · Surface roughness · Modeling · Abrasive particle

---

S. Singh · V. Gupta
Department of Mechanical Engineering, Thapar Institute of Engineering and Technology, Patiala 147004, India
e-mail: sachin.singh@thapar.edu

V. Gupta
e-mail: vishal.gupta@thapar.edu

M. R. Sankar (✉)
Department of Mechanical Engineering, Indian Institute of Technology Tirupati, Tirupati 517506, India
e-mail: evmrs@iittp.ac.in; ravisankarm@gmail.com

S. Das et al. (eds.), *Advances in Abrasive Based Machining and Finishing Processes*, Materials Forming, Machining and Tribology,
https://doi.org/10.1007/978-3-030-43312-3_8

Abbreviations

v	Volume of ferromagnetic particles
χ_m	Magnetic abrasive particles susceptibility
B	Magnetic flux density in the working zone
H	Magnetic field strength
μ_r	Relative magnetic permeability of pure iron
$[C]$	Derivative ofshape function matrix
μ_0	Magnetic permeability in vacuum
w	Volumetric ratio of iron in magnetic abrasive particles
n	Number of cutting edges of a magnetic abrasive particle taking part in finishing
N	Magnetic abrasives acting in the finishing region
v_p	Speed of magnetic abrasive particles
C	Constant of proportionality
t	Machining time
Δf	Force acting on a cutting edge of magnetic abrasive particle
H_w	Workpiece material hardness
θ	Half of asperity mean angle of abrasive cutting edges
R_a^0	Initial surface roughness of the workpiece
L_w	Workpiece length
q_i	Material removed from the 1 cm^2 area in the first minute
m	Constant (less than 1) depends on the productivity of the MAF process
$q_{\min}$	Minimum removal rate
μ_r	Relative permeability of MAPs
d	Diameter of abrasive particle
D_{map}	Diameter of MAP
F_z	Normal magnetic force
l_c	Length of cell
r_{tr}	Indentation projected area radius
$R_{\max}^0$	Initial surface roughness ($= 2R_a^0$)
R_{tr}	Radius of track
μ	Mean
σ	Standard deviation
k_w	Thermal conductivity of workpiece
$\{N\}$	Shape function matrix
v_b	Velocity of abrasive particles
μ_f	Coefficient of workpiece
R_w	Energy partition
σ_e	Effective stress

$\dot{\varepsilon}_e$	Effective strain
V_w	Volume of material under deformation
D_c	Constant
μ_{ab}	Magnetic permeability
μ_{fm}	Ferromagnetic phase magnetic permeability
V_{ab}	Abrasive phase volume fraction
V_{fm}	Ferromagnetic phase volume fraction
$V(x, y)$	Magnetic pole rotation speed
P_F	Instantaneous magnetic pressure the magnetic force
T_S	Number of test data
t_i	Target value
a_i	FFBP-NN predicted value
T_r	Total number of training sample
y_i	ANFIS output value in training

8.1 Introduction

Magnetic abrasive finishing (MAF) process was developed in East European countries due to the need for mechanization of the manual finishing of intricately shaped industrial parts [1]. Finishing is the last operation performed during the components manufacturing cycle. Surface roughness (R_a) greatly influences the functional properties such as fatigue, wear and friction of the component. Vibration of the rigid cutting tool during machining of the difficult to machine material (nickel, cobalt, titanium-based superalloys, etc.) and formation of cracks during machining of the brittle materials (ceramics, glasses, etc.) are major shortcomings of the conventional finishing processes. To achieve the demand of high surface finish with good accuracy simultaneously to minimize the machining cost with negligible surface defects during machining operation lead to the development of advanced finishing processes. Abrasive flow finishing (AFF), magnetic abrasive finishing (MAF), magnetorheological finishing (MRF) and magnetorheological abrasive flow finishing (MRAFF) processes are such recently developed advanced finishing processes. MAF process combines the magnetization behavior of the ferromagnetic particles (e.g., Fe) and machining behavior of the abrasive particles (e.g., Al_2O_3, SiC) to form a flexible finishing tool. MAF process is mainly used for finishing flat and cylindrical components. Material removal rate (*MRR*) and minimum R_a that can be achieved during MAF process is approximately 1 μm/min and 5 nm respectively [2]. Also, the low magnitude of finishing forces developed during the process minimizes the probability of generating defects such as microcracks during the finishing of brittle materials.

8.2 Primary Components of the Magnetic Abrasive Finishing Process

Some of the essential components of the MAF set-up are:

i. **Magnets**: Permanent or electromagnet magnets are used for generating the magnetic field. Magnetic lines of force generated by the magnets act as the finishing forces during the MAF process. Magnetic field intensity could be controlled in case of electromagnets by varying current, voltage and number of turns in the coil.
ii. **Ferromagnetic particles**: Ferromagnetic particles are required due to their magnetism properties in an external magnetic field. During the MAF process, ferromagnetic particles form a flexible magnetic abrasive brush (FMAB) along the magnetic lines between the north pole and the south pole of the magnets. The size of the ferromagnetic abrasive particles varies in between 50 μm to 400 μm. Size and type of the ferromagnetic particle influence the magnitude of the magnetic field density generated during the MAF process.
iii. **Abrasive particles**: Based on the requirement of the *MRR*, size of the abrasive particles are decided which generally lays between 1 μm to 10 μm. Abrasive particles can be silicon carbide (SiC), aluminium oxide (Al_2O_3), boron nitride, diamond, etc. Choice of the abrasive particles depends on the hardness of the workpiece to be finished. Ferromagnetic particle and abrasive particles are together known as magnetic abrasive particles (MAPs) can be of two types unbonded and bonded. Unbonded MAPs consists of a mechanical mixture of ferromagnetic particles and abrasive particles. While abrasive particles are held in ferromagnetic particles matrix formed by sintering or any other process in bonded MAPs [3].
iv. **Lubricant**: To enhance the flexibility of the FMAB sometimes lubricant is incorporated in the mixture of magnetic abrasive particles. Few authors added abrasive gel to the MAPs mixture. Lubricants increase the stability of the FMAB, decreases flown of MAPs from the working zone at high rotational speed (RS).

Mainly three types of energy are needed for the formation of FMAB (Fig. 8.1) [4]:

i. Magnetization energy, W_m for magnetizing the MAPs.
ii. Repulsion energy, W_f for countering the Faraday effect due to which there is a repulsion between MAPs.
iii. Tension energy, W_t for opposing the curved MAPs chains.

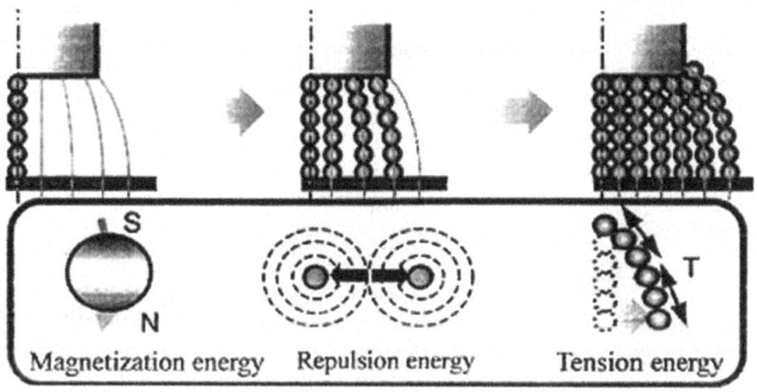

Fig. 8.1 Configurations of magnetic brush under the influence of various energies [4, Licence number: 4618170731940]

8.3 Types of Magnetic Abrasive Finishing Process

8.3.1 Flat Magnetic Abrasive Finishing Process

Finishing of large size flat workpiece is carried by flat magnetic abrasive finishing process (FMAF). As shown in Fig. 8.2a during FMAF process workpiece is placed between the magnet poles. A small gap (1mm –3 mm) is maintained for the FMAB which is formed as the magnetic field is applied during the FMAF process (Fig. 8.2b). As the magnetic field is applied ferromagnetic particles orient themselves along the magnetic lines of force due to dipole interaction. Abrasive particles get embedded between the ferromagnetic particles forming a finishing tool with random, multiple cutting edges. Finishing forces are transferred by the ferromagnetic particles to the abrasive particles for the removal of material during the FMAF process. As shown in Fig. 8.2c magnetic energies result in imparting the normal magnetic force (F_z) to the abrasive particles while the mechanical energy due to the rotation of FMAB generate tangential force (F_t). F_z helps the abrasive particle to indent into the workpiece surface forming a plastically deformed zone, which is later removed in the form of microchip due to F_t. F_z is influenced by several factors such as magnetic flux, size, shape, magnetic property and wt. % of the ferromagnetic particles, and working gap (WG). While, F_t is mainly influenced by the RS of the magnet.

8.3.2 Cylindrical Magnetic Abrasive Finishing Process

Cylindrical magnetic abrasive finishing (CMAF) process is used for finishing cylindrical rollers. As shown in Fig. 8.3a during the CMAF process workpiece (roller) is held by the spindle chuck between two magnets. Rotary motion is provided to the workpiece by the spindle, while the axial vibratory motion to the magnetic

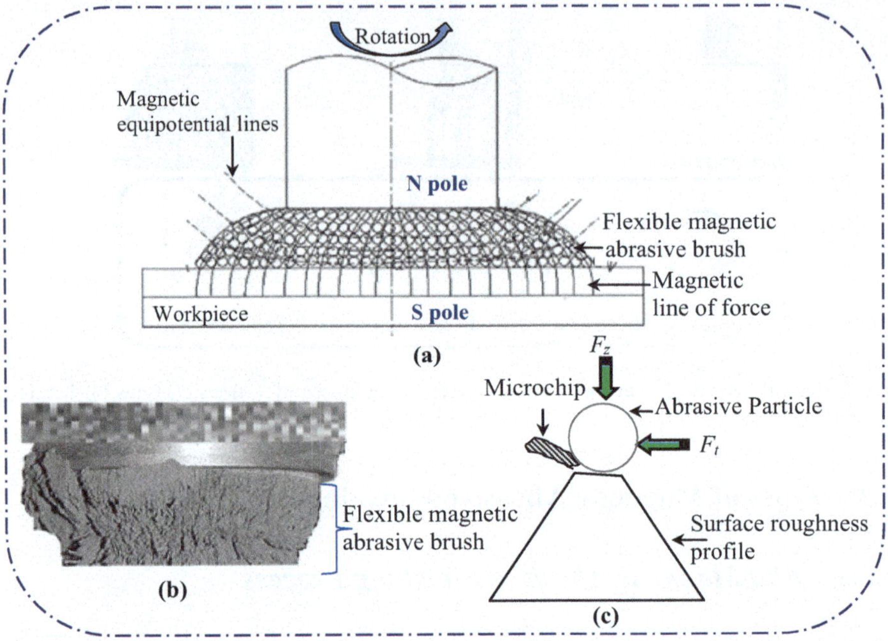

Fig. 8.2 **a** Schematic of plane magnetic abrasive finishing process, **b** image showing flexible magnetic abrasive brush [27, Licence number: 4618201082447], **c** schematic of abrasive particle indenting the roughness profile

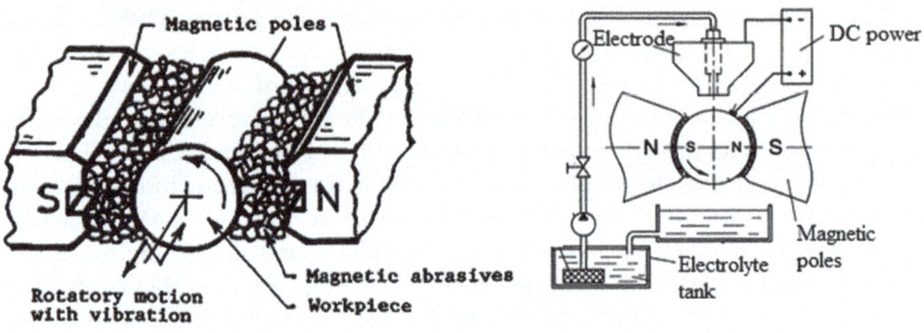

(a) (b)

Fig. 8.3 Schematic showing **a** cylindrical magnetic abrasive finishing process [1, Licence number: 4618201342816], **b** electrolytic magnetic abrasive finishing process [5, Licence number: 4618221231954] field is given by oscillating magnetic poles w.r.t. to the workpiece. FMAB acts as a deformable multi point finishing tool during the CMAF process. Flexibility of the FMAB can also be increased by adding solid lubricants such as zinc stearate. Thus, a finely finished surface on the workpiece is generated after the CMAF process.

8.3.3 Hybrid Magnetic Abrasive Finishing Process

To achieve high quality surface finish in combination with high efficiency several researchers tried to develop various hybrid MAF process. Their primary objective is to combine the advantages of the constituting processes and eliminating their disadvantages. It is found from the experimental studies hybrid MAF processes performed better than the MAF process. Some of the major MAF variants developed in literature are described in the following section.

8.3.4 Electrolytic Magnetic Abrasive Finishing Process

CMAF process finishing efficiency is enhanced by Yan et al. [5] by proposing the electrolytic magnetic abrasive finishing process (EMAF). Authors integrated the electrolyte supply system with the CMAF process set-up (Fig. 8.3b). During the finishing operation electrolyte fills the space between the workpiece and the electrode. Workpiece and electrode are connected to the anode and cathode of the DC power supply respectively. The dual action of electrolysis in combination with magnetic field results in accelerating the *MRR* and obtaining a superior finished surface compared with the MAF process. Taweel [6] employed the EMAF process for finishing 6061 Al/Al_2O_3 composite. Authors concluded that due to Lorentz force on the electrolytic ions, magnetic field lowers the electrolyte electric resistance. Thus, increasing current density which enhances *MRR*. Nickel-based superalloy is finished by Du et al. [7] with the help of EMAF process. During the EMAF process, care should be taken in matching the relationship between the electric field intensity and magnetic intensity. If electric field intensity is more, than the passive film is not completely removed by the abrasives. If magnetic field is more than the passive film is very thin and the role of electrolysis is not fully played. An uncontrolled electrolyte supply during finishing can cause surface defects such as pitting, sags and crests that can deteriorate the surface finish.

8.3.5 Vibration-Assisted Magnetic Abrasive Finishing

Few authors developed vibration assisted magnetic abrasive finishing process (VMAF) for enhancing the performance of the MAF process. As shown in Fig. 8.4a

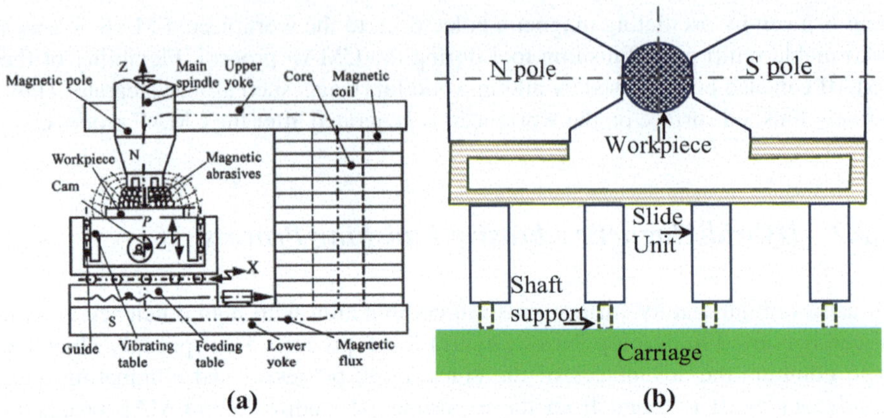

Fig. 8.4 Schematic of **a** vibration-assisted magnetic abrasive finishing process [10, Licence number: 4618240386941], **b** vibration assisted cylindrical–magnetic abrasive finishing process

during the VMAF process vibrations are given to the flat workpiece. Vibrations not only promote the abrasives relative motion but also cause the magnetic field to behave as dynamic. Thus, the magnitude of finishing forces generated during VMAF process is higher in comparison with MAF process. Natsume et al. [8] provided only the horizontal (X direction) vibration during the finishing by the MAF process. While Yin and Shinmura [9] modified the MAF process by providing vertical (Z direction) vibration with the help of a motorized cam system. Authors showed that developed set-up is particularly useful for deburring operation in addition to the efficient finishing of the workpiece. Later, authors [10] added horizontal and vertical vibration to obtain mixed mode (XZ direction) vibration of the working table. Authors showed experimentally that the developed set-up is particularly useful in nano finishing micro-curved surfaces of components. Authors [11] also provided vibrations in both the axis of the horizontal plane i.e., *X* and *Y* direction during finishing of the SS flat workpiece by VMAF process. Authors found the optimal combination of input parameters as 1 mm WG, weight of SiC: steel particles: machining fluid as 1:1.5:3, 16.67 Hz frequency of vibration, 500 rpm RS of magnet.

Judal et al. [12] developed a vibration assisted cylindrical–magnetic abrasive finishing (VAC–MAF) process for finishing cylindrical workpieces. As shown in Fig. 8.4b high speed lathe machine is modified into VAC–MAF process set-up. In combination with the rotating the workpiece, authors employed a vibration mechanism to vibrate the magnetic poles. Electromagnet arrangement is mounted on the slide unit. Such vibrations help in replenishing the worn out abrasive particles with the new ones during the finishing process. Hence, increasing ΔR_a and *MRR* during the VAC–MAF by 150 % and 100 % compared to the MAF process.

MAF process enjoys numerous advantages some of the features that provides MAF processes an edge over the other finishing process are:

(i) MAF process is economic as it doesn't require costly machines and tooling.

(ii) Finishing tool i.e., FMAB employed during the MAF process is self-sharpening and self-adaptable. Also, tool neither requires compensation nor dressing. Thus, increasing the productivity.
(iii) MAF process can successfully finish brittle materials such as ceramics and glass without inducing defects on their surfaces.
(iv) The magnitude of finishing forces generated during the MAF process can be controlled easily.

8.4 Experimental Investigation

Shinmura et al. [13] improved the finishing efficiency of the MAF process by developing a set-up that generates dynamic magnetic field during the MAF experiments. Authors found out that the finishing performance of the MAF process improves when time varying magnetic field is used in combination with the static magnetic field. Fox et al. [3] finished steel rollers to a R_a of 7.6 nm. Authors studied the effect of various MAF input parameter viz. bonded and unbonded magnetic abrasive particles, magnetic flux density, wt. % of lubricant, and axial vibration of the magnetic field on the R_a of the workpiece. Authors [14, 15] developed the MAF process for internally finishing straight pipes that are commonly found in the gas piping system. As the magnetic field is produced by the electromagnet, magnetic abrasive particles and iron abrasive particles conglomerate on the inner tube surface (Fig. 8.5a). As the tube rotates the relative motion between the tube inner surface and the abrasive achieve finished inner surface. Authors [15] achieved a minimum R_a of 0.2 μm in the least time by using 330 μm mean diameter iron particles, 50 μm mean diameter abrasive particles with 50 wt. % of each constituent in the mixture. Later, authors [16] developed the MAF set-up for finishing straight as well as bent pipes. As shown in Fig. 8.5b finishing unit consists of magnets that are connected to the arm of a 5 axis robot. The robot moves the finishing unit along the axis of the tube, maintaining a constant gap between the tube surface and the magnets. This helps in achieving nearly uniform R_a along the length of the bent tube. Yamaguchi and Shinmura [17] internally finished alumina ceramic tubes and achieved the minimum R_a of 0.02 μm. Authors also showed that MAF process produces a negligible effect on the residual stress of the finished surface. Yamaguchi et al. [19] extended the application of the MAF process by successfully finishing SS304 capillary tubes with the internal diameter of 800 μm and 400 μm. Authors developed a workpiece rotation system (Fig. 8.5c). Amount of the MAPs and magnitude of the magnetic field largely determines MAPs performance. Authors observed that in the workpiece region with weak magnetic field MAPs shows unstable behavior. Therefore, the developed system is not suitable for finishing long tubes due to the restrictions of the pole tip dimensions that result in weak magnetic forces in far off regions. Therefore, authors [18, 19] later developed a multi pole tip MAF set-up for finishing long capillary tubes. As shown in Fig. 8.5d multi pole tip system extends the finishing region along the axial direction

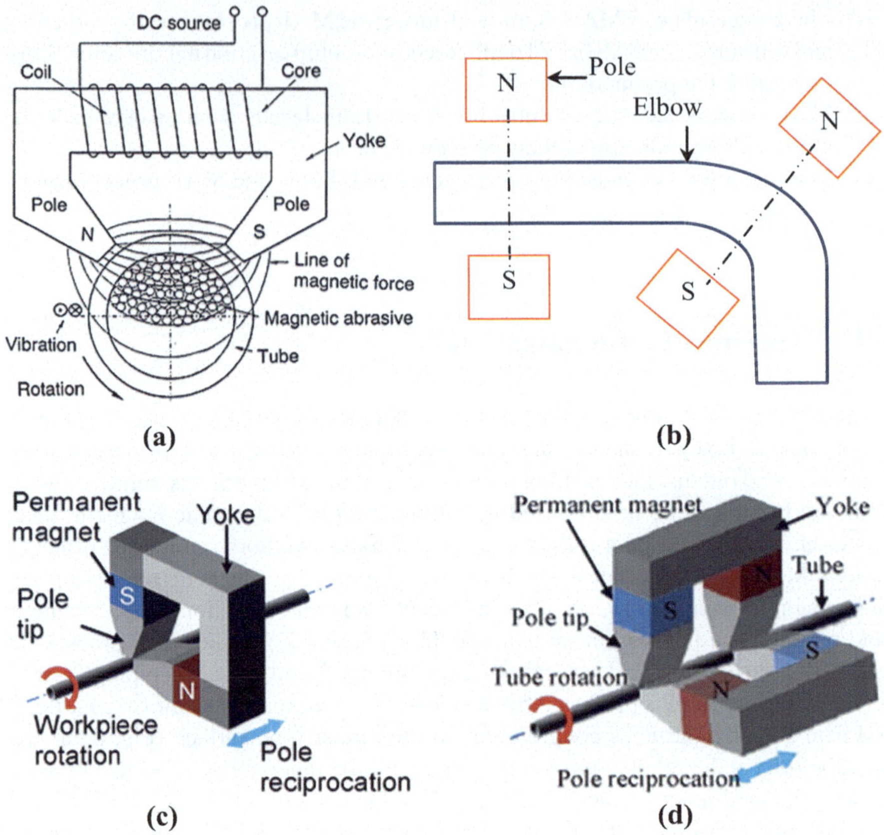

Fig. 8.5 Schematic showing **a** internal finishing of tube [14, Licence number: 4618250400487], **b** pole arrangement for finishing bend tube, **c** single pole-tip system [19, Licence number: 4618250561543], **d** multi pole-tip system [19, Licence number: 4618250561543]

of the tube. To further reduce the finishing time authors [20] modified the set-up by incorporating high speed (upto 30,000 rpm) spindle. It is necessary to continuously inject lubricant to reduce the adherence of the finishing medium with the workpiece surface at high RS.

Workpiece surface texture generated during the MAF process is studied by Yamaguchi and Shinmura [14]. Workpiece surface showed the dominant micro-scratches from the abrasive cutting edges. Authors reported that surface improvement during the MAF process takes place due to the material removal (*MR*) by the abrasive not only from the roughness peaks but valleys as well. Authors [21] showed that finishing performance of the abrasive particle is primarily dependent on the magnetic force and the friction force between the abrasive particle and the workpiece surface. Friction force greater than magnetic force eliminates the relative motion and there is no improvement in surface finish. Effect of MAF process input parameters (RS,

WG) on the output responses (MR, R_a) is reported by Jain et al. [22]. Authors experimentally found out that MR decreases with the increase in WG. Also, R_a improves with the increase in workpiece RS. Chang et al. [23] showed that employing harder ferromagnetic particles improves the MAF process finishing efficiency. Also, the size of the abrasive particle should be kept in a limit such that it doesn't make deep scratch marks on the workpiece surface. Abrasive particle bigger in size gives high MR but deteriorates the surface finish.

Singh et al. [24] carried out the detailed parametric study of the MAF process. Authors arranged voltage >WG> size of the abrasive particles >RS of the workpiece in significant order of their influence on the percentage change in surface roughness (% ΔR_a) of the workpiece. Best % ΔR_a is obtained by performing the MAF experiments at the voltage of 11.5 V, 1.25 mm WG, 180 rpm RS and #1200 abrasive particles. Yan et al. [25] demonstrated the capability of the MAF process by finishing tool steel workpiece machined by using electrical discharge machining (EDM) process. EDM process generates workpiece with a high amount of hard recast layer and cracks, thus resulting in a surface with high roughness. Authors reported that MAF process not only removed hard recast layer from the machined surface but also achieved a mirror like finished surface with the best surface finish of 0.04 μm (R_a). Authors achieved the minimum R_a by performing MAF experiments at 1 mm WG, 3 μm average SiC abrasive size, 320 μm average steel abrasive sizes, 0.75 T magnetic flux density, 0.6 m/s RS of the workpiece and 5 Hz is the vibration frequency of the workpiece.

Wang and Hu [26] found out that MAPs having mesh size between #30 and #50 results in maximum MRR. Although bigger size MAPs produces large magnetic force but the number of cutting edges taking part in shearing the roughness peaks decreases compared to the smaller size MAPs. Therefore, MRR decreases beyond the threshold MAPs size. Authors also demonstrated the effect of adding the finishing fluid (transformer oil) to the MAPs on the MRR. In case of dry finishing, MRR is decided by the workpiece material machinability. However, in case of adding finishing fluid chemical reaction at the workpiece surface interface and the finishing fluid is the critical factor influencing the MRR. Also, MRR increases by adding finishing liquid to the MAPs. Singh et al. [27] studied microscopic changes taking place in surface texture during the MAF process. Authors studied the wear pattern produced on the workpiece surface after finishing. The initial workpiece is prepared by performing the grinding operation. Therefore, grinding marks, pits and digs are visible in Fig. 8.6a. The final surface obtained after finishing contains circular lay due to the rotation of FMAB with sharp shearing marks (Fig. 8.6b). Thus, confirming that surface roughness peaks during MAF process is removed by micro-cutting and scratching. Sometimes, non-uniform magnetic field results in localized deep indentations of the abrasives or forming micro scratches with a width of less than 0.5 μm. Pashmforoush and Rahimi [28] showed that the mode of surface roughness peaks removal during processing of brittle materials can be micro cutting or micro fracture. Mode of material removal depends on the values of input parameters used for finishing. Material removal through micro cutting results in smoother surface compared to the surface obtained by micro fracture mechanism. Authors obtained a minimum R_a

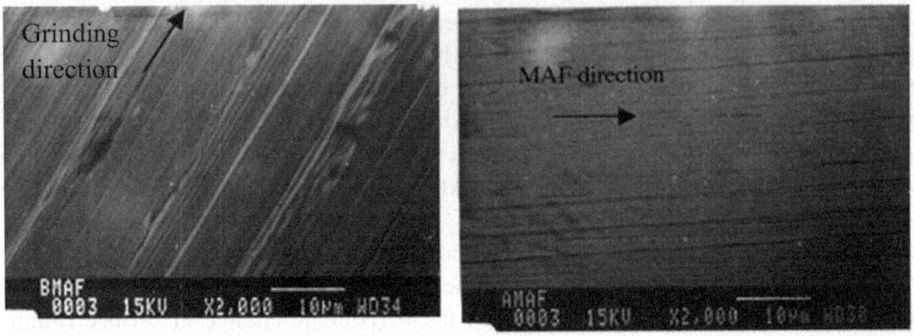

Fig. 8.6 Scan electron microscope images showing workpiece surface **a** initial after grinding, **b** finished (0.75 A current, 1.75 mm WG, #800 of abrasive particle and 9 no. of cycles) [27, Licence number: 4618201082447]

of 23 nm. R_a during finishing of brittle material is primarily influenced by abrasive mesh size.

Authors [29] carried out the detailed experimental study of the MAF process during finishing of steel (AISI 52100) workpiece. The developed regression eqn. of ΔR_a is given as [29]:

$$\Delta R_a = -674 + 5.48d_1 + 0.0628d_2 + 115d_3 + 15.8d_4 - 0.000018d_2^2 - 0.976d_1d_3 - 0.00203d_2d_4 - 2.41d_3d_4 \quad (8.1)$$

where, d_1, d_2, d_3 and d_4 are the voltage, abrasive size, electromagnet rpm, and wt. % of abrasive particles, respectively. ΔR_a is highly affected by abrasive size followed by their wt. %, electromagnet rpm and voltage. It is the magnetic force that decides the level of surface finish on the workpiece. Singh et al. [30] experimentally studied finishing forces generated during the MAF process. Authors planned the experiments using central composite rotatable design (CCRD) method. Regression equations after neglecting insignificant terms for normal magnetic force (F_m), tangential force (F_t) and change in surface roughness (ΔR_a) is given as [30]:

$$F_m = 48.0 + 12.0X_1 - 7.22X_2 + 6.38X_3 + 3.48X_1^2 + 2.51X_3^2 + 4.03X_3X_4 \quad (8.2)$$

$$F_t = 35.1 + 6.27X_1 - 6.05X_2 - 2.45X_4 - 1.04X_4^2 \quad (8.3)$$

$$\Delta R_a = 0.220 + 0.0250X_1 - 0.0392X_2 + 0.0175X_4 - 0.0250X_1X_2 - 0.0212X_2X_3 \quad (8.4)$$

where, X_1, X_2, X_3, X_4 and X_5 are the coded level values for current to the coil of electromagnet, WG, lubricant, RS of magnet and finishing time respectively. It was concluded that current and working are the most dominating MAF input parameters that affects the output responses. Authors also established relationship between F_m, F_t and ΔR_a as [30]:

$$\Delta R_a = (0.0200566)(F_m)^{0.583} \tag{8.5}$$

$$\Delta R_a = (0.008742)(F_t)^{0.885} \tag{8.6}$$

Lin et al. [31] finished SS 304 workpiece with the MAF process. Authors found out that 2.5 mm WG, 10 mm/min feed rate and 2 g of MAPs in the lubricant give the optimum surface finish. Girma et al. [32] concluded that bigger size MAPs are beneficial during finishing flat surfaces while smaller size MAPs provide better surface finish in case of cylindrical surfaces. Authors showed that using iron particle size 1.5–2 times the abrasive particle during MAF finishing provides best R_a. Authors [33] carried out the micro deburring and surface roughness improvement simultaneously on 0.1 mm diameter holes by the MAF process. In addition to the commonly studied MAF input parameters, the authors observed the role of coolant during the MAF process. Continuous supply of coolant provided better R_a compared to its use periodically. However, R_a obtained in both cases is better than the MAF process without coolant. Use of coolant during the MAF process decreases the adhesion of the finishing powder, which is responsible for the severe deterioration of surface finish. Kim and Kwak [34] enhanced the performance of the MAF process by installing a neodymium permanent magnet at the opposite face of the workpiece side (Fig. 8.7a). This results in a 30% increase of magnetic flux density which in turn increases the magnitude of finishing forces.

Im et al. [35] carried out the processing of 3 mm diameter steel rods by employing diamond abrasives. Authors got the best R_a and roundness as 0.06 μm and 0.12 μm respectively by using 1 μm diameter diamond abrasive paste as finishing medium.

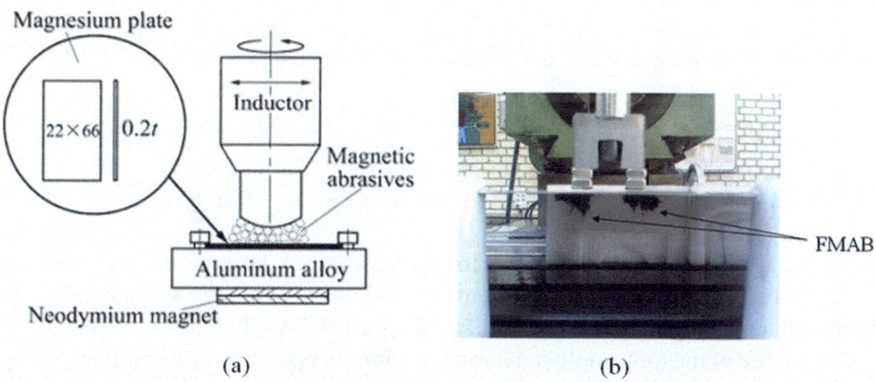

Fig. 8.7 Magnetic abrasive finishing process set-up developed for finishing **a** nonferrous materials [34, Licence number: 4618250788601], **b** large sheet of workpiece [38, Licence number: 4618250961826]

One of the shortcomings of the MAF process is flown away of the MAPs from the finishing region with a low magnitude of the magnetic force or at high RS. This hampers the efficiency of the MAF process. Intending to eliminate the above disadvantage Wang and Lee [36] developed magnetic finishing with gel abrasive (MFGA). Authors mixed ferromagnetic particles and the abrasives particles in the polymer (silicone) gel. Silicone gel not only prevents the flown of the abrasive particles but also resists the collapse of the magnetic brush. As a result, the reduction in R_a during the MFGA process is 3 times more as compared to the MAF process. Also, the authors demonstrated that the developed finishing medium possesses excellent recycling capability. Later, authors [37] carried out the experimental study of the MFGA process for finding the optimal combination of input parameters to give minimum R_a. Authors concluded that carrying out finishing with an abrasive gel containing harder abrasive particles, bigger in size and high concentration for a longer time promotes the finishing capability of MFGA process. Givi et al. [38] employed the MAF process for finishing of the aluminium sheets. Authors modified the MAF setup in such a way that the workpiece is in between the magnetic poles and FMAB brush (Fig. 8.7b). This arrangement compresses the abrasives into the workpiece surface rather than the magnetic poles which occur during the normal MAF process. Number of cycles of the workpiece across the FMAB brush and WG are the most influencing parameters. Regression eqn. for ΔR_a as found out by the authors is [38]:

$$\Delta R_a = 0.183 - 0.0238e_1 + 0.000073e_2 + 0.0104e_3 + 0.00525e_4 \tag{8.7}$$

where, e_1, e_2, e_3 and e_4 are the values for WG, RS of the magnet, wt. of MAPs and number of cycles respectively. Maximum ΔR_a of 0.267 μm is achieved by performing MAF experiments at 1.5 mm WG, 300 rpm, 2 g of MAPs and 6 cycles. Amineh et al. [39] machined aluminium alloy cylindrical workpiece by using wire electrical discharge machining (WEDM) process, later finished the same by MAF process. Due to the inherent characteristic of the WEDM process, it results in the generation of workpiece surface with hard recast layer. MAF process successfully removed the recast layer without generating any microcracks. The relation between ΔR_a, recast layer removal, *RL* and process input parameters are given as [39]:

$$\Delta R_a = 1.97 + 0.897 f_1 + 0.045873 f_2 - 0.224 f_3 - 0.0112 f_4 \tag{8.8}$$

$$RL = 10.2 + 14.7 f_1 + 0.380 f_2 - 2.83 f_3 - 0.0765 f_4 \tag{8.9}$$

where, f_1, f_2, f_3 and f_4 are the values for linear speed of workpiece, time, WG and mesh size respectively. Recast layer removal and improvement in surface finish are highly influenced by the abrasive size, linear speed followed by time and WG.

Cutting life of the tool employed during machining operation is also influenced by its surface finish. Authors [40] achieved a R_a of 25 nm on rake and nose while R_a of less than 50 nm on flank face of the tool. During turning of Ti64 alloy rods a finished tool enjoys twice the cutting life when compared to the unfinished tool. MAF process can also finish coated tools with minimal coating reduction [41]. Authors concluded that the use of small sized abrasive particles enhances the flexibility of the FMAB which helps the FMAB to deform properly along the tool surface. Magnetic force during the MAF process is affected by the permeability of MAPs, μ_m and is given as [41]:

$$\mu_m = \frac{\mu_{fm}\mu_{ab}}{\mu_{ab}V_{fm} + \mu_{fm}V_{ab}} \tag{8.10}$$

With an increase in the volume of abrasive particles, cutting edges of the MAPs increases but the permeability of MAPs given by Eq. 8.10 decreases and hence magnetic forces decline. Therefore, a proper balance is maintained between the volume fraction of the abrasive particles and ferromagnetic particles. Also, experimentally it is demonstrated that sintered MAPs finishing performance is better compared to the simply mixed MAPs [42].

Microscopic study of the finished workpiece surface is also carried out by Mulik and Pandey [43]. Authors carried out the finishing of the workpiece by using ultrasonic vibration assisted MAF (UAMAF) process. Authors provided the ultrasonic vibration of 5–12 μm amplitude to the workpiece. It is concluded by conducting XRD analysis that SiC abrasive particles are induced in the steel workpiece during the UAMAF process. This increases the microhardness of the workpiece from 730 to 960 HV. Extending their work further, authors [44] carried out the parametric study for finishing of steel (AISI 52100) workpiece by UAMAF process. Authors studied the role of ultrasonic vibrations pulse time in addition to the above mentioned input parameters. Machining marks that are not completely removed by the MAF process were successfully removed by UAMAF process. To demonstrate the finishing capability of UAMAF process authors compared the % ΔR_a obtained on the workpiece surface by MAF and UAMAF process (Table 8.1).

Mulik and Pandey [45] also modeled the normal finishing force, F_m and finishing torque, T_m generated during the UAMAF process. Magnitude of voltage and WG

Table 8.1 Comparative study of change in surface roughness obtained by MAF and UAMAF process [44]

Process parameters					Process	% ΔR_a	*Improvement ratio of UAMAF w.r.t. ΔR_a improvement in MAF (%)
V	Mesh	rpm	wt. %	T_{on}			
70	400	280	25	3	MAF UAMAF	43.99 56.13	27.60
70	800	280	25	3	MAF UAMAF	43.44 78.61	80.96
80	600	224	20	2	MAF UAMAF	24.34 54.49	123.87
90	800	280	25	3	MAF UAMAF	22.46 45.41	102.18

are the most significant factors influencing forces. Parametric eqn. of F_m and T_m as found out by the authors are [45]:

$$F_m = 14.8 + 0.0495g_1 - 0.814g_2 + 0.0597g_3 - 0.774g_4 + 0.196g_5 \quad (8.11)$$

$$T_m = 10.50 + 0.0718g_1 - 1.22g_2 + 0.0564g_3 - 1.04g_4 + 0.267g_5 \quad (8.12)$$

where, g_1, g_2, g_3, g_4 and g_5 are the values for voltage, rotation of electromagnet, wt. % of abrasive particles, finishing time, and pulse-on-time of ultrasonic vibrations. Harder the workpiece material smaller is the *MRR* of the MAF process. To address such issues Sihag et al. [46] developed chemical assisted MAF (CMAF) process and carried out the finishing of tungsten workpiece. During the CMAF process chemical slurry is used in addition with normal MAF process. Oxidizing agent in chemical slurry helps in the formation of an oxide layer on the workpiece which is easily sheared by the FMAB. Zhou et al. [47] showed that UAMAF process not only successfully removed the milling textured marks from the titanium workpiece but also induced residual compressive stress of 20 MPa. Finishing efficiency of the UAMAF process is 40% higher than the MAF process during finishing of the titanium workpiece. This is because, instantaneous pressure, P_0 due to the ultrasonic vibration adds in the rate of material removal, $R(x, y)$ by UAMAF process which is given as [47]:

$$R(x, y) = k(P_F + P_0)V(x, y) \quad (8.13)$$

Conventional MAF process suffers few major shortcomings such as the limited movement of abrasive under static magnetic field, non-uniform distribution of the MAPs that result in non-uniform surface finish. To address such issues authors [48] proposed the MAF process with low frequency alternating magnetic field. Alternating magnetic forces not only promotes the uniform dispersion of the MAPs but also enhance the stirring of the MAPs. Such advantages will promote the stability of the finishing tool during the MAF process. With the modified MAF process authors achieved a R_a of 4.38 nm from an initial R_a of 240.24 nm on SS plates.

8.5 Theoretical and Numerical Investigation

Shinmura et al. [1] did theoretical formulation for predicting the magnetic force generated on an abrasive in the magnetic field. Authors expressed the magnetic force F_x and F_y acting along x and y direction on an magnetic abrasive particle at position "A" shown in Fig. 8.8 as [1]:

$$F_x = \chi_m v H\left(\frac{dH}{dx}\right),\ F_y = \chi_m v H\left(\frac{dH}{dy}\right) \quad (8.14)$$

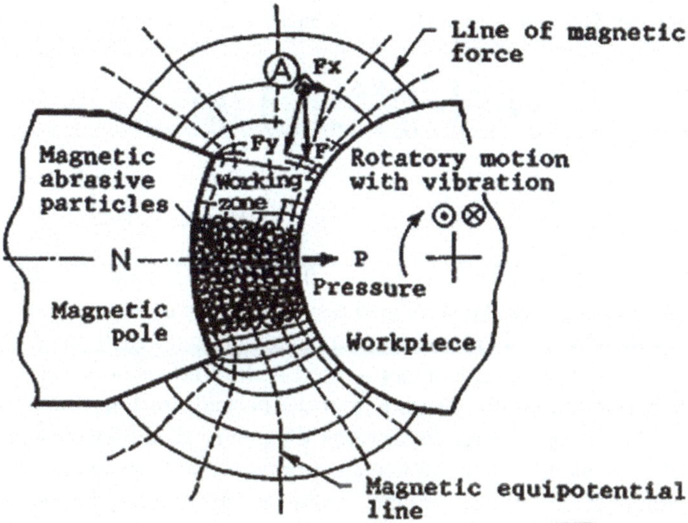

Fig. 8.8 Magnetic field distribution on a magnetic abrasive particle in magnetic field [1, License number: 4618201342816]

Also, by assuming shape of the magnetic abrasives as spherical, machining pressure between abrasive and the workpiece, P is [1]:

$$P = \frac{3\pi B^2 w}{4\mu_0}\left(\frac{(\mu_r - 1)}{3(\mu_r + 2) + \pi(\mu_r - 1)w}\right) \tag{8.15}$$

To avoid over finishing during the MAF process, authors [49] proposed a numerical model which predicts R_a as a function of time. Authors assumed uniform distribution of triangular initial surface roughness profile on the workpiece (Fig. 8.9a).

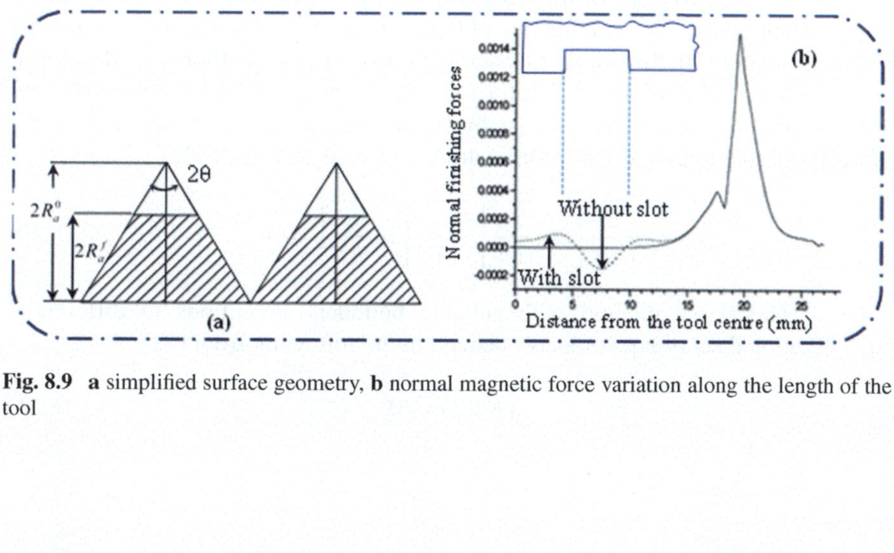

Fig. 8.9 **a** simplified surface geometry, **b** normal magnetic force variation along the length of the tool

Material removal, *MR* and final surface roughness R_a^f can be given as [49]:

$$MR = \left(C\frac{nN\Delta f v_p t}{H_w \pi \tan\theta L_w}\right)^{\frac{1}{4}} \left(R_a^0\right)^{-\frac{1}{4}} \quad (8.16)$$

$$R_a^f = R_a^0 - C^{\frac{1}{4}}\left(R_a^0\right)^{-1/8}(L_w)^{-5/4}\left(\frac{nN\Delta f v_p t}{H_w \pi L_w \tan\theta}\right)^{\frac{1}{4}} \quad (8.17)$$

Jayswal et al. [50] simulated the magnetic flux density generated during the MAF process. Authors compared the simulated results with the experimentally measured magnetic flux density and both are in good agreement with each other. During the experiments the slotted electromagnet is used and its effect on surface quality is studied. It is reported that the magnitude of the normal finishing forces generated during the MAF process increases towards the centre of the electromagnet with slot (Fig. 8.9b). Thus, the use of slotted electromagnet adds advantage to the MAF process. Baron et al. [51] developed the VMAF set-up for removing burrs from the edges of the drilled hole. MAF process not only successfully removed the burrs but also provided radiusing to the edges. Authors developed a theoretical equation for material removal, q in time, t as [51]:

$$q = q_i t^m \quad (8.18)$$

Also, the tool life, T of the MAPs can be given as [51]:

$$T = \sqrt[1-m]{m\frac{q_i}{q_{\min}}} \quad (8.19)$$

To reduce the experimental expenses, Jayswal et al. [52, 53] carried out the numerical study of the MAF process. Firstly, the authors developed a numerical model to predict the distribution of magnetic force using finite element (FE) model. Later, using the simulated values of the magnetic force as input, the theoretical model is proposed that predicts R_a as a function of flux density, WG, abrasive size and number of electromagnet slot. To reduce the computational expenses authors [52] developed the axisymmetric FE model of the process. After some simplified assumptions governing equation of the MAF process by considering magnetic potential, ϕ as the primary variable in cylindrical coordinates (r and z) is given as [52]:

$$\frac{1}{r}\frac{\partial}{\partial r}\left[r\mu_r\frac{\partial\phi}{\partial r}\right] + \frac{\partial}{\partial z}\left[\mu_r\frac{\partial\phi}{\partial z}\right] = 0 \quad (8.20)$$

Using Galerkin's method with suitable boundary conditions to differential Eq. 8.20, global set of equations is obtained in the following from [52]:

$$[K]\{\phi\} = 0 \quad (8.21)$$

where global coefficient matrix, $[K]$ is obtained by summing elemental matrix, $[K]^e$ [52]:

$$[K] = \sum [K]^e \tag{8.22}$$

$$[K]^e = \int_{\Omega^e} \mu_r [C]^{eT} [C]^e 2\pi r \, dr \, dz \tag{8.23}$$

Authors considered the surface roughness profile uniform with triangular nature. Also, to simplify the model workpiece surface is decomposed in small square (Fig. 8.10a). MAPs are assumed to be arranged in circular tracks in FMAB as shown in Fig. 8.10b.

Surface roughness, $R^n_{\max(i,j)}$ after 'n'th revolution in a cell (i, j), during the MAF process is given as [52]:

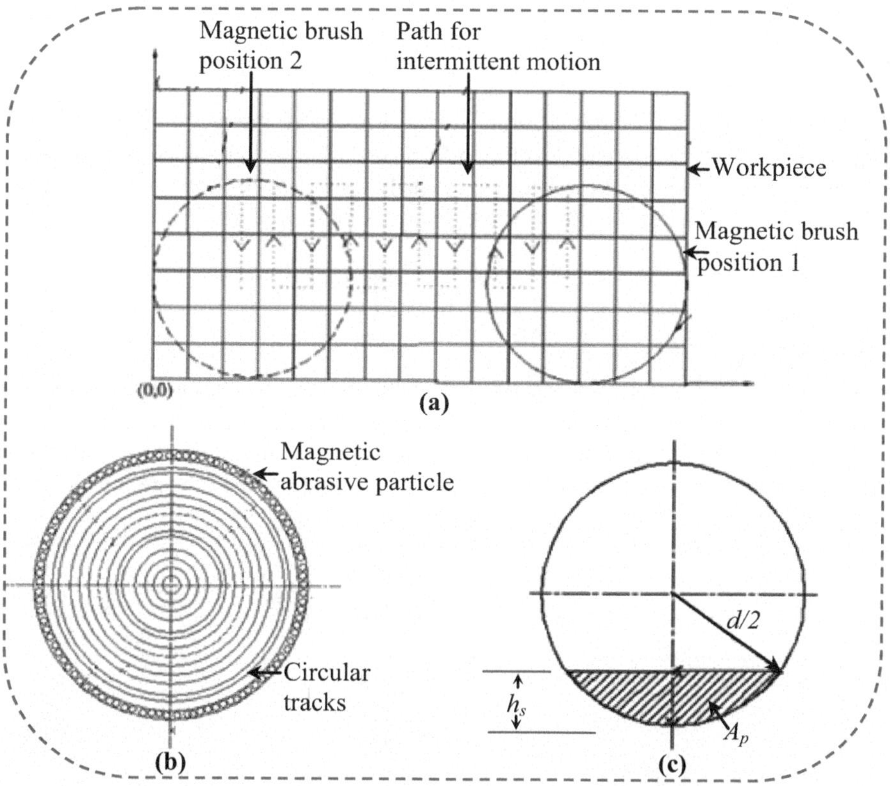

Fig. 8.10 Schematic showing **a** movement of tool along y-direction [52, License number: 4618251432167], **b** circular tracks formed during the magnetic abrasive finishing process, **c** abrasive particle depth of penetration and area of penetration [52, License number: 4618200104090]

$$R^n_{\max(i,j)} = R^0_{\max} - \left(\left(R^0_{\max} - R^{(n-1)}_{\max(i,j)} \right)^2 + \left(\frac{\Delta V^n_{(i,j)} R^0_{\max} D_{\mathrm{map}}}{\pi r_{tr} l_c R_{tr} n_a} \right) \right)^{1/2} \quad (8.24)$$

where, volume of material removed in nth revolution, $\Delta V^n_{(i,j)}$ is given as [52]:

$$\Delta V^n_{(i,j)} = \frac{2 A_{p(tr)} hs^n_{(tr)} l_c \pi R_{(tr)} n_a}{R^0_{\max} D_{\mathrm{map}}} \quad (8.25)$$

Also, as shown in Fig. 8.10c depth of indentation of the MAPs, $hs^n_{(tr)}$ and projected area, $A_{p(tr)}$ in nth revolution in trth track is given as [52]:

$$hs^n_{(tr)} = \frac{d}{2} - \sqrt{\left(\frac{d}{2}\right)^2 - \frac{F_z}{H_w \pi}} \quad (8.26)$$

$$A_{p(tr)} = \frac{(d)^2}{4} \left[\cos^{-1}\left(1 - \frac{2hs^n_{(tr)}}{d}\right) \right] - \left(\left(\frac{d}{2} - hs^n_{(tr)} \right) \sqrt{hs^n_{(tr)} \left(d - hs^n_{(tr)} \right)} \right) \quad (8.27)$$

Further, to predict R_a more accurately at various MAF input conditions, authors [54] modified their previously developed R_a model. Authors considered the surface roughness profile following the Gaussian distribution. Considering $f(x)$ as value for surface profile ordinate at position x, roughness is generated as [54]:

$$f(x) = \frac{1}{\sigma \sqrt{2\pi}} \exp\left[-\frac{1}{2} \left(\frac{x-\mu}{\sigma} \right)^2 \right] \quad (8.28)$$

Kumar and Yadav [55] developed FE model to predict the temperature generated during the MAF process. Authors concluded that magnetic pressure and the rpm of the FMAB should be carefully chosen to avoid high rise in temperature during the MAF process. Governing equation for 2-D axisymmetric steady state temperature distribution is [55]:

$$k_w \left\{ \frac{1}{r} \frac{\partial}{\partial r} \left(r \frac{\partial T}{\partial r} \right) + \frac{\partial}{\partial z} \left(\frac{\partial T}{\partial z} \right) \right\} = 0 \quad (8.29)$$

Elemental eqn. by using Galerkin's method is [55]:

$$[K_t]^e \{t\}^e = \{f\}^e \quad (8.30)$$

where, elemental coefficient matrix, $[K_t]$ and right-side force vector, $\{f\}$ is given as [55]:

$$[K_t]^e = k_w \int\limits_{D^e} [C]^{eT}[C]^e r dr dz \quad (8.31)$$

$$\{f\}^e = \int\limits_{S} \{N\}^T q_w r dS \quad (8.32)$$

Also, heat flux entering into the workpiece, q_w is [55]:

$$q_w = R_w \times P \times \mu_f \times v_b \quad (8.33)$$

Authors [56] also simulated the temperature generated at the interface of the FMAB and the workpiece using the FE based ANSYS software. Predicted temperature range lay in between 34 °C and 51 °C which might not hamper the finishing efficiency of FMAB brush. Authors compared the simulated temperature results with the experimentally measured values and are found in good agreement. Finishing of para/diamagnetic material workpiece is difficult by the MAF process due to the low magnetic field. Kala and Pandey [57] developed double disc magnetic abrasive finishing (DDMAF) process for increasing the magnetic field, hence finishing forces during the processing of para/diamagnetic material workpiece. DDMAF process set-up consists of electromagnet arrangement on both sides of the workpiece. Authors showed by performing the simulation that magnetic flux density in the case of a single electromagnet is mostly concentrated inside the electromagnet (Fig. 8.11a). While in case of double magnet it is uniformly distributed across electromagnet, WG and workpiece (Fig. 8.11b). Experimental study showed that WG is the significant parameter during finishing of hard materials like stainless steel. While finishing of soft material like copper is dominated by the RS.

Later, extending their work further authors developed a tool for internally finishing tubes with an internal diameter less than 80 mm. Authors carried out a simulation

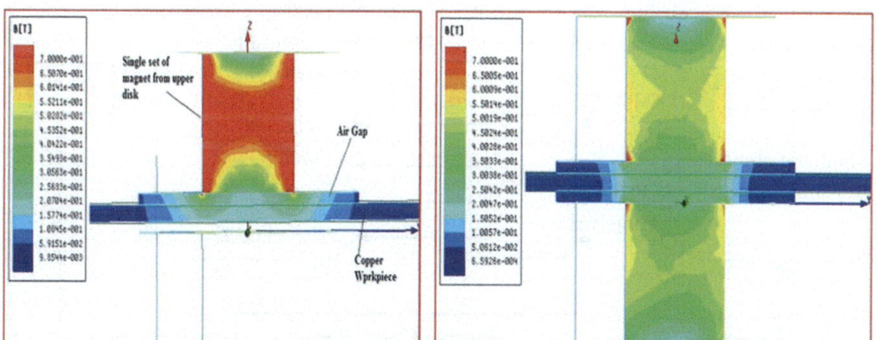

(a) (b)

Fig. 8.11 Simulation results showing magnetic flux density for **a** single electromagnet, **b** double electromagnet [57, License number: 4618260129213] study on the magnetic field produced by the designed tool using the Maxwell software. The tool is designed to produce a high magnetic field in the WG. Authors also carried out the detailed experimental study with the developed tool and given the parametric eqn. for % ΔR_a as [58]:

$$\%\Delta R_a = -49 + 0.201h_1 - 83.0h_2 + 3.3h_3 + 0.0796h_4 - 0.000236h_1^2 + 72.2h_2^2 + 0.107h_1h_2 - 0.00286h_1h_3 - 0.00242h_3h_4 \quad (8.34)$$

where, h_1, h_2, h_3 and h_4 are the values for RS, magnetic flux, wt. % of abrasive particles and mesh size respectively. Maximum % ΔR_a of 89.6% is achieved at 0.8 T magnetic flux density, 500 rpm, 20 wt. % of abrasive particles with #1200 mesh size.

8.6 Soft Computing

Hung et al. [59] tried to develop MAF process into an adaptive control system with the help of soft computing technique. Authors proposed statistical assisted fuzzy nets in-process surface finish prediction (S-FN-IPSFP) system for the MAF process (Fig. 8.12). Developed system predicts R_a as the output which is a function of input parameters viz. spindle speed, frequency, current, and weight of abrasive particles with 97% accuracy. Developed model is based on fuzzy-net theory following the five step approach that are:

i. Decompose input space into fuzzy regions,

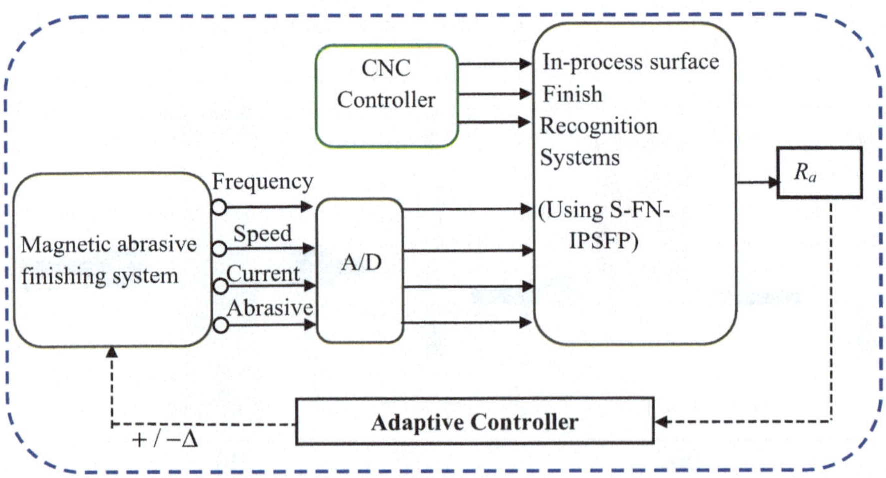

Fig. 8.12 Experimental set up of statistical-assisted fuzzy nets in-process surface finish prediction system

ii. Generate fuzzy rules,
iii. Avoid conflicting rules,
iv. Statistical assisted fuzzy-nets rule base,
v. Determine a mapping based on the fuzzy rule base.

Oh and Le [60] developed an in process monitoring system for the MAF process by monitoring acoustic emission signals (AES) and force signals (FS). Also energy released, AE_{rms} during the MAF process can be given as [60]:

$$AE_{rms} = D_c\{\sigma_e \times \dot{\varepsilon}_e \times V_w\}^{1/2} \tag{8.35}$$

Later, using the AE signals and FS signals as individually as input for the artificial neural networks (ANNs) authors predicted the R_a. It was concluded from the study that using combined data (AE + FS) as input, ANNs (Fig. 8.13) provided least error between predicted and experimentally measured compared to the individually used AE based and FS based networks.

Djavanroodi [61] used artificial neural network for modeling the MAF process. Effect of current, workpiece velocity and finishing time is predicted on $\Delta R_{a.}$ Tangent sigmoid (*Tansig*(z)) transfer function with weighted sum of the input, z is used in ANN and is given as [61]:

$$Tansig(z) = 2(1 + \exp(-2z)) - 1 \tag{8.36}$$

Experimentally obtained and predicted by ANN results are in good agreement, with current as the significant factor affecting ΔR_a. Teimouri and Baseri [62] developed two models feed forward back-propagation neural network (FFBP-NN) and

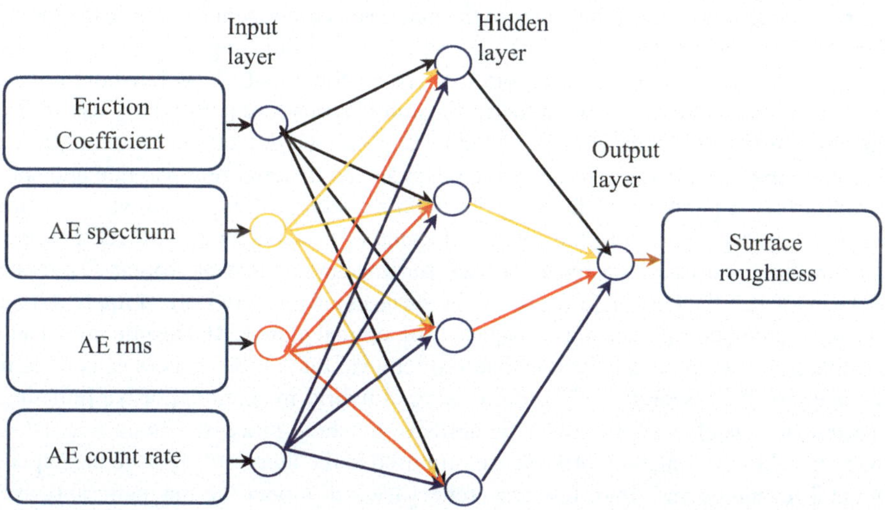

Fig. 8.13 Schematic of acoustic emission based artificial neural network

adaptive neuro-fuzzy inference system (ANFIS) for prediction of % ΔR_a as a function of MAF input parameters. Performances of the developed models are evaluated by measuring mean absolute error, *MSE* and root mean square error, *RMSE* given as [62]:

$$MSE = \frac{1}{T_s}\sum_{i=1}^{T_s} |t_i - a_i|^2 \tag{8.37}$$

$$RMS = \left(\frac{1}{T_r}\sum_{i=1}^{T_r} (t_i - y_i)^2\right)^{1/2} \tag{8.38}$$

Comparing the experimental and predicted values of % ΔR_a authors concluded that ANFIS system models the MAF system more accurately compared to the FFBP-NN model. Later, by using the objective function i.e., % ΔR_a developed by the ANFIS, authors used simulated annealing (SA) and particle swarm optimization (PSO) algorithms to optimize the value of various MAF input parameters to get maximum % ΔR_a. Moosa [63] also used the ANFIS for predicting and optimizing the ΔR_a obtained on the workpiece surface during the MAF process. Authors used membership function as Gaussian and obtained the predicted results within 3% deviation with experimentally obtained values.

8.7 Summary

Magnetic abrasive finishing process is an economical finishing process useful for finishing as well as deburring the edges of the machined components. Flexible magnetic abrasive brush generated during the magnetic abrasive finishing process eliminates the need for rigid, expensive and vibration free machine tools. Low finishing forces generated during the magnetic abrasive finishing process are particularly useful for finishing brittle materials without inducing any surface defects on the workpiece. The magnitude of finishing forces which decides the material removal rate and surface finish is dependent on the magnetic field density, size, type and wt. % of the magnetic abrasive particles, the rotational speed of the magnet etc. Magnetic abrasive finishing process is successfully used for finishing of ceramic rollers [2], inner surfaces of capillary tubes [19, 20], inner surfaces of pipes in critical applications like semi-conductor plants and aerospace [16], optical glass [28]. Despite numerous advantages of the process, the low material removal rate of the process is a concern for many of the researchers. Therefore, several hybrid magnetic abrasive finishing process such as electrolytic magnetic abrasive finishing process, vibration assisted magnetic abrasive finishing process are proposed in the literature. To save the experimental resources and time, few researchers tried to model the magnetic abrasive finishing process by using the finite element method and soft computing techniques.

Predicted results by the developed models are in fair agreement with the experimental results.

Although magnetic abrasive finishing process is studied by several researchers but still it is infancy in many respects some of them are:

(i) Generation of non-uniform magnetic during magnetic abrasive finishing field causes the non-uniform surface finish on the workpiece surface.
(ii) It is reported in the literature sometimes during magnetic abrasive finishing process localized large magnetic pressure creates deep scratches/cuts [27]. Such phenomenon deteriorates the surface finish.
(iii) Still, the magnetic abrasive finishing process needs to be developed commercially by increasing its productivity in terms of components finished per unit time.
(iv) Few researchers did modeling of the magnetic abrasive finishing process but there are still a considerable number of assumptions such as considering the spherical shape of magnetic abrasive particles, constant WG, initial surface roughness profile is either triangular or Gaussian in nature etc. that causes the predicted results to deviate from the experimental results.
(v) Developing an adaptive control system that can predict the relation between process input parameters and its output responses. Later, automatization of the process still needs to be carried out.

Such gaps should be addressed properly to develop magnetic abrasive finishing process to its full potential.

References

1. Shinmura T, Takazawa K, Hatano E, Matsunaga M, Matsuo T (1990) Study on magnetic abrasive finishing. CIRP Ann 39(1):325–328
2. Komanduri R (1998) Magnetic field assisted finishing of ceramics—Part III: on the thermal aspects of magnetic abrasive finishing (MAF) of ceramic rollers
3. Fox M, Agrawal K, Shinmura T, Komanduri R (1994) Magnetic abrasive finishing of rollers. CIRP Ann 43(1):181–184
4. Mori T, Hirota K, Kawashima Y (2003) Clarification of magnetic abrasive finishing mechanism. J Mater Process Technol 143:682–686
5. Yan BH, Chang GW, Cheng TJ, Hsu RT (2003) Electrolytic magnetic abrasive finishing. Int J Mach Tools Manuf 43(13):1355–1366
6. El-Taweel TA (2008) Modelling and analysis of hybrid electrochemical turning-magnetic abrasive finishing of 6061 Al/Al_2O_3 composite. Int J Adv Manuf Technol 37(7–8):705–714
7. Du ZW, Chen Y, Zhou K, Li C (2015) Research on the electrolytic-magnetic abrasive finishing of nickel-based superalloy GH4169. Int J Adv Manuf Technol 81(5–8):897–903
8. Natsume M, Shinmura T, Sakaguchi K (1998) Study of magnetic abrasive machining by use of work vibration system (characteristics of plane finishing and application to the inside finishing of groove). Trans Jpn Soc Mech Eng 64(627C):4447–4452
9. Yin S, Shinmura T (2004) Vertical vibration-assisted magnetic abrasive finishing and deburring for magnesium alloy. Int J Mach Tools Manuf 44(12–13):1297–1303

10. Yin S, Shinmura T (2004) A comparative study: polishing characteristics and its mechanisms of three vibration modes in vibration-assisted magnetic abrasive polishing. Int J Mach Tools Manuf 44(4):383–390
11. Lee YH, Wu KL, Jhou JH, Tsai YH, Yan BH (2013) Two-dimensional vibration-assisted magnetic abrasive finishing of stainless steel SUS304. Int J Adv Manuf Technol 69(9–12):2723–2733
12. Judal KB, Yadava V, Pathak D (2013) Experimental investigation of vibration assisted cylindrical–magnetic abrasive finishing of aluminum workpiece. Mater Manuf Processes 28(11):1196–1202
13. Shinmura T, Hatano E, Takazawa K (1986) The development of magnetic-abrasive finishing and its equipment by applying a rotating magnetic field. Bull JSME 29(258):4437–4443
14. Yamaguchi H, Shinmura T (1999) Study of the surface modification resulting from an internal magnetic abrasive finishing process. Wear 225:246–255
15. Shinmura T, Yamaguchi H (1995) Study on a new internal finishing process by the application of magnetic abrasive machining: internal finishing of stainless steel tube and clean gas bomb. JSME Int J Ser C Dyn Control Robotics Design Manuf 38(4):798–804
16. Yamaguchi H, Shinmura T, Kobayashi A (2001) Development of an internal magnetic abrasive finishing process for nonferromagnetic complex shaped tubes. JSME Int J Ser C 44(1):275–281
17. Yamaguchi H, Shinmura T (2004) Internal finishing process for alumina ceramic components by a magnetic field assisted finishing process. Precis Eng 28(2):135–142
18. Yamaguchi H, Kang J, Hashimoto F (2011) Metastable austenitic stainless steel tool for magnetic abrasive finishing. CIRP Ann 60(1):339–342
19. Kang J, Yamaguchi H (2012) Internal finishing of capillary tubes by magnetic abrasive finishing using a multiple pole-tip system. Precis Eng 36(3):510–516
20. Kang J, George A, Yamaguchi H (2012) High-speed internal finishing of capillary tubes by magnetic abrasive finishing. Procedia CIRP 1:414–418
21. Yamaguchi H, Shinmura T (2000) Study of an internal magnetic abrasive finishing using a pole rotation system: discussion of the characteristic abrasive behavior. Precis Eng 24(3):237–244
22. Jain VK, Kumar P, Behera PK, Jayswal SC (2001) Effect of working gap and circumferential speed on the performance of magnetic abrasive finishing process. Wear 250(1–12):384–390
23. Chang GW, Yan BH, Hsu RT (2002) Study on cylindrical magnetic abrasive finishing using unbonded magnetic abrasives. Int J Mach Tools Manuf 42(5):575–583
24. Singh DK, Jain VK, Raghuram V (2004) Parametric study of magnetic abrasive finishing process. J Mater Process Technol 149(1–3):22–29
25. Yan BH, Chang GW, Chang JH, Hsu RT (2004) Improving electrical discharge machined surfaces using magnetic abrasive finishing. Mach Sci Technol 8(1):103–118
26. Wang Y, Hu D (2005) Study on the inner surface finishing of tubing by magnetic abrasive finishing. Int J Mach Tools Manuf 45(1):43–49
27. Singh DK, Jain VK, Raghuram V, Komanduri R (2005) Analysis of surface texture generated by a flexible magnetic abrasive brush. Wear 259(7–12):1254–1261
28. Pashmforoush F, Rahimi A (2015) Nano-finishing of BK7 optical glass using magnetic abrasive finishing process. Appl Opt 54(9):2199–2207
29. Mulik RS, Pandey PM (2011) Magnetic abrasive finishing of hardened AISI 52100 steel. Int J Adv Manuf Technol 55(5–8):501–515
30. Singh DK, Jain VK, Raghuram V (2006) Experimental investigations into forces acting during a magnetic abrasive finishing process. Int J Adv Manuf Technol 30(7–8):652–662
31. Lin CT, Yang LD, Chow HM (2007) Study of magnetic abrasive finishing in free-form surface operations using the Taguchi method. Int J Adv Manuf Technol 34(1–2):122–130
32. Girma B, Joshi SS, Raghuram MVGS, Balasubramaniam R (2006) An experimental analysis of magnetic abrasives finishing of plane surfaces. Mach Sci Technol 10(3):323–340
33. Ko SL, Baron YM, Park JI (2007) Micro deburring for precision parts using magnetic abrasive finishing method. J Mater Process Technol 187:19–25
34. Kim SO, Kwak JS (2008) Magnetic force improvement and parameter optimization for magnetic abrasive polishing of AZ31 magnesium alloy. Trans Nonferrous Metals Soc China 18:s369–s373

35. Im IT, Mun SD, Oh SM (2009) Micro machining of an STS 304 bar by magnetic abrasive finishing. J Mech Sci Technol 23(7):1982–1988
36. Wang AC, Lee SJ (2009) Study the characteristics of magnetic finishing with gel abrasive. Int J Mach Tools Manuf 49(14):1063–1069
37. Wang AC, Tsai L, Liu CH, Liang KZ, Lee SJ (2011) Elucidating the optimal parameters in magnetic finishing with gel abrasive. Mater Manuf Process 26(5):786–791
38. Givi M, Tehrani AF, Mohammadi A (2012) Polishing of the aluminum sheets with magnetic abrasive finishing method. Int J Adv Manuf Technol 61(9–12):989–998
39. Amineh SK, Tehrani AF, Mohammadi A (2013) Improving the surface quality in wire electrical discharge machined specimens by removing the recast layer using magnetic abrasive finishing method. Int J Adv Manuf Technol 66(9–12):1793–1803
40. Yamaguchi H, Srivastava AK, Tan MA, Riveros RE, Hashimoto F (2012) Magnetic abrasive finishing of cutting tools for machining of titanium alloys. CIRP Ann 61(1):311–314
41. Yamaguchi H, Srivastava AK, Tan M, Hashimoto F (2014) Magnetic abrasive finishing of cutting tools for high-speed machining of titanium alloys. CIRP J Manuf Sci Technol 7(4):299–304
42. Liu ZQ, Chen Y, Li YJ, Zhang X (2013) Comprehensive performance evaluation of the magnetic abrasive particles. Int J Adv Manuf Technol 68(1–4):631–640
43. Mulik RS, Pandey PM (2010) Mechanism of surface finishing in ultrasonic-assisted magnetic abrasive finishing process. Mater Manuf Process 25(12):1418–1427
44. Mulik RS, Pandey PM (2011) Ultrasonic assisted magnetic abrasive finishing of hardened AISI 52100 steel using unbonded SiC abrasives. Int J Refract Metal Hard Mater 29(1):68–77
45. Mulik RS, Pandey PM (2012) Experimental investigations and modeling of finishing force and torque in ultrasonic assisted magnetic abrasive finishing. J Manuf Sci Eng 134(5):051008
46. Sihag N, Kala P, Pandey PM (2015) Chemo assisted magnetic abrasive finishing: experimental investigations. Procedia CIRP 26:539–543
47. Zhou K, Chen Y, Du ZW, Niu FL (2015) Surface integrity of titanium part by ultrasonic magnetic abrasive finishing. Int J Adv Manuf Technol 80(5–8):997–1005
48. Wu J, Zou Y, Sugiyama H (2015) Study on ultra-precision magnetic abrasive finishing process using low frequency alternating magnetic field. J Magn Magn Mater 386:50–59
49. Kim JD, Choi MS (1995) Simulation for the prediction of surface-accuracy in magnetic abrasive machining. J Mater Process Technol 53(3–4):630–642
50. Jayswal SC, Jain VK, Dixit PM (2004) Analysis of magnetic abrasive finishing with slotted magnetic pole. In: AIP conference proceedings, vol 712, no 1. AIP, pp 1435–1440
51. Baron YM, Ko SL, Park JI (2005) Characterization of the magnetic abrasive finishing method and its application to deburring. In: Key engineering materials, vol 291. Trans Tech Publications, pp 291–296
52. Jayswal SC, Jain VK, Dixit P (2005) Modeling and simulation of magnetic abrasive finishing process. Int J Adv Manuf Technol 26(5–6):477–490
53. Jayswal SC, Jain VK, Dixit PM (2005) Magnetic abrasive finishing process—a parametric analysis. J Adv Manuf Syst 4(02):131–150
54. Jain VK, Jayswal SC, Dixit PM (2007) Modeling and simulation of surface roughness in magnetic abrasive finishing using non-uniform surface profiles. Mater Manuf Process 22(2):256–270
55. Kumar G, Yadav V (2009) Temperature distribution in the workpiece due to plane magnetic abrasive finishing using FEM. Int J Adv Manuf Technol 41(11–12):1051–1058
56. Mishra V, Goel H, Mulik RS, Pandey PM (2014) Determining work-brush interface temperature in magnetic abrasive finishing process. J Manuf Process 16(2):248–256
57. Kala P, Pandey PM (2015) Comparison of finishing characteristics of two paramagnetic materials using double disc magnetic abrasive finishing. J Manuf Process 17:63–77
58. Verma GC, Kala P, Pandey PM (2017) Experimental investigations into internal magnetic abrasive finishing of pipes. Int J Adv Manuf Technol 88(5–8):1657–1668
59. Hung CL, Ku WL, Yang LD (2010) Prediction system of magnetic abrasive finishing (MAF) on the internal surface of a cylindrical tube. Mater Manuf Process 25(12):1404–1412

60. Oh JH, Lee SH (2011) Prediction of surface roughness in magnetic abrasive finishing using acoustic emission and force sensor data fusion. Proc Inst Mech Eng Part B J Eng Manuf 225(6):853–865
61. Djavanroodi F (2013) Artificial neural network modeling of surface roughness in magnetic abrasive finishing process. Appl Sci Eng Technol 6(11):1976–1983
62. Teimouri R, Baseri H (2013) Artificial evolutionary approaches to produce smoother surface in magnetic abrasive finishing of hardened AISI 52100 steel. J Mech Sci Technol 27(2):533–539
63. Moosa AA (2014) Utilizing a magnetic abrasive finishing technique (MAF) via adaptive nero fuzzy (ANFIS). Al-Khwarizmi Eng J 10(2):49–56

Chapter 9
Magnetic Field Assisted Finishing Processes

Arpan Kumar Mondal, Atul Singh Rajput, Deopkant Prasad, and Dipankar Bose

Abstract In today's manufacturing industry surface roughness is highly concerned. Poor surface quality leads to many problems like malfunctioning, large amount of wear, low life span. In case of fluid flow through pipes poor surface quality causes higher friction loss. Lots of efforts were made to improve the surface quality of a product but by conventional processes high level of surface finish are not achievable. By means of hybrid technologies it is feasible to achieve surface roughness up to nano level. The most controlled and effective way to produce surface roughness up to nano meter by using magneto rheological fluid (MRF). With the help of MRF high grade surface finish can achieve with convenient tolerances and without ruining the surface contour. Working principle and their advantages of MR fluid and rotational magneto-rheological abrasive flow finishing is discussed in this chapter.

Keywords Magnetorheological fluid · Nano finishing · Hybrid surface finishing process · R-MRAFF

9.1 Introduction

Finishing of a product is very essential process in manufacturing to obtain a good surface texture, with a good surface quality the value of a product enhances.

Finishing mainly classified in two categories:

A. K. Mondal (✉) · A. S. Rajput · D. Prasad · D. Bose
Mechanical Engineering Department, NITTTR, Kolkata, Block-FC, Sector-III, Salt Lake City, Kolkata 700106, India
e-mail: arpan@nitttrkol.ac.in

A. S. Rajput
e-mail: singhatul55@gmail.com

D. Prasad
e-mail: deokant0512@gmail.com

D. Bose
e-mail: dbose@nitttrkol.ac.in

S. Das et al. (eds.), *Advances in Abrasive Based Machining and Finishing Processes*, Materials Forming, Machining and Tribology,
https://doi.org/10.1007/978-3-030-43312-3_9

- Conventional finishing processes (Grinding, Lapping, Honing etc.)
- Advanced finishing processes (Abrasive flow finishing, Magnetorheological finishing etc.)

Surface finishing is an essential stage in the production of all constituents that need high precision and close tolerances for execution their jobs as a distinct constituent or as a portion of an assembly. These constituents have extensive variabilities in terms of dimension, shape, material properties, etc. Moreover, latest materials having greater properties are being established unceasingly to encounter the necessities of diverse fields, such as medical, aerospace, optics, automobile, electronics, etc. As an outcome, incessant development of different machining and finishing methods are essential to process or fabricate such materials. Numerous conventional processes are broadly used for finishing. Though, they are not effective enough when it comes to finishing very soft materials that are used in optics, to achieve nanometer-level (or, in some cases, Angstrom) surface roughness value, to finish multifaceted freeform surfaces, etc. Just advancing these processes will not function due to the formerly stated necessities, which are not easy to achieve. Hence, development of newfangled finishing processes is the lone solution to the current necessities. Numerous new and unconventional finishing processes have been developed for different types of workpiece materials. Precise surface finish is required for today's manufacturing industries. From last few decades lots of efforts have been made to achieve high level of surface finish [1–4]. Earlier lots of conventional methods were there for finishing process such as grinding, honing, lapping etc. but the limitation of these conventional processes were that they are not applicable on complex structures and new materials. To overcome such kind of problems hybrid finishing were introduced.

One of the methods by which surface finish up to the value of nanometer can be achieved is by introduction of magnetorheological finishing process. First time the effort in the field of magnetorheological finishing was made in late 80s by Kordonski when he invented the first magnetorheological finishing machine [5].

9.2 Magnetorheological Fluid

Magnetorheological fluid established finishing processes depends on the properties of MR fluid. The primary exploration and advancement of MR fluids and setup is given to Rabinow in the late 1940s at the US National Bureau of Standards [6]. Basically MR fluid consists of a termination of non-colloidal, magnetically influenced particles in a non-magnetic medium.

Magnetorheological fluid composed of silicon carbide abrasives and carbonyl iron powder circulated in the viscoplastic base of carrier fluid like grease and mineral oil; rheological property of MR fluid changes when it comes under external magnetic field influence as shown in Fig. 9.1. When MR fluid comes in the region of extrinsic magnetic field columnar structure were formed by carbonyl iron particles due to dipole effect as shown in Fig. 9.2. This agile action of MR fluid is applied to precisely

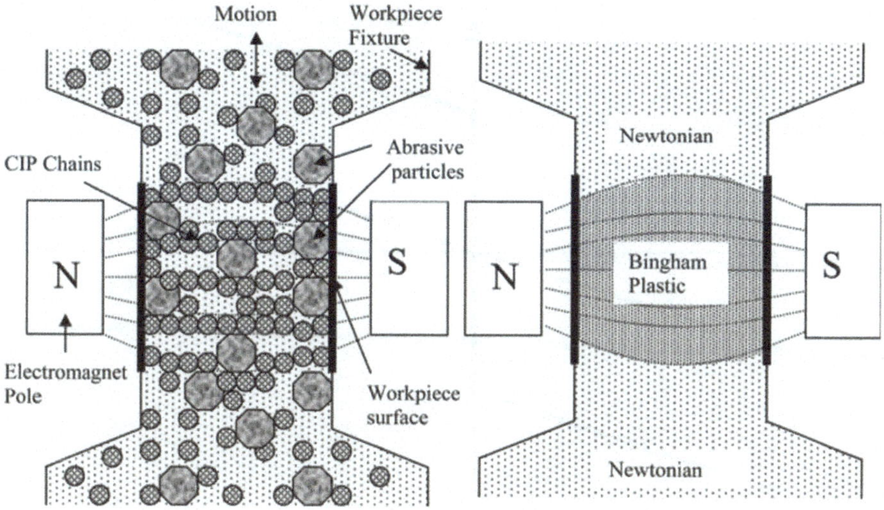

Fig. 9.1 Alteration in rheological properties of MR fluid in finishing process [11]

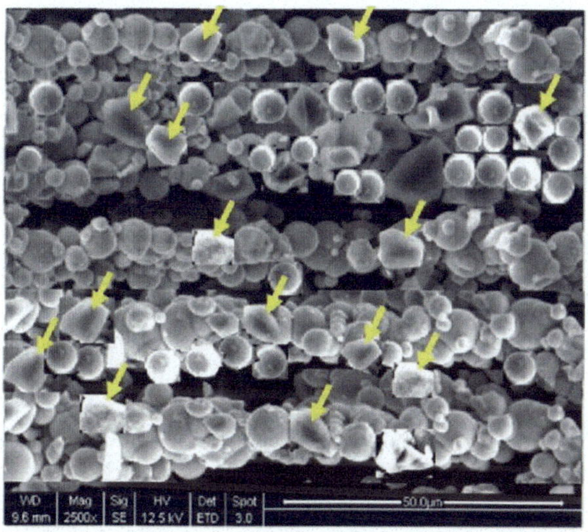

Fig. 9.2 Formation of columnar structure in MR fluid [19]

regulate the finishing forces, thus roughness of the surface. It is also termed as smart fluid because it changes its properties on presence of external magnetic field, fluid properties changes to bingham plastic from Newtonian [7]. The shear stress is represented by [8].

$$\tau = \tau_0(H)sgn(\gamma) + \eta\gamma,^{\cdot} \quad |\tau| > |\tau^{\cdot}| \tag{9.1}$$

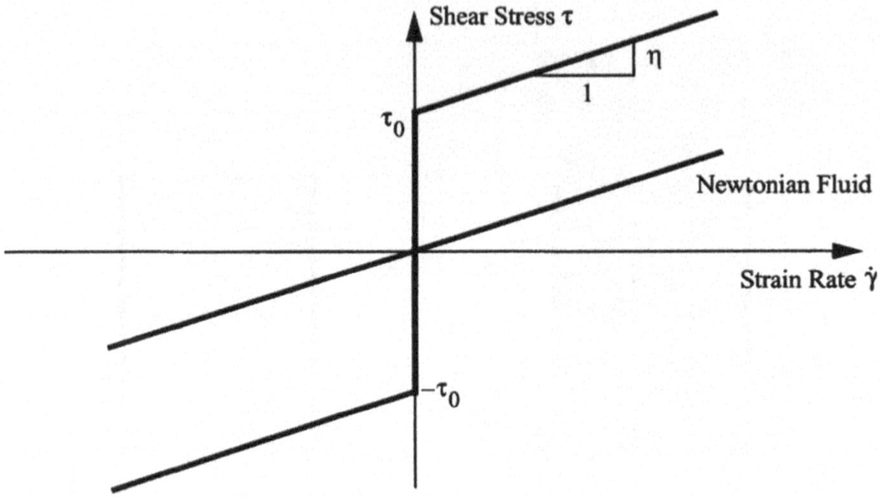

Fig. 9.3 Bingham plasticity model [4]

$$\tau = 0,\ |\tau| < |\tau^{\cdot}| \tag{9.2}$$

where, τ_0 is yield stress due to magnetic field, $^{\cdot}\gamma$ is rate of change of shear strain; H is magnitude of the external magnetic field; and η is plastic viscosity which is independent of yield, that is slope of calculated shear stress against rate of change of shear strain as shown in Fig. 9.3.

Ginder et al. [9] developed mathematical and theoretical models of a magnetorheological fluid phenomenon that account for magnetic nonlinearity and saturation effects in particular. The magneto-static strength of the interparticle and the resulting shear stress was measured in this analysis. They performed FEA calculations to determine the effect of shear stress in MR fluids on magnetic nonlinearity and saturation. The maximum shear stress of the particles increases from their results as a square of the particle saturation magnetization.

Seok et al. [10] in his work tribological properties of MR fluid in a surface finishing process derived a mathematical model and observed that MR fluid shows solid as well as fluid like properties in external fluid which impacted on the material removal rate as described in equation.

$$\frac{dz}{dt} = \alpha\left(\mu_s P V_p + \frac{\eta}{H}(V_p)^2\right) \tag{9.3}$$

where, $\frac{dz}{dt}$ is the rate of change work piece height with respect to time or material removal rate; α is proportionality constant between the MRR and the shear done at the work piece surface; μ_s is kinetic friction coefficient; P is normal pressure enforced; V_p is velocity of work piece; Viscosity of MR fluid is denoted by η; H is the height MR fluid ribbon.

9.3 Working Principle of R-MRAFF

In rotational magnetorheological fluid flow finishing process with the help of magnetorheological fluid (MR-fluid) which works as finishing material, surface finish upto nano level can be acquired [11]. Magnetorheological fluid is a non Newtonian fluid which is comprises of carbonyl iron particles (CIP), carrier fluid abrasive changes its viscosity when it comes under the influence of some external field and behaves as Bingham fluid [12]. MR-fluid forms a columnar structure in external magnetic field. Back and forth motion of the MR-fluid is provided by the reciprocation of pistons in R-MRAFF process and rotational motion is provided to the magnets of certain magnetic field. In between magnets cylindrical workpiece made of steel is placed as shown in Fig. 9.4.

In Fig. 9.4 shows the working principle of R-MRAFF process. Axial forces (F_a) generated by the back and forth motion of the MR-fluid causes the cutting action. By the extrusion pressure some radial forces (F_r) is also generated on the workpiece. Centrifugal forces (F_c) and magnetic force (F_m) generated because of the magnetic field and rotation of magnets respectively. Hence the total force normal to the workpiece is shown in Eq. (9.4) [13] (Fig. 9.5).

$$F_n = F_r + F_c + F_m \tag{9.4}$$

Das et al. [14] performed the experiment on nano surface finishing of stainless-steel tubes using R-MRAFF process in which they used work piece of stainless steel. In the experiment it was observed that the rotation of magnet provide better surface profile as compared to the case where magnet was stationary. The reason behind this

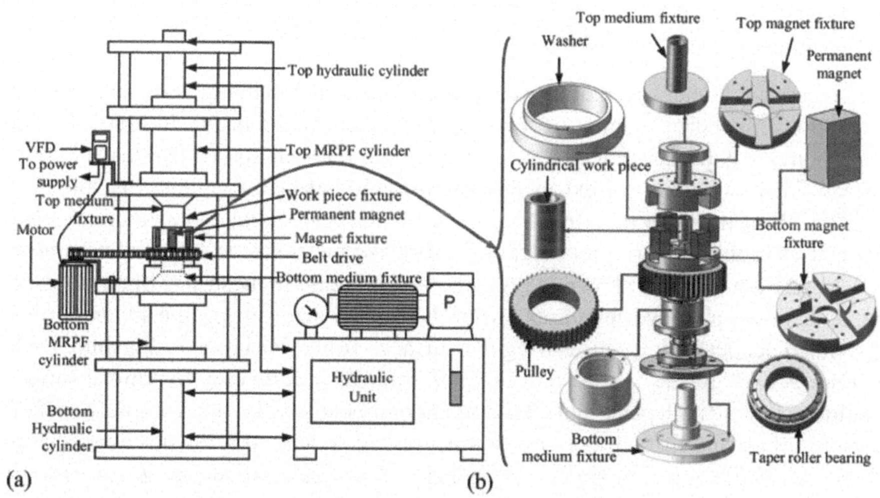

Fig. 9.4 Schematic diagram of R-MRAFF process [12]

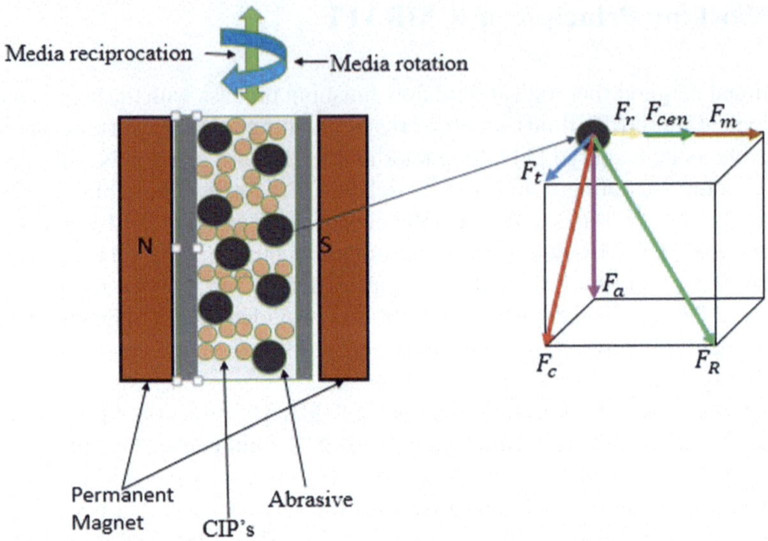

Fig. 9.5 Forces in the R-MRAFF system operating on abrasive

conclusion was the effective length of an abrasive during the finishing process was higher than the former one. With the help of rotation of magnets, abrasive followed a helical path which increases its effective length. Process parameters of R-MRAFF process are.

- Rotational speed of the magnet.
- Extrusion pressure.
- Abrasive mesh size.
- Finishing cycle.

Effect of rotational speed of magnets on surface is shown in Fig. 9.6, with increase in rotational speed of magnet over its optimal speed the capacity to hold abrasive by carbonyl iron particles decreases as centrifugal force increases that leads to poor surface finish. Variation of extrusion pressure with surface roughness is shown in Fig. 9.7. With increase in extrusion pressure axial and radial forces increases and that causes higher material removal rate. Abrasive mesh size is a crucial parameter in R-MRAFF process as shown in Fig. 9.8, with increase in mesh size of an abrasive, diameter of an abrasive increases. After 120 mesh size during the indentation it penetrated deeper which causes irregular surface profile. Variation of finishing cycle over surface roughness is shown in Fig. 9.9. Initially it quite easy to remove surface roughness having high peaks but after optimal number of cycles the workpiece surface become soften and further increase in finishing cycle lead to scratches over surface which causes decrease in surface roughness. And obtained surface finish (Ra) = 16 nm.

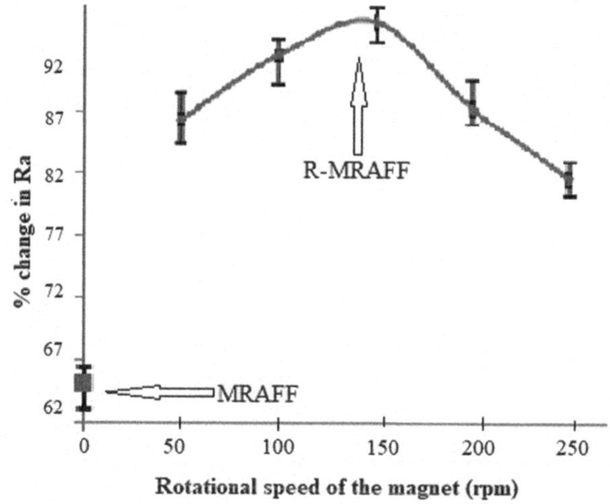

Fig. 9.6 Alteration of RPM of magnet with percentage change in surface roughness (Ra) at P = 375 bar, N = 1000 cycles, M = 180 [14]

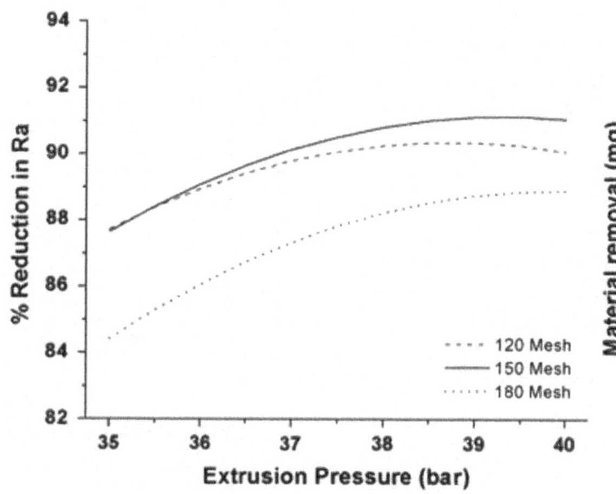

Fig. 9.7 Alteration of extrusion pressure with percentage change in surface roughness (Ra) with at N = 1000 cycles, S = 150 RPM [14]

9.4 Mathematical Modelling of the R-MRAFF Process

A mathematical model can be developed for calculating the material removal and surface roughness during the R-MRAFF process.

Following assumptions were made for developing the mathematical model

- Workpiece material is assumed to be uniform throughout and having same properties in all directions.
- Uniform roughness profiles are present on the material.
- The peaks and valleys are spread over the entire surface and the peaks are considered to be triangular as shown in Fig. 9.10.

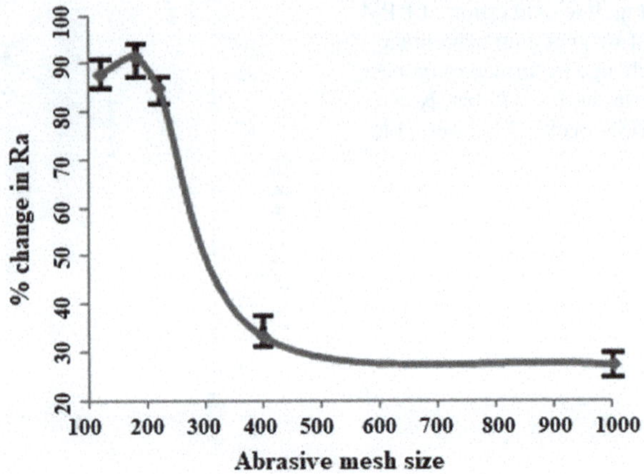

Fig. 9.8 Alteration of abrasive mesh size with percentage change in surface roughness (Ra) P = 375 bar, N = 1000 cycles, S = 100 RPM [14]

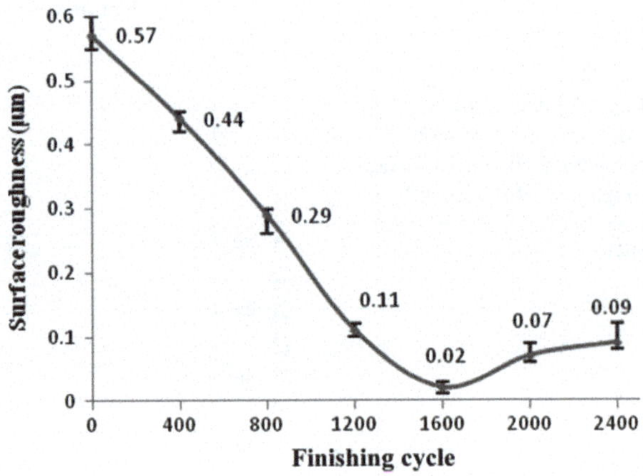

Fig. 9.9 Alteration of number of finishing cycles with percentage change in surface roughness (Ra) at P = 375 bar, S = 150 RPM, M = 180 [14]

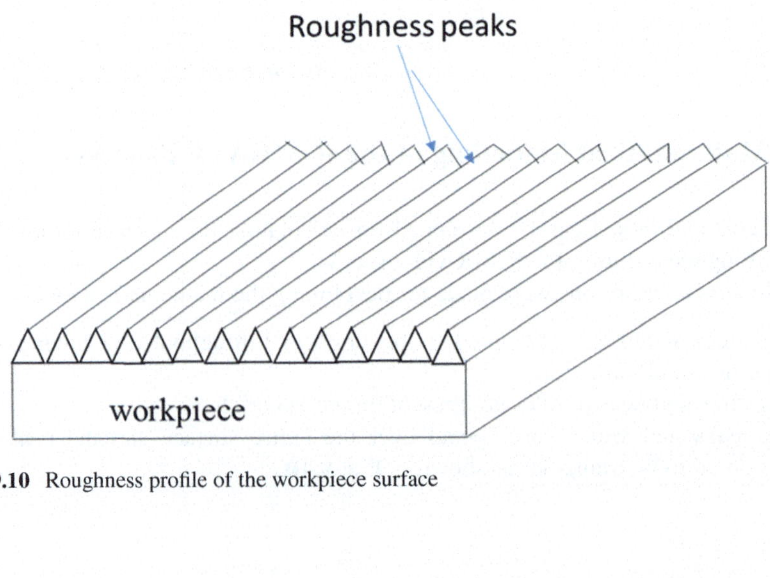

Fig. 9.10 Roughness profile of the workpiece surface

- Carbonyl iron particles and abrasives are considered to be perfect spheres of same diameters.
- The strength of the magnetic field remains constant with respect to time during the finishing process.
- Magnetic field intensity at a point remains constant throughout the finishing process.
- The distance between two abrasive particles remains constant, i.e. there is no relative movement between them.
- Carbonyl iron particles present in the magnetorheological fluid rearrange themselves in a chain like structure around abrasive particles, when they come under the influence of external field.
- Abrasive particles get trapped within the iron chains and are held tightly.
- Normal forces acting on the abrasives helps them to penetrate into the workpiece surface and axial forces shear away the roughness peaks.
- Peaks that are sheared away during the finishing process do not affect the process capability as the material removal rate is very low.

9.4.1 Analysis of Forces

Carbonyl iron particles present in the MR fluid acquire dipole moment in the presence of external magnetic field. When this dipole exceeds a particular value of thermal energy, carbonyl iron particles forms a chain in the direction of magnetic field. In the finishing area carbonyl iron particles pushes abrasive particles onto the surface of the workpiece along the line of magnetic field by magnetic force (F_m) given by the Eq. (9.5).

$$F_m = m\mu_0\chi_m H\nabla H \tag{9.5}$$

where,

m = Mass of a magnetic carbonyl iron particle (kg)
χ_m = Mass-susceptibility of the carbonyl iron particle ($\frac{m^3}{Kg}$)
H = Magnetic field intensity $\left(\frac{A}{m}\right)$
∇H = Gradient of the magnetic field intensity and
μ_0 = Permeability in free space $\left(4^{\pi} \times 10^{-9}\frac{Wb}{Am}\right)$

As the magnets are placed adjacent to each other so, a curved path is followed by the line of magnetic force near the workpiece surface and hence, there are two components of magnetic force acts on the workpiece surface, one acts normal to the workpiece surface known as normal component (F_{mn}) other acts along the circumference of the workpiece known as tangential components (F_{mt}).

A radial force (F_r) acts on the abrasive particle. This force is a function of medium viscosity and applied extrusion pressure. An axial force (F_a) acts on the abrasive particle, this force is generated due to reciprocating motion of the MR fluid during finishing operation and is proportional to hydraulic extrusion pressure (P). Simulation of MR fluid can be done for calculating the radial and axial forces acting on the workpiece. Rotating magnets generates rotational magnetic field, which in turn rotates the MR fluid inside the workpiece and due to this rotation a centrifugal force ($F_{cen} = mr\omega^2$) acts on the abrasive particle normal to the workpiece surface. So, there are three forces (F_{mn}, F_r and F_{cen}) which acts normal to workpiece surface and are responsible for indenting the abrasives on the workpiece.

$$F_{\text{indentation}} = F_{mn} + F_r + F_{\text{cen}} \tag{9.6}$$

Rotary motion of the magnet generates a force on the abrasive particle tangential to the fluid rotation, known as tangential cutting force, $F_t = 2\,m\,(\omega \times V)$, where is angular velocity and is linear velocity of the particle. Shearing of roughness peaks is done by the resultant of three forces F_{mt}, F_a and F_t. Total cutting force F_c is given by,

$$F_c = F_t + F_a + F_{mt} \tag{9.7}$$

This resultant cutting force along with axial force forces abrasive particles to move in a helical path. However the path followed will not be exactly helical as the MR fluid is in semisolid state. Helix angle of the helical path is given by the Eq. (9.8) and total of the helical path is given by the Eq. (9.9),

$$\theta = \tan^{-1}\frac{V_a}{V_t} \tag{9.8}$$

$$s = h_w\sqrt{\left(\frac{V_t}{V_a}\right)^2} \tag{9.9}$$

Normal forces given in Eq. (9.6) acting on abrasive particles pushes them onto the workpiece surface and produces indentations. Cutting forces given in Eq. (9.7) removes the roughness peaks. Depth of indentation can be calculated by the given Eq. (9.10).

$$d = \frac{D_g}{2} - \frac{1}{2}\sqrt{D_g^2 - D_i^2} \tag{9.10}$$

where,

d = depth of indentation,
D_g = diameter of abrasive grain, and
D_i = diameter of indentation.

Diameter of indentation D_i can be calculated using known values of BHN of workpiece using given Eq. (9.11).

$$BHN = \frac{F_{indentation}}{\frac{\pi D_g}{2}\left(D_g - \sqrt{D_g^2 - D_i^2}\right)} \tag{9.11}$$

9.4.2 Material Removal Model

Abrasives present in the MR fluid remove the material from the work piece surface by shearing action and the size of these chips are very small. Since abrasives indent on the surface so they remove volume equal to the indented groove. As assumed a triangular roughness the surface area to be finished is large as compared to the abrasive particles so, we can take a small portion as a cell for computation and the same can be applied to the whole surface. Volume of material removed through a cell by a single abrasive is given by the Eq. (9.12).

$$\Delta V = A_p \times l_t \tag{9.12}$$

A_p is the sheared area in the helical track and l_t length of contact of abrasive with the surface of the cell is given by the Eq. (9.13).

$$l_t = l \times nf \tag{9.13}$$

the contact length (BC) of a single profile can be obtained as Eq. (9.14)

$$l = 2d \tan\theta \tag{9.14}$$

where d is the depth of penetration of the abrasive and 2 is the mean angle of surface asperity number of profile is given by Eq. (9.15)

$$n_f = \frac{l_c}{l_b} \tag{9.15}$$

l_c is length of cell as shown in Fig. 9.11 and l_b base length of single profile.

$$l_b = 2R_{\max} \tan\theta \tag{9.16}$$

$R_{\max}$ = initial or maximum value of surface roughness Therefore;

$$l_t = l \times n_f = l \times \frac{l_c}{l_b} = \frac{d \times l_c}{R_{\max}} \tag{9.17}$$

Therefore, volume of material removed by a single abrasive

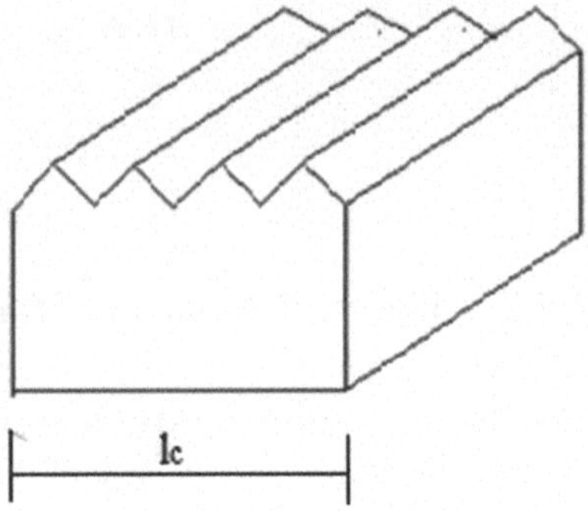

Fig. 9.11 Length of cell

$$\Delta V = \frac{A_P \times d \times l_c}{R_{\max}} \tag{9.18}$$

Number of abrasives passing through a cell in each stroke is given by Eq. (9.19).

$$n = \frac{length\ of\ helical\ track}{distance\ between\ two\ abrasives} \tag{9.19}$$

Distance between two abrasives is given by Eq. (9.20).

$$l_a = \frac{h_w\sqrt{\left(\frac{V_t}{V_a}\right)^2 + 1}}{l_a} \tag{9.20}$$

Total volume removed through a cell in a single stroke is shown in Eq. (9.21)

$$V = \frac{A_p \times d \times l_c \times h_w\sqrt{\left(\frac{V_t}{V_a}\right)^2 + 1}}{R_{\max} \times l_a} \tag{9.21}$$

9.4.3 *Surface Roughness Model*

From the geometry of the figure as shown in the Fig. 9.12. Let R_1 is the Roughness after one stroke. Volume of material removed in one stroke is given by Eq. (9.22).

$$\Delta V = \frac{1}{2} \times AD \times BC \times r \tag{9.22}$$

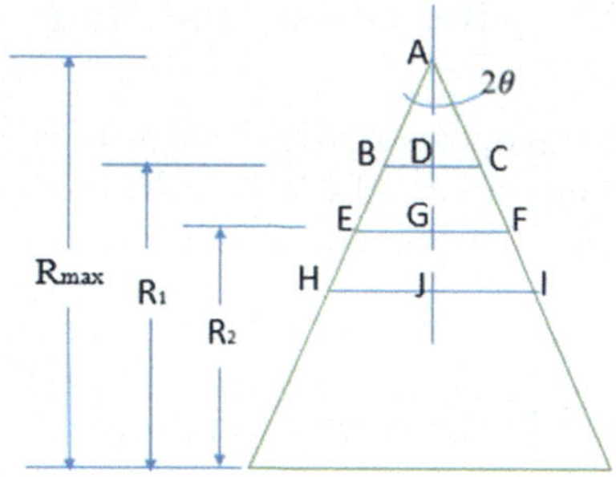

Fig. 9.12 A single surface profile

r is the average width of cut, it is equal to the diameter of the projected area of penetration of the abrasive grain

$AD = R_{\max} - R_1$ and
$BC = 2(R_{\max} - R_1)\tan\theta$ Therefore $\Delta V = (R_{max} - R_1)^2 \tan\theta \times r$

Total volume of material removed through a cell $\Delta V_c = \Delta V \times n_f$
We know $n_f = \frac{l_c}{2\times R\max \times tan\theta}$
Volume of material removed through a cell in first stroke can be written as

$$\Delta V_{c1} = \frac{(R\max - R_1)^2 \times r \times l_c}{2R\max} \tag{9.23}$$

This volume removed from the cell in one stroke is equal to the volume removed in one stroke as discussed earlier. Equating the two equations we get,

$$R_1 = R_{max} - \left(\frac{2R\max_{\times V}}{r \times l_c}\right)^2 \tag{9.24}$$

R_2 is the roughness after second stroke, BCFE is the cross sectional area of the material removed in the second stroke.

$$BCFE = \left[(R\max - R_2)^2 - \left((R\max - R_1)^2\right] \tan\theta \tag{9.25}$$

Equating this to volume of material removed per stroke V, we get Eq. (9.26)

$$R_2 = R\max - \left[(R\max - R_1)^2 + ((2V\ R\max)/(rlc))\right]^{(1/2)} \tag{9.26}$$

It can be generalised that after n stroke the final surface roughness is

$$R_n = R\max - \left[(R\max - R_n - 1)^2 + ((2V\ R\max)/(rlc))\right]^{(1/2)} \tag{9.27}$$

9.5 Simulation Modelling

The purpose of this segment is to analyse the fluid flow variation with change in compositions of CIP by calculating the axial stresses and radial stresses due to the impact of abrasive on surface of the work piece with numerical analysis.

9.5.1 Governing Equation

The mathematical depiction of the movement of the MR fluid is described in continuity Eq. (9.28) and momentum Eq. (9.29) in combination with a suitable rheological constitutive Eq. (9.15).

$$\frac{\partial u_i}{\partial x_i} = 0 \tag{9.28}$$

$$\rho\left[\frac{\partial u_i \partial u_i}{\partial x_i}\right] = \frac{\partial p}{\partial x_i} + \frac{\partial}{\partial x_j}\left[\mu\dot{\gamma}\left(\frac{\partial u_i}{\partial x_j} + \frac{\partial u_j}{\partial x_i}\right)\right] \tag{9.29}$$

Here, density of the MR-fluid is denoted by ρ, pressure is denoted by p and velocity in i and j direction is denoted by u_i, u_j respectively. In Herschel Bulkley model for Bingham fluid as shown in Fig. 9.13 describes the behavior of Bingham fluid, when the value of n is equals to 1 then the non-Newtonian fluid behaves like a Bingham fluid.

Shear stress of the Bingham fluid is defined as in Eq. (9.30).

$$\tau^{--} = \tau_0^{--} + \eta D^{--} \tag{9.30}$$

where, τ_0 is the yield shear stress and D is a tensor which denotes rate of deformation and denoted by Eq. (9.31).

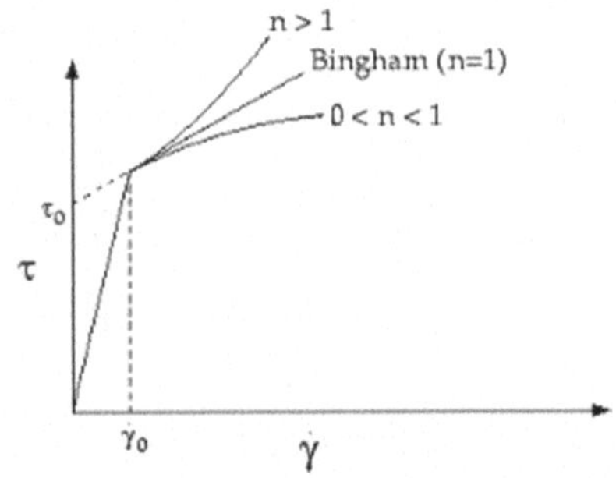

Fig. 9.13 Herschel-Bulkley model

$$\bar{\bar{D}} = \left(\frac{\partial u_i}{\partial x_j} + \frac{\partial u_j}{\partial x_i}\right) \tag{9.31}$$

Viscosity of Bingham fluid is defined in Eq. (9.6) where η is the viscosity of the Bingham fluid, $\dot{\gamma}$ is strain rate as defined in Eq. (9.32), K is viscosity of the Bingham fluid and n is power law index which is in the case of Bingham fluid is equals to 9.1.

$$\eta = \frac{\tau_0}{\dot{\gamma}} + K\dot{\gamma}^{n-1} \tag{9.32}$$

$$\dot{\gamma} = \sqrt{\frac{1}{2}\bar{\bar{D}} : \bar{\bar{D}}} \tag{9.33}$$

K and τ_0 be contingent on the concentration of MR fluid and magnitude of magnetic field described in Eqs. (9.34) and (9.35) [15].

$$K = -202.64 + 11.09A + 9.54B + 5C - 110.23D - 0.38AB - 0.53AC + 19.23\,CD \tag{9.34}$$

$$\begin{aligned}\tau_0 = {} & -176.13 + 9.11A + 14.58B + 2.19C - 218.46D - 0.34AB \\ & - 0.4AC - 0.57BC + 17.15CD + 0.39C^2 + 170.42H^2\end{aligned} \tag{9.35}$$

where, A is concentration of CIP, B is concentration of SiC, C is concentration of AP3 grease in MR fluid and D is magnitude of magnetic field. Magnetohydrodynamics is implemented in fluent to achieve required magnetic field. The link amongst the external magnetic field and fluid stream field depends upon the two basic concept first one is the electric field produced due to the movement of electrons in conductive materials in the influence of some external magnetic field, and the second one is Lorentz force produced by the electric and magnetic field. Electromagnetic fields are defined by Maxwell's equation 9.36–9.39.

$$\Delta \cdot B = 0 \tag{9.36}$$

$$\Delta \cdot E = -\frac{\partial B}{\partial t} \tag{9.37}$$

$$\Delta \cdot D = q \tag{9.38}$$

$$\nabla \cdot \boldsymbol{H} = \boldsymbol{J} + \frac{\partial \boldsymbol{D}}{\partial t} \tag{9.39}$$

where, magnetic field is defined by *B*, electric field is defined by *E*, induction field for magnetic field is defined by *H*, induction field for electric fields is defined by *D*, electric charge density is defined by q and electric current density is defined by *J*. The induced magnetic and electric fields *H* and *D* are shown in Eqs. (9.40)–(9.41).

$$H = \frac{1}{\mu} B \tag{9.40}$$

$$D = \in E \tag{9.41}$$

where, μ is magnetic permeability and *is* electric permittivity. From the law of Ohm and the equation of Maxwell, the magnetic induction equation is derived. The equation delivers the relationship amongst the magnetic field and fluid flow field. In general, Ohms law that defines the current density j is given by Eq. (9.42).

$$j = \sigma E \tag{9.42}$$

where, σ *is* electrical conductivity. For fluid velocity field U in a magnetic field B, Ohm's law takes the form as shown in Eq. (9.43).

$$j = \sigma[E + U \times B] \tag{9.43}$$

From the law of Ohm and the equation of Maxwell, the induction equation can be derived as in Eq. (9.44) [16].

$$\frac{\partial B}{\partial t} + (U.\nabla)B = \frac{1}{\mu\sigma}\nabla^2 B + (B.\nabla)U \tag{9.44}$$

Modelling of surface texture is carried out after investigating the surface roughness profile data obtained from the surface irregularity evaluating device. Ra value of the surface roughness was found to be 0.6 μm. It is considered that the diameter of the abrasive particle is greater than the gap between the two successive crests of surface roughness. Normal force required for indenting the abrasives into the workpiece is obtained from the simulation [17]. Using this normal force and by knowing the BHN of the workpiece material we can find out the diameter of indentation by using Eq. (9.45).

$$D_i = \sqrt{D_g^2 - \left(D_g - \frac{2 \times 10^{-6} F_n}{9.81 H_{BHN} \pi D_g}\right)} \tag{9.45}$$

Again from the Fig. 9.14 using geometry the depth of indentation can befound out by using Eq. (9.46).

$$d = \frac{D_g}{2} - \frac{1}{2}\sqrt{D_g^2 - D_i^2} \tag{9.46}$$

Following steps are considered for simulating the surface roughness model.

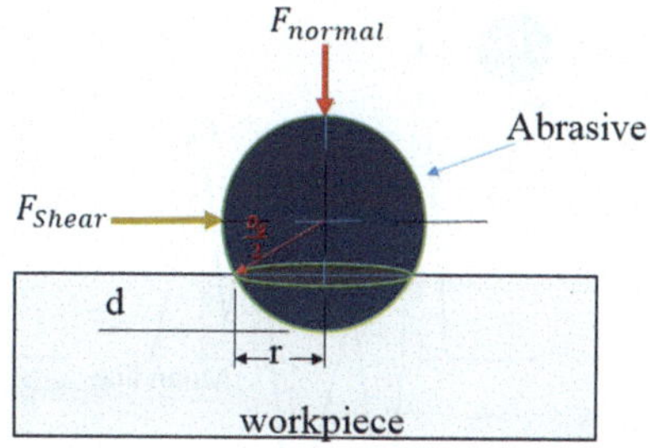

Fig. 9.14 Indentation by abrasive on work piece

Step 1. The initial surface roughness statistics obtained from the surface irregularity evaluating device are stockpiled in an array. Crests of the surface texture are identified if the following Eq. (9.47) is satisfied.

$$C_{i-1}\, C_i \geq C_{i+1} \tag{9.47}$$

where C_{i-1}, C_i and C_{i+1} are (i−1)th, ith and (i + 1)th crests respectively.

Step 2. It is essential that all the crests identified should undergo cutting action at each stroke, if a point is not identified as crest its altitude remains same for the next stroke. It is also considered that at any stroke (n) shearing occurs from a specific crest if Eq. (9.48) is satisfied.

$$C_{i,n} > (C_{i,max} - d) \tag{9.48}$$

Step 3. In the next stroke (n + 1) the new altitude of ith crest $C_{i,n+1}$ will be shown in Eq. (9.49).

$$C_i, n + 1 = (C_{i,max} - d) \tag{9.49}$$

where $C_{i,max}$ is maximum crest height in the former stroke (nth stroke). New crest height obtained after shearing by one abrasive has been shown in Fig. 9.15.

Step 4. After each recapitulation all the crests which were parted initially into two arrays containing crests C_i and valleys V_i are organized together into an array $Y_{i:j}$.

Step 5. Most important thing is that the roughness value is computed about the mean line of the surface texture. Mean line is the line which divides the texture so that area above and below the line are equal. After each recapitulation abrasive particles shears away materials from the crests present above the mean line. Hence, the position of the mean line changes ΔY and after each recapitulation new mean line of the updated surface texture is evaluated and consequently all the data $Y_{i,n}$ are

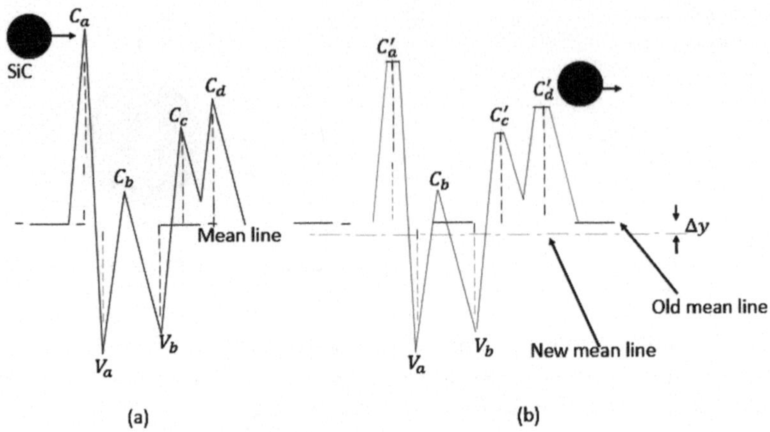

Fig. 9.15 Impact of single abrasive on surface roughness

updated and given in Eq. (9.50).

$$Y_{i,n} + 1 = Y_i - \Delta Y \tag{9.50}$$

Step 6. The centerline average surface roughness value R_a of the surface texture can be evaluated using surface texture data points Y_i from the Eq. (9.51).

$$R_a = \frac{\sum_{i=1}^{n} |Y|_i}{n} \tag{9.51}$$

where, n is the data points number.

The formation of velocity circulation in Zone 1 is shown in Fig. 9.16. The axial velocity profile plot in Zone 2 is shown in Fig. 9.17, specifies that nearby axis of work piece fixture there is a compact core section where the MR fluid flows with highest velocity and the reduction in velocity is non-linearly, in the radial direction from the center of the work piece fixture where velocity becomes zero. This approves the MR fluid flow circulation of the non-Newtonian Bigham plastic fluid [18]. Effect of magnetic field over velocity profile is shown in Figs. 9.18 and 9.19.

Due to an upsurge in the CIP concentration in the MR fluid, the persuaded Strength of magnetic field also increased. The induced magnetic field simulation with the help of magnetohydrodynamics. The radial variation of inlet velocity and outlet velocity of different compositions of CIP in MR fluid in Zone 2 is shown in Figs. 9.20 and 9.21 respectively.

The average values of axial and radial stresses generated by the action of single abrasive over different compositions of CIP are listed in Table 9.1. It can be observed that with increase in composition of CIP, radial and axial stresses also get increased. The cause for this kind of behavior is described below. CIP is the key constituent of the MR fluid, and offers a noteworthy influence to the physical characteristic of MR

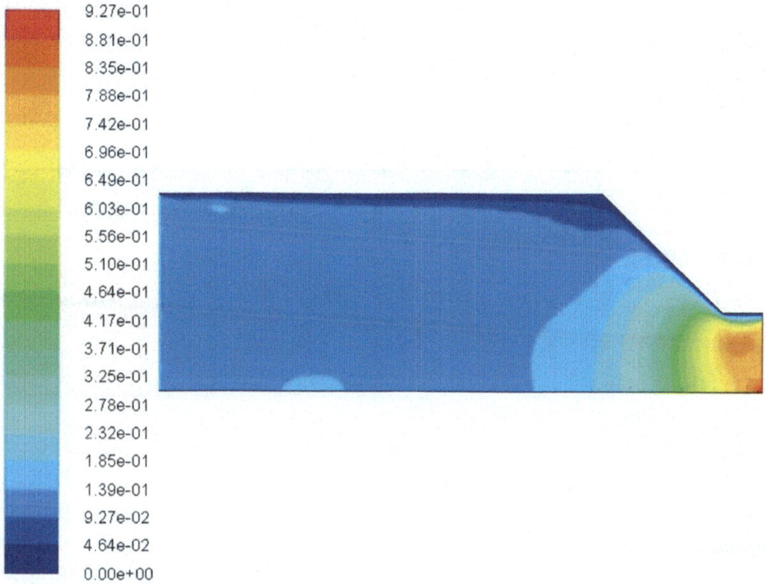

Fig. 9.16 Velocity profile for zone 1 (15% CIP with 10% SiC)

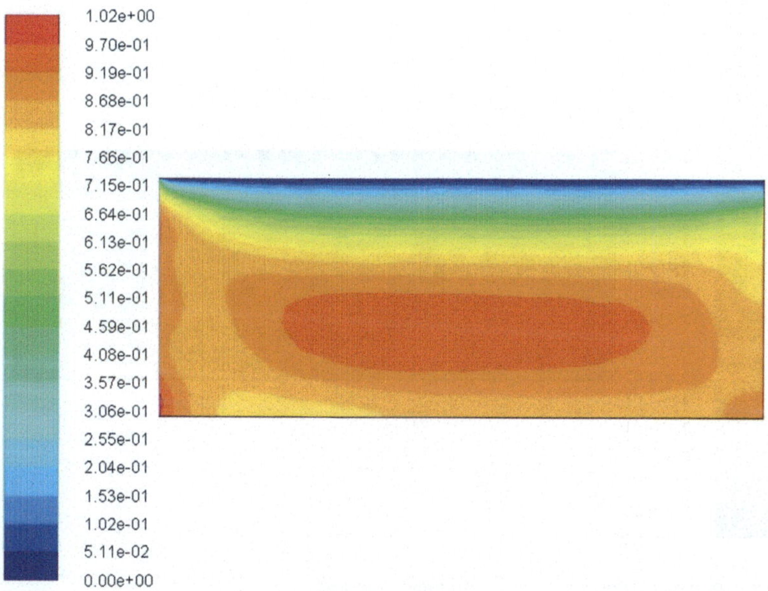

Fig. 9.17 Velocity profile for zone 2 (15% CIP with 10% SiC)

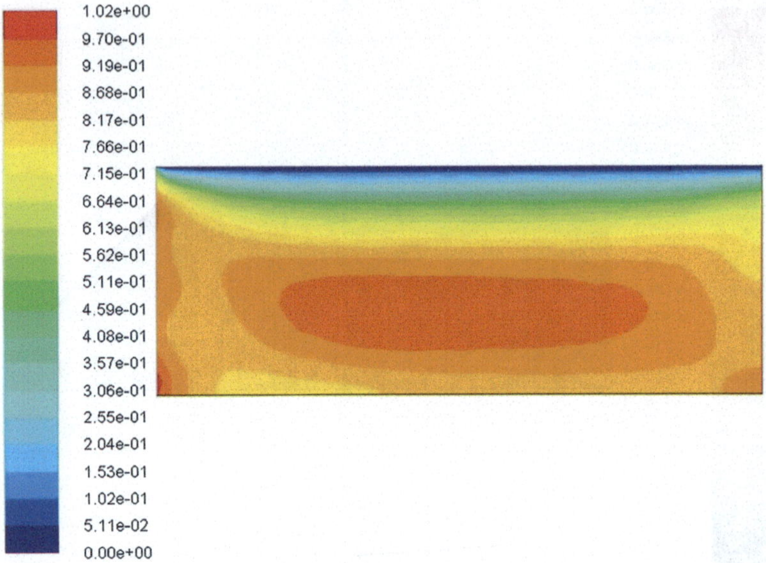

Fig. 9.18 Velocity profile for zone 2 with magnetic field

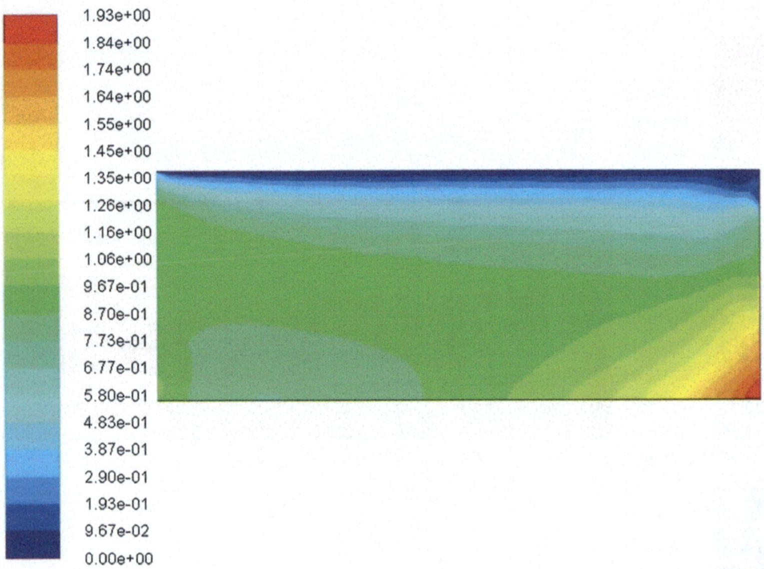

Fig. 9.19 Velocity profile for zone 2 with magnetic field

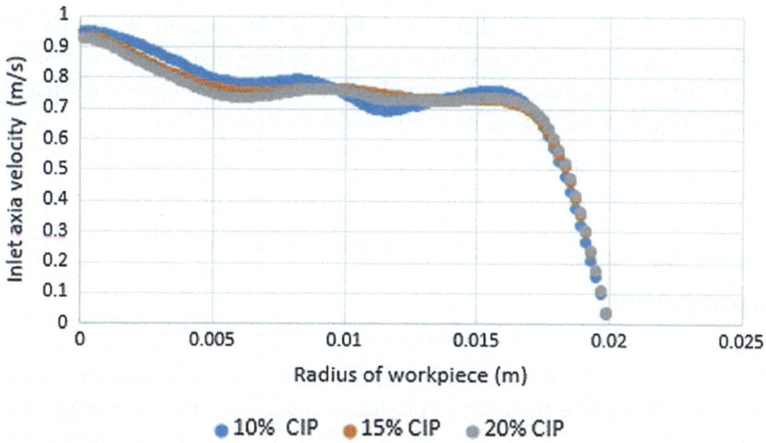

Fig. 9.20 Velocity profile for zone 1 (15% CIP with 10% SiC)

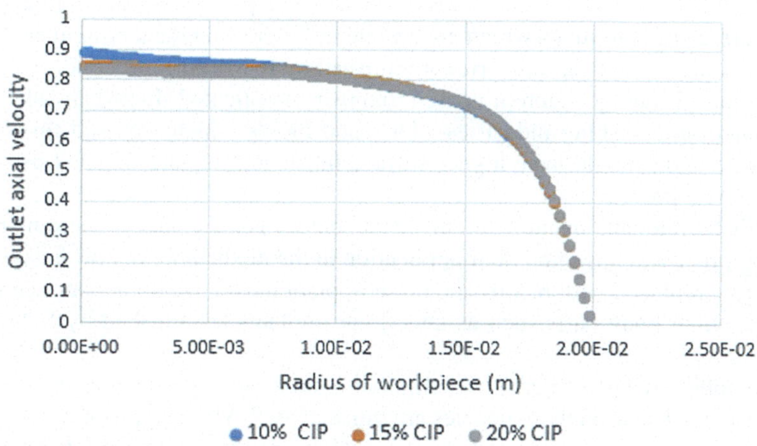

Fig. 9.21 Velocity profile for zone 2 (15% CIP with 10% SiC)

Table 9.1 Axial stresses applied by different CIP concentration

CIP (% vol.)	Avg. axial stress (KPa)	Avg. radial stress (Pa)	Radial force (nN)	Surface roughness (μm)
10	12.498	15.889	4.494	0.032
15	14.978	23.782	6.729	0.027
20	21.229	41.658	11.795	0.019

fluid. In Table 9.1, it is observed that with increase in concentration of CIP, axial velocity in zone 2 get decreased as shown in Fig. 9.20 Likewise, as the volume fraction of CIP increases, body centered tetragonal crystal structure is predicted because of that large yield stress was noticed. For the similar cause the MR fluid imposes higher radial and axial stresses near the wall if fixture with increase in composition of CIP as shown in Table 9.1.

9.6 Applications

R-MRAFF process finds its applications in the areas where surface finish plays an important role. Reduction in surface roughness means savings of energy, low heat generation and less wear and tear of parts. Following are some of the areas where R-MRAFF process is used.

- Medical Industry—In medical industry R-MRAFF process is used to finish internal surface of bent tubes which are an integral part of medical equipment. Heart valves used in medical industries are finished by R-MRAFF process.
- Aerospace Industry—Here it is used to finish complicated shaped airfoil.
- Die and mold making industries—Dies and molds which are used for making intricate parts should have higher surface finish so, they are finished using this R-MRAFF process.
- Prosthetic implants manufacturers—21st century is the century where manufacturing processes have find their application in the medical field. now a days body parts are replaced with metals and these implant must have higher surface finish. These small body parts such as knee joint implants are finished by R-MRAFF process.
- Automobile manufacturers—In automobile industry very fine nozzles with high surface finish is used these nozzles are finished by R-MRAFF process. Also there are many more components such as gear, bearings etc. are used where relative between parts take place. These parts are finished by R-MRAFF process for higher surface finish (Fig. 9.22).

9.7 Summary

R-MRAFF is a magnetic field assisted finishing process, which uses magnetic force for finishing the workpiece surfaces. This process has the capability to reduce surface roughness up to nanometer range. This process finds its application in many important areas which were not processed by conventional processes. Complicate intricate surfaces can be polished. Close tolerances can be achieved by polishing the workpiece with this process, which will increase interchangeability of parts and hence productivity will increase. This process produces high quality surfaces so,

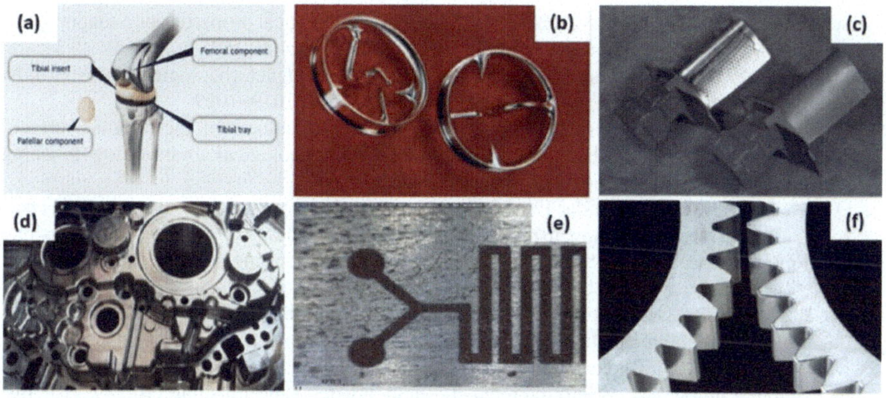

Fig. 9.22 Application of R-MRAFF process **a** Knee joint implant, **b** Heart valves, **c** Air foil, **d** Engine crank case die, **e** Complex microchannel, and **f** Automotive gear

friction between parts reduces, energy losses reduce, and less wear and tear takes place. This process generates crosshatched pattern on the surface of the workpiece which increases oil (lubricate) retention capabilities of the workpiece. However this process can be hybridized with some other processes for achieving surface roughness in the atomic level.

References

1. Golini D, Kordonski WI, Dumas P, Hogan SJ (1999) Magnetorheological finishing (MRF) in commercial precision optics manufacturing. In: Optical manufacturing and testing III, vol 3782. International Society for Optics and Photonics, pp 80–92
2. Golini D (1999) Precision optics manufacturing using magnetorheological finishing (MRF). In: Optical fabrication and testing, vol 3739. International Society for Optics and Photonics, pp 78–86
3. Arrasmith SR, Kozhinova IA, Gregg LL, Shorey AB, Romanofsky HJ, Jacobs SD, Golini D, Kordonski WI, Hogan SJ, Dumas P (1999) Details of the polishing spot in magnetorheological finishing (MRF). In: Optical manufacturing and testing III, vol 3782. International Society for Optics and Photonics, pp 92–101
4. Menapace J, Penetrante B, Miller P, Parham T, Nichols M, Peterson J, Golini D (2002) Combined advanced finishing and uv-laser conditioning for producing uv-damage-resistant fused silica optics. In: Optical fabrication and testing, page OMB4. Optical Society of America
5. Harris DC (2011) History of magnetorheological finishing. In: Window and dome technologies and materials XII, vol 8016. International Society for Optics and Photonics, pp 80160 N
6. Rabinow J (1948) The magnetic fluid clutch. Electr Eng 67(12):1167
7. Menapace JA, Dixit SN, Génin FY, Brocious WF (2004) Magnetorheological finishing for imprinting continuous-phase plate structures onto optical surfaces. In: Laser-induced damage in optical materials: 2003, vol 5273. International Society for Optics and Photonics, pp 220–231
8. Yang G, Spencer BF Jr, Carlson DJ, Sain MK (2002) Large-scale MR fluid dampers: modeling and dynamic performance considerations. Eng Struct 24(3):309–323
9. Ginder JM, Davis CL, Elie LD. Rheology of magnetorheological fluids: models and measurements. Int J Modern Phys B 10(23–24):3293–3303

10. Seok JW, Lee SO, Jang K-I, Min B-K, Lee SJ (2009) Tribological properties of a magnetorheological (MR) fluid in a finishing process. Tribol Trans 52(4):460–469
11. Jha S, Jain VK (2004) Design and development of the magnetorheological abrasive flow finishing (MRAFF) process. Int J Mach Tools Manuf 44(10):1019–1029
12. Das M, Jain VK, Ghoshdastidar PS (2011) The out-of-roundness of the internal surfaces of stainless steel tubes finished by the rotational-magnetorheological abrasive flow finishing process. Mater Manuf Process 26(8):1073–1084
13. Das M, Jain VK, Ghoshdastidar PS (2009) Parametric study of process parameters and characterization of surface texture using rotational-magnetorheological abrasive flow finishing (R-MRAFF) process. In: ASME 2009 international manufacturing science and engineering conference. American Society of Mechanical Engineers, pp 251–260
14. Das M, Jain VK, Ghoshdastidar PS (2010) Nano-finishing of stainless-steel tubesusing rotational magnetorheological abrasive flow finishing process. Mach Sci Technol 14(3):365–389
15. Das M, Jain VK, Ghoshdastidar PS (2012) Computational fluid dynamics simulation and experimental investigations into the magnetic-field-assisted nano-finishing process. Proc Inst Mech Eng Part B J Eng Manuf 226(7):1143–1158
16. Gedik E, Kurt H, Recebli Z, Balan C (2012) Two dimensional CFD simulation of magnetorheological fluid between two fixed parallel plates applied external magnetic field. Comput Fluids 63:128–134
17. Rajput AS, Prasad D, Mondal AK, Bose D (2019) 2D Computational fluid dynamics analysis into rotational magnetorheological abrasive flow. Advances in materials and manufacturing engineering: proceedings of ICAMME, p 67
18. Chhabra RP, Francis Richardson J (2011) Non-Newtonian flow and applied rheology: engineering applications. Butterworth-Heinemann
19. Nagdeve Leeladhar, Sidpara Ajay, Jain VK, Ramkumar J (2018) On the effect of relative size of magnetic particles and abrasive particles in MR fluid-based finishing process. Mach Sci Technol 22(3):493–506
20. Jolly MR, David Carlson J, Munoz BC (1996) A model of the behaviour of magnetorheological materials. Smart Mater Struct 5(5):607

Chapter 10
Technological Advances and Challenges in Chemical Mechanical Polishing

Samad Nadimi Bavil Oliaei, Muslim Mukhtarkhanov, and Asma Perveen

Abstract It has been already several decades since chemical mechanical polishing (CMP) process has been deployed as a planarization technique for the fabrication of integrated circuit (IC) in the semiconductor industries. CMP is considered to be a wet polishing technique that has the capability to generate ultrafine surfaces for numerous materials using the combined effect of chemical and mechanical interactions. As CMP involves both mechanical and chemical actions, the process efficiency of CMP also varies with the parameters involved with mechanical and chemical aspects. This chapter presents an overview of CMP technology, working principles, its recent advancement status in terms of green slurry development, abrasives development, new polishing method, research trends and challenges.

Keywords Wafer · Surface · Planarization · Abrasives · Polishing

10.1 Introduction

With the tremendous growth of electronic industries in the past few decades, it has become possible to achieve electronic devices with reduced cost and increased performance due to the fabrication of ultra large scale integrated (ULSI) circuits containing 10^8 or even more devices on chip within semiconductors substrates or wafer. As per the forecast of Moore, the growth of the number of components in the semiconductor chip has roughly multiplied by two every two years over the last few decades. At the same time, advanced fabrication technology also causes the device-features dimension to shrink by 13% per year. Eventually, reduced features size results in overall reduction of device size with increased packing density but with reduced cost of function. Moreover, device speed increases and power consumption reduces

S. Nadimi Bavil Oliaei
Department of Mechanical Engineering, Çankaya University, Ankara, Turkey

M. Mukhtarkhanov · A. Perveen (✉)
Department of Mechanical and Aerospace Engineering, School of Engineering and Digital Sciences, Nazarbayev University, Nur-Sultan, Kazakhstan
e-mail: asma.perveen@nu.edu.kz

S. Das et al. (eds.), *Advances in Abrasive Based Machining and Finishing Processes*, Materials Forming, Machining and Tribology,
https://doi.org/10.1007/978-3-030-43312-3_10

due to the reduced feature size. In addition, the complex nature of chip design and fabrication continues to increase along with miniaturization and integration. Process requirements of this kind of wafer fabrication include high degree of repeatability, as well as uniformity for higher process yields. One of the common challenges faced by fabrication technology arises from the demand for global planer surfaces associated with lithography process in back end process integration. Chemical Mechanical Polishing (CMP) has become an inevitable technology to achieve this smooth surface topography of the multiple layers of interlevel dielectric involved [1, 2].

Chemical mechanical Polishing process was initially deployed for the purpose of oxide polishing and later for tungsten polishing in the year of 1986 and 1988, respectively. After few decades of development by IBM, CMP has become one of the paramount processes in the thin film fabrication which aids in smooth as well as damage-free surface generation. As the name implies, CMP removes material with the combined effect of chemical and mechanical actions. CMP fabricates global as well as local surfaces by removing material in micro scale, nanoscale or even atomic scale, which also satisfies the requirements sets by the lithography process. Due to its contribution towards thinning and flattening of thin films, it has become established as one of the leading planarization process in the Integrated circuits(IC) manufacturing industry [3]. Chip manufacturing plant consists of the front-end and the back-end process. While the front-end process manufactures the circuit elements, the back-end process combines them into IC. Both of these processes are in need of CMP process to create flat and thin structures of inter-level dielectric (ILD, SiO_2) or to remove additional materials to create inlaid metal or isolation trench structure. The ILD CMP is usually implemented for aluminum metallization. Other than these two processes, chip interconnect functioning as back-end of the line process(BEOL) also contributes significant role in the development of reduced size of features and increased speed of the devices other than product reliability for large volume production [4, 5]. CMP is also useful in BEOL process and aids in stacking layer upon layer for interconnect materials [6]. With time, demand for increased wafer size, smaller feature size and higher integration has been increasing rapidly. While the level of interconnection and wafer diameter increases, CMP has been facing challenges such as Nano-level planarity, sub Nano-level roughness and reduced surface/subsurface damage on the wafer.

CMP is also found to play critical roles in other emerging technologies like copper damascene patterns polishing, low-K dielectrics as well as shallow trench isolation. Moreover, the introduction of Cu and W in Ultra-large-scale integrated (ULSI) circuits also need CMP process to create inlaid interconnected structure. The complexity associated with concurrent or sequential removal of a wide range of materials eventually necessitates the deeper understanding of the process principles in order to facilitate the optimal process design, control and increased process yield [7]. While, the understanding of micro/nano/atomic scale removal due to combined mechanical and chemical action is essential, additionally mechanism involving global planer surface generation by local material removal for large dimension wafers is also equally important.

10.2 Principles of CMP

CMP is based on a very simple concept of polishing the surface of the workpiece with tiny abrasives and chemicals until achieving a flat, smooth and flawless surface. In this process, the aim is to achieve an acceptable global and local planarization [8], the former being more influential on the yield of device fabrication. The origin of CMP dates back to ancient times. However, its usage in the semiconductor industry by IBM company dates back to the 1980s [9]. CMP is one of the essential steps of ultra-precision machining for different crystals [10]. Despite the developments made in the tooling and polishing capabilities of the CMP, less attention has been paid to the mechanisms involved in this process. Therefore, understanding the underlying mechanism of CMP could be helpful for further development of this process and making the process more efficient. Because of its complex nature, due to the interactions between chemistry and mechanics of the process and diversity in the factors involved in this process, CMP is considered as a challenging technology.

In order to realize a CMP process, four different configurations can be used as illustrated in Fig. 10.1 [11]. Among these configurations, the rotary type with a reciprocating mode is widely used. The schematic drawing of CMP is demonstrated in Fig. 10.2. In this process, both the rotating platen and the workpiece (wafer) carrier rotates in the same direction with a synchronized reciprocating radial motion of the carrier. It has been mentioned by Runnels [12] that there is a separation of about 10 μm between the pad and workpiece which is filled by a slurry during the CMP process. This separation is generally attributed to the pressure increase in the slurry due to the relative motion between platen and workpiece which results in a hydrodynamic lift force. This hydrodynamic lift force which balances the downforce is responsible for the film separation. The formation of hydrodynamic film separation has been verified by measuring the vertical displacement of the workpiece carrier by Nakamura et al. [13].

In a study conducted by Cook et al. [14], a mechanism other than hydrodynamic film CMP has been considered which is called asperity contact mechanism. According to this mechanism, the material removal occurs when the workpiece contacts the asperities of the polishing pad, which is very similar to the mechanical polishing techniques such as lapping.

In the CMP process, two fundamental mechanisms of material removal are combined. One of them is related to the removal by the abrasion due to a mechanical contact between abrasive particles and the workpiece while the other is based on chemical corrosion. In this process, chemical reactions soften the surface of the workpiece, then this soften layer is removed through a mechanical removal process by abrasive particles. For example, for the case of dielectric CMP process, the reaction between silicon oxide surface with OH^- from the slurry solution occurs and generates silicon hydroxide ($Si(OH)_4$). The hydration reaction occurring during the process is as follows [15]:

$$(SiO_2)_x + 2H_2O \rightarrow (SiO_2)_{x-1} + Si(OH)_4 \quad (10.1)$$

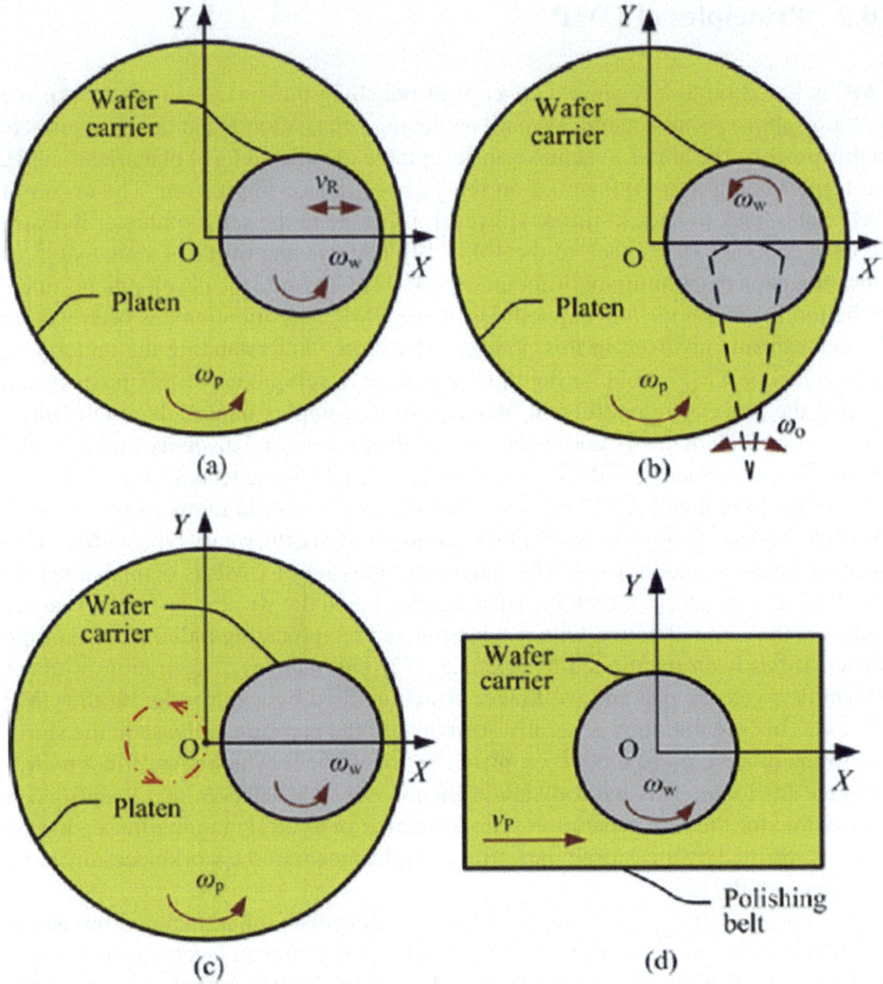

Fig. 10.1 Illustration of different CMP configurations: **a** Rotary type-reciprocating mode CMP **b** rotary type-Oscillatory mode CMP, **c** Orbital CMP, **d** Linear CMP [11]

The soften $Si(OH)_4$ is disintegrated by mechanical abrasion with the help of abrasive particles in the slurry.

CMP basically is intended to do two different tasks simultaneously. The first expected task is material removal, while the second one is the planarization, where a smooth, uniform and almost defect-free surface is generated at the end of this process. Therefore, to have a high performance CMP process, it is necessary to consider two different aspects of the process. The first aspect is related to the material removal mechanism involved in micro/nano scale realized by a combination of mechanical and chemical activities. The second important aspect is to understand how a global planar surface is achieved by local material removal [11].

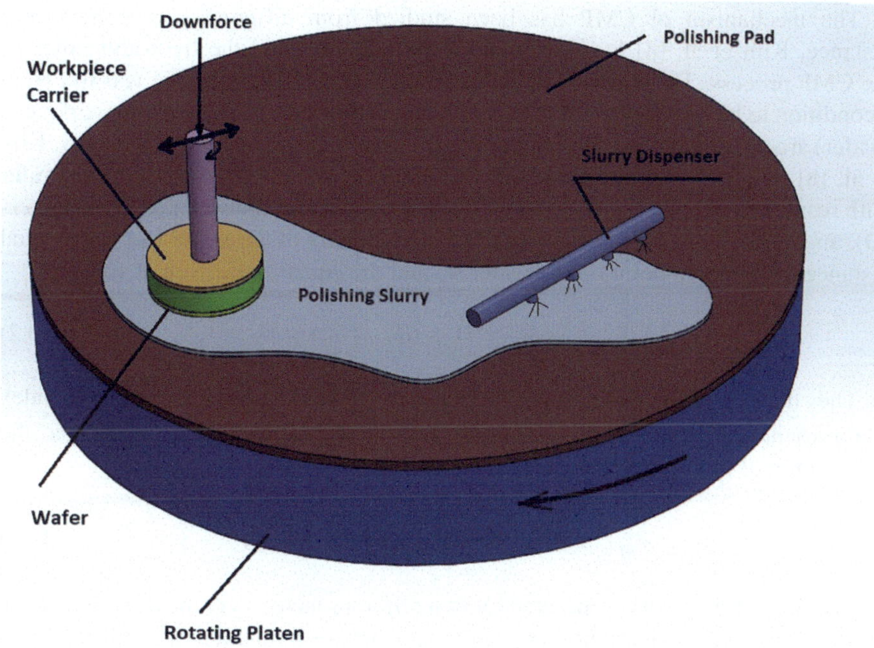

Fig. 10.2 Illustration of the CMP process

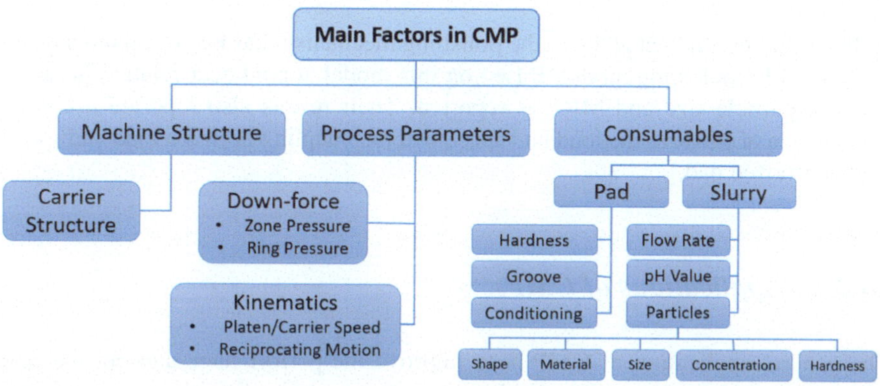

Fig. 10.3 Main factors involved in a CMP process

Several parameters are involved in CPM process including the pressure distribution over the surface of the workpiece, relative velocity distribution, size, type and concentration of the abrasive particles in the slurry, flow characteristics and chemical state of the slurry, thermal management, pad conditioning state and other mechanical and chemical factors. Figure 10.3 illustrates the main factors involved in a CMP process.

The mechanism of CMP has been studied from different perspectives. For instance, Kim et al. [8] have analyzed the kinematics and the hydrodynamics of the CMP process. The study of the kinematic condition of the device was to obtain a condition to have a uniform removal of material from the surface of the workpiece (wafer) area. By excluding the effect of oscillatory motion in the CMP process, Kim et al. [8] obtained a relationship for the relative velocity of a point on the wafer with respect to the pad ($\vec{v}_{p/w}$) as a function of the distance between rotation centers (D), angular velocity of the pad (ω_P), angular velocity of the wafer (ω_w) and radial distance to the given point on the wafer (r) and the equation is given as follows:

$$\vec{v}_{p/w} = \vec{\omega}_p \times \vec{D} + (\vec{\omega}_p - \vec{\omega}_v) \times \vec{r} \tag{10.2}$$

They defined a dimensionless parameter $\zeta = \frac{r_w}{D}(1 - \frac{\omega_w}{\omega_p})$ called a kinematic index to represent the kinematic aspects of the CMP process. Using this parameter, the distribution of the velocity is determined as follows:

$$v_{p/w} = \vec{\omega}_p D\sqrt{(\rho\zeta)^2 + 2\rho\zeta\cos\varphi + 1} \tag{10.3}$$

The flow of the slurry is affected by two different factors, i.e. the direction of the friction force and the distribution of velocity, which are a function of ζ values. Based on the tribological aspects of friction and velocity, their interpretations revealed that a non-uniform material removal would happen due to a large velocity deviation within the wafer area (large ζ value), resulting from local deviations of the friction force [8].

In a study by Zhou et al. [16], the polishing mechanism has been explained using the Nano-film polishing model. Based on this model, a nonlinear relation between abrasive particle size and MRR is reported. Their results also reported a critical dependence of MRR to the load and speed and the coupling between load and speed for the grooved pad.

10.2.1 Applications of CMP

The general application of CMP is to improve the performance and cut the cost of the semiconductor materials. With the developments made in the semiconductor industry, there is a continuous need to build transistors and interconnectors smaller and smaller for the new generation of devices. In addition, the achievable precision and surface finish requirements for CMP is also becoming narrower. The advantages offered by CMP as a precision polishing technique provides a platform for chip manufacturers to shrink their devices by producing defect-free wafers with an acceptable level of flatness and smoothness. Additionally, it was the emergence of CMP that made it possible to use copper interconnectors in the semiconductor industry. Therefore, CMP is considered as one of the challenging steps and enabling technology in lithography based production processes. Since in lithography, the devices are being manufactured layer by layer, the CMP process may be repeated several times throughout the overall manufacturing cycle of a device.

CMP has found widespread industrial applications, including microelectronics, MEMS-based sensors and actuators, micro-opto-electro-mechanical systems (MOEMS), fabrication of high-voltage high-power transistors, surface acoustic wave (SAW) devices to optical coatings [17], giant-large scale integrated circuits (GLSI) [18], etc. Especial applications include the use of CMP for polishing of silicon wafers [19], polishing of silicon dioxides and silicon nitrides [20] and polymer layers [21], polishing of sapphire in the semiconductor industry as a substrate for epitaxial growth of Gallium Nitride [22] and polishing of Silicon Carbide substrates, processing of conductive polycrystalline silicon as a fill material for through silicon vias (TSV) [23] and the thinning of Silicon on Sapphire (SOS) and Silicon on Insulator (SOI).

10.3 Abrasives for CMP

The slurry in the CMP process consists of two different phases. i.e. liquid and solid phases, even though most recent applications based on abrasive free slurries are also being proposed [24]. The solid phase of slurry is composed of abrasive materials such as ceria, aluminum oxide, silicon carbide, etc. The slurry may contain only one type of abrasive particles, which is called single abrasive slurry (SAS) or it may contain a mixture of abrasive particles known as a mixed abrasive slurry (MAS). Abrasives are among the most important key elements of the CMP process which largely influence the polished surface quality and material removal rate (MRR) [25, 26] and largely affect the performance of the CMP process [15]. Perhaps silica abrasives are among the most frequently deployed abrasives in the CMP process. Silica abrasives can be produced by two common synthesizing techniques.

In a study by Zhu et al. [27] different abrasive particles such as alumina, monocrystalline diamond, and polycrystalline diamond are used to study the material removal rate and surface quality of Sapphire in an aqueous slurry with different pH values. It has been observed that α-alumina abrasives promote higher polishing rates because of surface hydration and chemical-mechanical reaction between α-alumina and sapphire. The sol-gel technology has been used by Xu et al. [28] to prepare a mixed abrasive slurry for CMP of a sapphire wafer. In their study, diamond abrasive particles are used as hard abrasives, while high-reactivity silica is used as the soft abrasive. MAS is shown to be capable of increasing the material removal rate by more than 52.6% compared to SAS. They also reported a reduced surface roughness of 21.6% when using MAS. The use of ellipsoidal rod-shaped silica Nano-composite abrasives produced by the Chromium ion/PEG200 induced method for CMP of sapphire substrate is reported by Dong et al. [29]. An increased MRR and reduced surface roughness have been demonstrated, which was credited to an optimal combined effect of chemical and mechanical interaction. Xu et al. [30] reported the use of irregular flower-shaped silica abrasives in the CMP process of sapphire wafers. Their results revealed a remarkable increase of 217.4% in the material removal rate compared to spherical silica for an abrasive content of 6%. Comparable surface roughness has been obtained when using flower-shaped abrasives compared to spherical silica. The study of Liu and Lei [31] uses an Nd^{3+}-doped colloidal spherical SiO_2 composite abrasives synthesized by seed-induced growth in the CMP process of sapphire. A lower average surface roughness values and higher material removal rates are obtained compared to pure silica abrasives. The seed-induced growth method is also used by Lei and Tong [25] to produce LA-doped colloidal silica composite abrasives for the CMP process. The abrasives are shown to have good dispersibility and inerratic spherical shape. When applied to the CMP of sapphire, lower Ra values are reported along with improved material removal rate compared to pure spherical silica particles. In another study by Yin et al. [32], MgO doped colloidal SiO_2 abrasives are used for sapphire substrate polishing. Again, a higher MRR and lower Sq (square roughness) values are reported compared to pure spherical silica particles.

The density of the abrasive particles performs a key role in a successful CMP process. Since the change of polishing rate is proportional to the change of density of the slurry, therefore to have consistent polishing rates, it is necessary to have accurate control of the density of the slurry. Mixed abrasive slurries are also used in the CMP process. In the study of Jindal et al. [33], MAS.

Which is formed by dispersing alumina and silica particles in deionized water is used for copper damascene polishing. It is demonstrated that the performance of the CMP process has been improved significantly. The polishing rate in CPM is also subjected to the particle size distribution (PSD). The PSD consists of the major population of abrasive particles that contribute to the primary polishing of the wafer. Generally, in CMP a narrow PSD is preferred to have better control on the process [34].

Zhou et al. [16] have researched the impact of nano-scale colloidal abrasive size on the MRR and surface finish of thermally-grown silicon dioxide on p-type single crystal silicon (100). The mean diameter of the abrasive nano-particles is selected in the range of 10-140 nm. A strong dependency of MRR to the abrasive particle size has been reported. Their results revealed that an abrasive particle size of 80 nm is the optimum particle size with respect to the surface finish and MRR. As a rule of thumb, particles with zeta values larger than 30 are considered colloidally stable [34]. In order to assess the stability of the abrasive particles in the slurry, Zeta potential is used.

10.4 Recent Advancement in CMP

The main application of Chemical Mechanical Polishing (CMP) is to obtain planar, smooth and defect-free surfaces for manufacturing semiconductors. Since device dimensions continue to shrink, the CMP mechanism is becoming more complex and the research area is widening with every year passes. Since its first use in IBM in 1986 [11], the CMP process remains to include two stages of treatment: softening of material's surface by chemicals (slurry); abrasive particles removal of softened parts through mechanical force [35]. Hence, this section reviews few recent advancements in CMP concerning new types of slurries, abrasives and methods. Table 10.1 summarizes a list of the mentioned innovations. Recent advancements in CMP can be grouped into three categories.

Table 10.1 Recent advancements in CMP

Type	Content	Application	Advancement	Reference and year
Slurry	SiO_2 Nano spheres, H_2O_2, malic and citric acids	Semiconductors	Environment friendly	[36] 2016
Slurry	Silica, hydrogen peroxide and chitosan oligosaccharide	Polishing copper	Environment friendly	[37] 2018
Slurry	Non-noble metal materials ($Fe\text{-}N_x/C$)	Optoelectronics devices	Catalyst free of platinum to improve the MRR of sapphire	[38] 2015
Slurry	Hybrid polymers	Chemical mechanical planarization	Lowering topographical variations and surface roughness	[39] 2015
Abrasive	Silica nanoparticles	Thin films	High quality surfaces	[40] 2018
Abrasive	Polyelectrolyte–Al_2O_3/SiO_2 composite nanoparticle	Sapphire planarization	Dispersibility and superior polishing performance	[41] 2019
Method	Multilayer ice bonded abrasive polishing (IBAP)	Polishing of Ti-6Al-4 V alloy	Morphology and finish	[42] 2017
Method	Polishing inner surfaces of objects printed in 3D printers	Additive manufacturing	Polishing and removal of semi-welded particles on the interior surfaces of components	[43] 2017
Method	Electro-mechanical polishing	SiC surface	Pad-free treatment	[44] 2017
Method	Negative pressure cavity jet polishing (NPCJP)	Optical processing	Efficiency and surface quality	[45, 46] 2017

10.4.1 Applying New Efficient and Environment Friendly Slurries

Exploration of new types of slurry that are both efficient and environment friendly has been a focus for many inventors and scientists. For example, it is well-known that mercury cadmium telluride (HgCdTe or MCT) semiconductors have been polished using corrosive and toxic chemical reagents recently. However, a new type of slurry that consists of mainly SiO_2 nano-spheres, H_2O_2, malic and citric acids proved to have excellent characteristics in terms of impact on the environment. Polished surface has a roughness of nearly $R_a = 0.45$ nm, and peak-to-valley (PV) value of 4.74 nm respectively on MCT wafers according to Zhang et al. [36].

Another example of applying environment-friendly slurry is using silica, hydrogen peroxide and chitosan oligosaccharide for polishing copper. Polishing copper is associated with using strong acids and hazardous chemicals. This invention is attributed to Zhang et al. [37]. Polished surface achieved average surface roughness R_a of 0.444 nm and peak-to-valley value of 5.468 nm respectively after the process. The effect of using above mentioned substances is that hydrogen peroxide oxidizes Cu surface and CuO, $Cu(OH)_2$ are formed. After that chitosan oligosaccharide is ionized releasing H^+ which dissolves CuO and $Cu(OH)_2$. As a result, chelating of Cu^{2+} ions takes place through chitosan oligosaccharide molecules. At the end, silica nano-spheres act as remover for adsorbed layer, leaving ultra-smooth surface of copper untouched. X-ray photoelectron spectra and electrochemical measurements can be deployed to reveal average surface roughness as well as peak-to-valley values.

Now a day sapphire, gallium nitride (GaN) and silicon carbide (SiC) are considered to be the main candidates as single crystal substrate material for next generation optoelectronics devices. In order to achieve improved material removal rate for sapphire, Xu et al. [38] have proposed to develop non-noble metal materials ($Fe\text{-}N_x/C$) such as platinum-free catalyst. CMP process can achieve smooth sapphire surface containing periodically organized step structure along with improved MRR as high as 38.43 nm/min. This MRR is nearly 15.44% larger than the case when no catalyst is used.

Lowering topographical variations and surface roughness are one of the main concerns in CMP. The invention of slurries containing hybrid polymers made it possible to successfully tackle this issue [39]. At the same time, this approach does not compromise frictional characteristics and removal rates since slurries consisting of only silica or ceria nanoparticles that have similar or lower qualities. In order to obtain such microgels, colloidal particles of a cross-linked, thermally responsive polymer were used for preparation of organic-inorganic composites. Polymeric particles with embedded nanoparticles, hybrid PNIPAM-polysiloxane particles and composite of an inorganic metal-oxide (MO) such as CeO_2 and TiO_2were formed as a result.

10.4.2 Using New Types of Abrasive Materials

Rapidly growing interest in developing new nanoparticles that can act as abrasive materials to create high quality surfaces for thin films has become a trend recently. Process requirements of abrasive slurry set by advanced CMP in order to achieve higher wafer surface quality can be met by mesoporous silica abrasive particularly with lower hardness along with lower young modulus. Ryu et al. [40] have developed a way of making uniform wrinkled silica nanoparticles (WSNs) without using a pot. Uniform particles can be obtained through rapid cooling of reactants at the convenient time resulting in separation of nucleation and growth stages. In addition, WSNs show better properties as an abrasive material compared to nonporous silica nanospheres as well as fumed silica-based products in CMP processes.

Wang and Lei [41] have developed a novel type of nanoparticle which is able to provide outstanding dispersibility and greater polishing performance and it contains mainly polyelectrolyte–Al_2O_3/SiO_2 composite. The use of abrasive polyelectrolyte-Al_2O_3/SiO_2 increased the material removal rate by 30% compared to pure Al_2O_3. Polyelectrolytes being a polymer with a number of ionizable groups in the side chain have certain benefits compared to nonionic polymers because of its characteristics related to higher hydrophilicity and stability. Alumina was covalently bonded together with a silica bifunctional molecule, which also was adsorbing polyelectrolytes by electrostatic interaction. Being well dispersed, these particles increase removal rate causing the acceleration of mechanical process. This kind of abrasive penetrates the wafer relatively in less depth and very smooth surface emerges after the end of the process. All of the mentioned characteristics of novel material meet those high requirements set for sapphire planarization.

10.4.3 New Polishing Methods

Ultra-precision surface generation processes like lapping and polishing has attracted huge attention from researcher due to the growing needs of nano-level surface on a range of product exploited in different industries such as semiconductor, medical or aerospace. For example, a novel multilayer icebonded abrasive polishing (IBAP) tool was proposed by Ramesh Babu [42]. This tool was developed for a polishing sample of Ti-6Al-4 V alloy in multiple stage based on a systematic study where required number of layers, thickness of each subsequent layer, and type, size, concentration of abrasives particles in each layer can be determined. In order to perform ultrafine polishing of Ti-6Al-4 V alloy, this IBAP tool comes with three layers and each layer is different with different thickness and concentration.

- Bottom layer, made of soft 3 μm Al_2O_3 abrasive with concentration of 5%;
- Middle layer, made of 8 μm moderately hard SiC with concentration of 10%;
- Top layer, made of 15 μm hard B_4C with concentration of 30%.

According to the assessed morphology and finish, the three-layered IBAP showed 81% improvement compared to a single-layered IBAP tool. Obtained ultrafine surfaces have 37 nm finish and almost free of scratch surface of the alloy.

In addition, the recent advancement in additive manufacturing (AM) has brought about new challenges like polishing inner surfaces of complicated shape objects printed in 3D printers. Selective laser melting (SLM) and electron beam melting (EBM) are two most important AM process that faces key challenges due to the roughness issue in the interior surface of built part that is complex in nature. If we consider engines and other fluid systems, there are two main concerns related to 3D printed parts: occurrence of semi-welded particles and undesirable surface roughness. Both issue can cause complications for engines either by blocking or hindering fluid flow. Research study done by Mohammadian et al. [43] demonstrated that it is feasible to combine chemical-abrasive flow polishing technique to completely remove semi-welded particles present on the interior surfaces of the components, designed for the aerospace industry, and considerably improve surface roughness as well as texture. By comparison, of different testing features for SLM printing, it was discovered that the most effective way of polishing is when applying a flow that contains both abrasives and chemicals. When using this technique, the results revealed reduction of R_a value by 45% for a build orientation of 15°. Both schematic and real illustration of the CMP process set-up are presented in Fig. 10.4.

New advancements have been reported in electro-mechanical polishing too. A new method developed by Murata et al. [44] proposed a pad-free treatment. Polishing pad appears to be an unnecessary accessory for this technique since the surface of the single-crystalline is electrochemically oxidized and CeO_2 is deployed to remove oxides from the polyurethane–CeO_2 core–shell particles. Generally, due to rapid wear caused by friction, the resin pad needs periodic replacement, which leads to an increase of process cost. A schematic of the core-shell particles is shown in Fig. 10.5. The core is made of polyurethane and the role of the shell is given to soft CeO_2 particles, which constitutes an abrasive layer. Since CeO_2 particles are soft

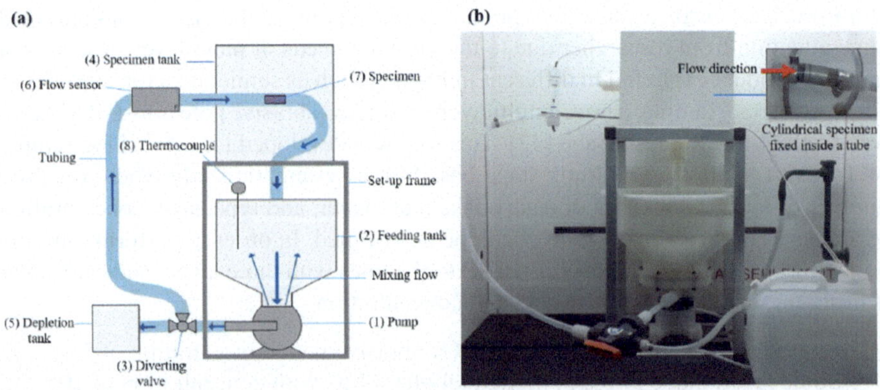

Fig. 10.4 Schematic **a** and real **b** illustration of the chemical abrasive polishing set-up [43]

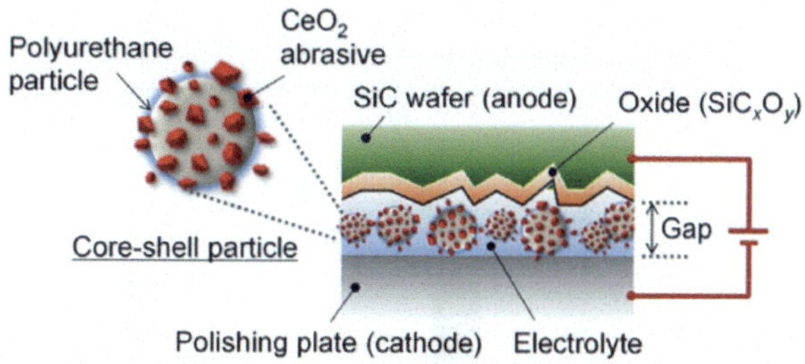

Fig. 10.5 Schematic view of the proposed electrochemical mechanical polishing (ECMP) method [44]

compared to SiC surface, the conventional way of CMP process cannot be realized without electrolysis. Consequently, the surface continuously experiences anodization that creates many scratches and rough surface. Through optimization of electrolysis and periodical application of bias voltage, the oxide film disappears from the surface and the average roughness reaches as high as 0.5 nm. While comparing with the traditional polishing process that comes with colloidal SiO_2 slurry, this proposed technique demonstrates remarkable polishing productivity even without the help of a polishing pad.

Abrasive water jet technology was first introduced as a well drilling technology. Since then many types of abrasive jet technologies emerged in other fields, including CMP. Abrasive jet polishing (AJP) is a new ultra-precision optical processing technology which comprises of complex fluid jet of water and abrasives [45]. Higher flexibility, better accuracy, user-friendly control, and reduced cost are the main advantages of AJP. The basic principle of abrasive water jet polishing (AWJP) is shown in Fig. 10.6. Well proportioned micro-abrasive is first mixed with water. Then high-pressure pump transfers the polishing liquid to the nozzle forming high-speed abrasive jet. The flow moves along workpiece surface tangentially generating strong shear impact and performing micro-removal of the material [46]. Chen et al. [45] has introduced an advancement to AWJP by creating a new negative pressure

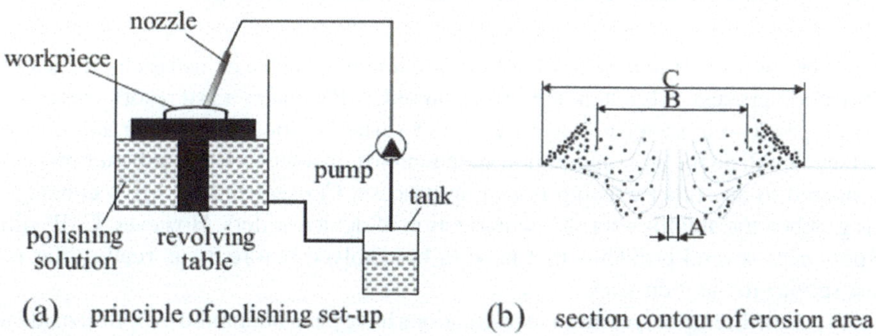

Fig. 10.6 Principle of Abrasive water jet polishing **a** polishing arrangement and **b** section contour of erosion area [45]

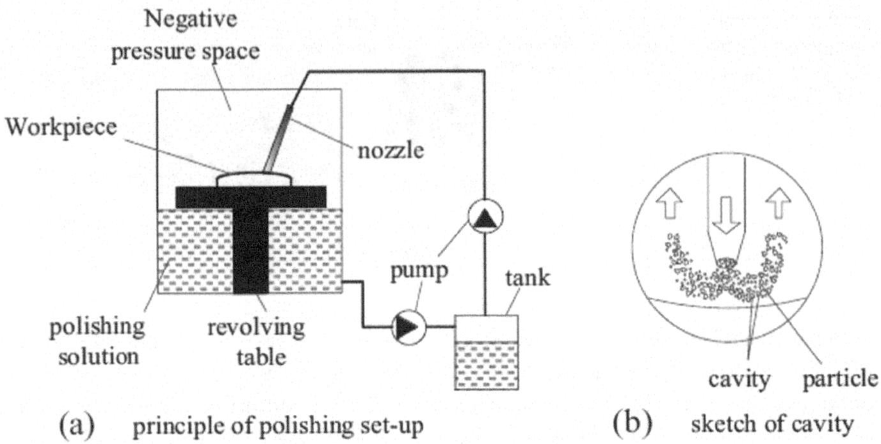

Fig. 10.7 Principle of negative pressure cavity jet polishing with **a** polishing arrangement and **b** sketch of a cavity [45]

cavity jet polishing (NPCJP). The application of AWJP along with negative pressure cavity technology allows diminishing air disturbance and improves polishing accuracy. A closed polishing container provides a negative pressure space. Differential pressure allows the abrasive liquid to be ejected from the storage tank with high speed creating a high-speed jet. The presence of negative pressure enhances the cavitation effect causing more material removal and improvement of surface quality. Figure 10.7 provides a schematic of NPCJP method. As a result, this innovation may ensure superiority of NPCJP over conventional AWJP in terms of efficiency and surface quality.

10.5 Development Trend and Research Challenges

It has been almost 30 years; CMP has been implemented in the industries for the wafer fabrication process. Although, the invention of CMP came as a last resort, currently, this process can fabricate damage free wafer surfaces along with planer and mirror surface. It will not be an exaggeration statement to consider CMP as far better process compared to any other existing polishing process. Consequently, CMP technology has grabbed the attention for the planarization of device wafers. However, CMP still experiences several problems that need to be resolved before it can really offer its best service to the industries.

As per the literature, research investigations have been attempted to optimize ideal metal CMP slurry in order to achieve effective polishing of metallic layers, with optimum MRR as well as uniformity and reduced surface defects, where the variation of additive types, their concentration and pH of the slurry were considered mainly. Effective polishing also depends on other parameters such as oxidizing agents, chelating agent, corrosion agents and other chemical additives. However, recent environmental concern associated with the CMP process requires reduced slurry consumption. Additionally, post CMP cleaning process also exploits CMP and contributes towards chemical contamination due to the involvement of slurry. Therefore, large amount of process waste associated with CMP appears to have negative environmental effects. In order to address this hurdle, research efforts to develop next generation CMP slurry free of corrosion inhibitors and abrasive for post CMP cleaning process has been under investigation currently which will also help removing organic Cu residue on the wafers. Nevertheless, CMP slurry experiences challenges when it comes about new materials needs, complex integration schemes, or cleaning requirements. Better understanding of the CMP slurry fundamentals using new strategies such as 3D pH-potential diagram, or other chemistry studies may assists in verifying the device yield of products [47].

By introducing copper interconnect instead of Al interconnects, and damascene process, RC delay of the device can be reduced. It has been also predicted that further reduction in RC delay can be achieved by deploying ultra low-k materials. ITRS roadmap 2012 reported on the integration of low-k materials (such as dielectric value of 2) into IC by 2025, which can replace existing oxide, ILD layer [48]. As a result, it will reduce capacitance loss and increase the signal transmission rate. However, these low-k materials faced challenges due to their relatively soft and weak nature compared to metal. Therefore, buckling and crushing failure become inevitable for both single and double damascene structure containing low k materials. Apart from that, post CMP surface quality and planarity will be affected extensively due to the different value of mechanical properties and polishing rates of Cu and low-k materials. Hence, to address these issues connected with non-uniform polishing phenomenon, and complex materials structures, researches on the development of new technology/process that can accommodate stress free CMP and low downward force CMP are indispensable [11, 49]. Moreover, the potential of CMP in the fabrication of micro-electro-mechanical systems (MEMS), shallow trench isolation, back-side polishing needs to be explored in the future as well. MEMS topography comes in larger dimension usually along with the step height of 10 microns and features size of few hundreds' microns. Longer polishing time associated with MEMS enhances issues such as pattern density and process instability. Moreover, there is a growing trend for direct wafer bonding which joins anything with anything literally. However, challenge arises as surface roughness requirements for such situation is below 1 nm in order to achieve good bond [50].

With technological advancement, IC experiences the trend towards reduced size, lower cost and higher integration. As per the ITRS report (Table 10.2), half pitch of metal will be further reduced to 5 nm by year 2025. Along with the reduction of feature size, the focus depth of the lithography process also needs to be shortened. Ultimate chip failure occurs due to non-uniform lithography width caused by non-uniformity of the wafer surface. In addition, transistors that can fit on single chip of ULSI reached to over 1 billion. As with the increasing number of transistors per chip, an increased number of interconnector layers is necessary. Multi-level interconnectors are already in place for increasing the connection efficiency 0.65 nm node requires

Table 10.2 Interconnect CMP demand from International Technology Roadmap for Semiconductors (ITRS) [48]

Year	Metal 1 wiring half pitch (nm)	Number of metal levels	Interval metal insulator effective dielectric constant k
2012	32	12	2.82–3.16
2013	27	13	2.55–3.00
2014	24	13	2.55–3.0
2015	21	13	2.55–3.00
2016	19	13	2.40–2.78
2017	17	14	2.40–2.78
2018	15	14	2.40–2.78
2019	13	14	2.15–2.46
2020	12	14	2.15–2.47
2025	7	16	1.6–2.00

9–10 interconnect layers, and when the size goes below 45 nm, this number exceeds 10. Eventually, node of 32 nm requires 12 interconnect layers while node of 22 nm demands 13 layers of interconnector. Therefore, further challenges impose to CMP process along with an increased number of interconnect layers which causes the accumulation of non-uniformity. Moreover, the wafer dimension has been increasing from 200 to 300 mm and recently to 450 mm in order to enhance production efficiency as well as to reduce chip cost. Although, current CMP technology in semiconductor industries is sufficient for wafer size of 300 mm, achieving global planarity across the whole wafer surface for larger size wafer will impose additional challenges for current CMP technology [11].

Polishing efficiency varies or deteriorates with the polishing time. Therefore, polisher conditioning is implemented as countermeasure, which shortens the life cycle of polishing pad. On the other hand, polisher conditioning also makes it challenging to identify the end of polishing with the aid of polishing time as reference. In order to address this issue, it is necessary to develop higher efficiency as well as long-lasting polishing pads. Other influential parameters to be considered are slurry, pad materials, polishing pressure, removal rate and polishing pass. Deeper understanding of these parameters can aid to improve the CMP efficiency. Additional global planarization across the entire wafer surface as well as detection of end point will be quite challenging for further improvement of the CMP process [51].

As CMP also causes material removal by abrasive action, material removal rate varies with the hardness level of concerned materials. Workpiece consisting of more than one type of materials suffers different rate of material removal rate. Usually softer materials abrade faster than harder materials causing a generation of a concave surface. Therefore, composite materials suffer from concavity and convexity, which poses a challenge on the CMP process.

10.6 Summary

CMP has been extensively used in the semiconductor and Integrated circuit manufacturing Industries and has remained a hot research topic until recently. Even though CMP has been in the market for over decades, this process is still considered a complex one due to the involvement of both mechanical and chemical aspects, which are also affected by several factors. These factors that include carrier structure, polishing pad, slurry, and other process parameters affect the quality of final surface polishing. Abrasive used along with slurry has a paramount effect on the process performance and larger abrasive particle causes mechanical abrasion while chemical interaction is dominated in nano-scale range. Advancement in CMP has occurred mainly in three major areas i.e. environmentally friendly efficient slurry, new types of abrasives and new polishing methodology. With the reduction of feature size in the IC device, an increment of wafer dimensions and usage of low k materials, CMP faces increasing number of process challenges. Therefore, further precision technology with pressure control provision and end point detection technology may have a substantial bearing on the improvement of CMP process control.

References

1. Moore GE (1965) Cramming more components onto integrated circuits. McGraw-Hill, New York, USA
2. Chang CY, Sze SM (1996) ULSI technology. McGraw-Hill
3. Kahng AB, Samadi K (2007) CMP fill synthesis: A survey of recent studies. IEEE Trans Comput Aided Des Integr Circ Syst 27(1):3–19
4. Ryan JG, Geffken RM, Poulin NR, Paraszczak JR (1995) The evolution of interconnection technology at IBM. IBM J Res Dev 39(4):371–381
5. Bai P, Auth C, Balakrishnan S, Bost M, Brain R, Chikarmane V, Heussner R, Hussein M, Hwang J, Ingerly D (2004) A 65 nm logic technology featuring 35 nm gate lengths, enhanced channel strain, 8 Cu interconnect layers, low-k ILD and 0.57/spl mu/m/sup 2/SRAM cell. In: IEDM technical digest. IEEE international electron devices meeting, 2004. IEEE, pp 657–660
6. Zantye PB, Kumar A, Sikder A (2004) Chemical mechanical planarization for microelectronics applications. Mater Sci Eng R Rep 45(3–6):89–220
7. Landis H, Burke P, Cote W, Hill W, Hoffman C, Kaanta C, Koburger C, Lange W, Leach M, Luce S (1992) Integration of chemical-mechanical polishing into CMOS integrated circuit manufacturing. Thin Solid Films 220(1–2):1–7
8. Kim HJ, Kim HY, Jeong HD, Lee SH, Dornfeld D (2003) Kinematic analysis of chemical mechanical polishing and its effect on polishing results. In: Key Engineering Materials. Trans Tech Publications Ltd, pp 229–234
9. Oliver MR (2013) Chemical-mechanical planarization of semiconductor materials, vol 69. Springer Science & Business Media
10. Huang C, Zhou H, Zhu Y, Xia C (2019) Effect of chemical action on the chemical mechanical polishing of β-Ga2O3 (100) substrate. Precision Eng 56:184–190
11. Zhao D, Lu X (2013) Chemical mechanical polishing: theory and experiment. Friction 1(4):306–326
12. Runnels SR (1994) Feature-scale fluid-based erosion Modeling for chemical-mechanical polishing. J Electrochem Soc 141(7):1900–1904

13. Nakamura T, Akamatsu K, Arakawa N (1985) A bowl feed and double sides polishing for silicon wafer for VLSI. Bull JSPE 19(6):120
14. Cook L, Wang J, James D, Sethuraman A (1995) Theoretical and practical aspects of dielectric and metal CMP. Semicond Int 18(13):141–143
15. Rudawska A (2018) Abrasive technology: characteristics and applications. BoD–books on demand
16. Zhou C, Shan L, Hight JR, Danyluk S, Ng S, Paszkowski AJ (2002) Influence of colloidal abrasive size on material removal rate and surface finish in SiO2 chemical mechanical polishing. Tribol Trans 45(2):232–238
17. Gaidarzhy A, Imboden M, Mohanty P, Rankin J, Sheldon BW (2007) High quality factor gigahertz frequencies in nanomechanical diamond resonators. Appl Phys Lett 91(20):203503
18. Zhou J, Niu X, Cui Y, Wang Z, Wang J, Wang R (2019) Study on the film forming mechanism, corrosion inhibition effect and synergistic action of two different inhibitors on copper surface chemical mechanical polishing for GLSI. Appl Surf Sci 144507
19. Zhao Y, Chang L (2002) A micro-contact and wear model for chemical–mechanical polishing of silicon wafers. Wear 252(3–4):220–226
20. Oh M-H, Singh RK, Gupta S, Cho S-B (2010) Polishing behaviors of single crystalline ceria abrasives on silicon dioxide and silicon nitride CMP. Microelectron Eng 87(12):2633–2637
21. Neirynck JM, Yang G-R, Murarka SP, Gutmann RJ (1996) The addition of surfactant to slurry for polymer CMP: effects on polymer surface, removal rate and underlying Cu. Thin Solid Films 290:447–452
22. Xu W, Cheng Y, Zhong M (2019) Effects of process parameters on chemical-mechanical interactions during sapphire polishing. Microelectron Eng 216
23. Pirayesh H, Cadien K (2015) Chemical mechanical polishing in the dry lubrication regime: Application to conductive polysilicon. J Mater Process Technol 220:257–263
24. Kamigata Y, Kurata Y, Masuda K, Amanokura J, Yoshida M, Hanazono M (2001) Why abrasive free Cu slurry is promising? MRS Online Proc Lib Arch 671
25. Lei H, Tong K (2016) Preparation of La-doped colloidal SiO2 composite abrasives and their chemical mechanical polishing behavior on sapphire substrates. Precision Eng 44:124–130
26. Lei H, Liu T, Xu L (2019) Synthesis of Sm-doped colloidal SiO_2 composite abrasives and their chemical mechanical polishing performances on sapphire substrates. Mater Chem Phys 121819
27. Zhu H, Tessaroto LA, Sabia R, Greenhut VA, Smith M, Niesz DE (2004) Chemical mechanical polishing (CMP) anisotropy in sapphire. Appl Surf Sci 236(1–4):120–130
28. Xu Y, Lu J, Xu X (2016) Study on planarization machining of sapphire wafer with soft-hard mixed abrasive through mechanical chemical polishing. Appl Surf Sci 389:713–720
29. Dong Y, Lei H, Liu W, Chen Y (2019) Preparation of ellipsoidal rod-shaped silica nanocomposite abrasives by Chromium ion/PEG200 induced method for sapphire substrates chemical mechanical polishing. J Alloy Compd 777:1294–1303
30. Xu L, Lei H, Wang T, Dong Y, Dai S (2019) Preparation of flower-shaped silica abrasives by double system template method and its effect on polishing performance of sapphire wafers. Ceram Int 45(7):8471–8476
31. Liu T, Lei H (2017) Nd3 + -doped colloidal SiO_2 composite abrasives: Synthesis and the effects on chemical mechanical polishing (CMP) performances of sapphire wafers. Appl Surf Sci 413:16–26
32. Yin D, Niu X, Zhang K, Wang J, Cui Y (2018) Preparation of MgO doped colloidal SiO_2 abrasive and their chemical mechanical polishing performance on c-, r-and a-plane sapphire substrate. Ceram Int 44(12):14631–14637
33. Jindal A, Hegde S, Babu S (2002) Chemical mechanical polishing using mixed abrasive slurries. Electrochem Solid-State Lett 5(7):G48–G50
34. Babu S (2016) Advances in chemical mechanical planarization (CMP). Woodhead Publishing
35. Kim HJ (2018) Abrasive for chemical mechanical polishing. abrasive technology: characteristics and applications. 183

36. Zhang Z, Wang B, Zhou P, Guo D, Kang R, Zhang B (2016) A novel approach of chemical mechanical polishing using environment-friendly slurry for mercury cadmium telluride semiconductors. Sci Rep 6:22466
37. Zhang Z, Cui J, Zhang J, Liu D, Yu Z, Guo D (2019) Environment friendly chemical mechanical polishing of copper. Appl Surf Sci 467:5–11
38. Xu L, Zou C, Shi X, Pan G, Luo G, Zhou Y (2015) Fe-Nx/C assisted chemical–mechanical polishing for improving the removal rate of sapphire. Appl Surf Sci 343:115–120
39. Gupta V, Kumar A, Coutinho C, Mudhivarthi S (2015) U.S. Patent No. 9,120,952. Washington, DC: U.S. Patent and Trademark Office
40. Ryu J, Kim W, Yun J, Lee K, Lee J, Yu H, Kim JH, Kim JJ, Jang J (2018) Fabrication of uniform wrinkled silica nanoparticles and their application to abrasives in chemical mechanical planarization. ACS Appl Mater Interfaces 10(14):11843–11851
41. Wang T, Lei H (2019) Novel polyelectrolyte–Al 2 O 3/SiO 2 composite nanoabrasives for improved chemical mechanical polishing (CMP) of sapphire. J Mater Res 34(6):1073–1082
42. Rambabu S, Ramesh Babu N (2018) Empirical approach to develop a multilayer icebonded abrasive polishing tool for ultrafine finishing of Ti-6Al-4 V alloy. Mater Manuf Process 33(4):359–366
43. Mohammadian N, Turenne S, Brailovski V (2018) Surface finish control of additively-manufactured Inconel 625 components using combined chemical-abrasive flow polishing. J Mater Process Technol 252:728–738
44. Murata J, Yodogawa K, Ban K (2017) Polishing-pad-free electrochemical mechanical polishing of single-crystalline SiC surfaces using polyurethane–CeO2 core–shell particles. Int J Mach Tools Manuf 114:1–7
45. Chen F, Miao X, Tang Y, Yin S (2017) A review on recent advances in machining methods based on abrasive jet polishing (AJP). Int J Adv Manuf Technol 90(1–4):785–799
46. Fang H, Guo P-J, Yu J-C (2004) Research on material removal mechanism of fluid jet polishing. Opt Tech 2
47. Lee D, Lee H, Jeong H (2016) Slurry components in metal chemical mechanical planarization (CMP) process: A review. Int J Precision Eng Manuf 17(12):1751–1762
48. Information (2012) http://www.itrs.net/Links/2012ITRS/Home2012.htm
49. Singer P (2003) Low-pressure CMP developed for 300 mm ultralow-k. Semicond Int 26(12):30–30
50. Banerjee G, Rhoades RL (2008) Chemical mechanical planarization historical review and future direction. ECS Trans 13(4):1–19
51. Doi T, Marinescu ID, Kurokawa S (2011) Advances in CMP polishing technologies. William Andrew

Chapter 11
Finishing of Curved Surface by Rotary Abrasive Float Polishing

Alakesh Manna

Abstract The traditional polishing process can produce very good smooth surface but the efficiency of this process is very low for polishing of ductile, hard and brittle materials. Again, it is very difficult to generate mirror finish on irregular and curved surface workpiece. The principle issues are geometrical shape, deviations in dimension throughout the surface and no repeatability of the procedure while polishing these surfaces, therefore, it is essential to introduce a new polishing method which can overcome the aforesaid problem. Keeping in view, a rotary abrasive float polishing (RAFP) set-up has been design, fabricated and utilized for polishing of curved surface specimens. This chapter presents the properties of Alloy 61S workpiece material and fabrication of curved surface specimens by machining of Alloy 61S on CNC Vertical Milling centre ANFORD ECO VP 850. This chapter also presents the feasibility and applications of the novel polishing technique RAFP. Different concentrations abrasive slurry were prepared and used to finish the curved surface of Alloy 61S specimen. The effects of RAFP process parameters e.g. abrasive concentration, rpm, time duration on reduction in surface roughness heights ΔR_a and ΔR_t have been analyzed. The experiments were carried out based on Taguchi design of experiments and optimized the RAFP process parameters for surface roughness heights. Experimental results reveal 66.49% reduction in the surface roughness height, R_a (μm). The scanning Electron microscope (SEM) study also reveals the improvement in machined surface.

Keywords Curved surface · CNC milling · RAFP · Surface finish · SEM images

A. Manna (✉)
Departmental of Mechanical Engineering, Punjab Engineering College (Deemed to be University), Sector-12, Chandigarh 160012, India
e-mail: kgpmanna@rediffmail.com

S. Das et al. (eds.), *Advances in Abrasive Based Machining and Finishing Processes*, Materials Forming, Machining and Tribology,
https://doi.org/10.1007/978-3-030-43312-3_11

11.1 Introduction

With the advancement, the application of curved surface in enterprises required unpleasantness and shape precision. Again these are prerequisites for curved surfaces. The curved surfaces have been used in numerous engineering applications, for example, frameworks that requires steady and high temperature safe optics which can continue to any obscure vulnerabilities. Again, fabrication of high quality precision curved surface is becoming more important in defense industry. The needs for freeform surface fabrication are widely recognized in the optics, aerospace, biomedical and automotive industries. But fabrication of freeform surfaces are very difficult and expensive because of involvement of advance multi-axis dedicated machining centre and allied processes. Forming and cleaning of complex optical surfaces have dependably been a test to the optical creation industry. The high precision diamond turning is one of the processes can be used to create super finish curved surfaces, however it is a costly process. Machining of curved and irregular surfaces are difficult and costly besides this tight tolerance are utmost important in modern era. Precision machining of complex and complicated shapes and/or sizes, machining of unable to reach surfaces at micro or Nano levels with narrow tolerances demands to the development and application of new advanced nonconventional machining process. The application of well-known conventional machining processes such as WEDM, EDM, and ECM etc. are not capable to produce mirror finished surfaces. Even these processes have there on limitations based on predominance of selected processes and produces poor surface finish. Subsequently an endeavour has been made to recognize a financially cost effective finishing process for finishing of Al sample by designing rotary abrasive float polishing setup.

Several researches have already worked on abrasive flow finishing processes and explained about the consequences and effects of process parameters like number of cycles, extrusion pressure and abrasive concentration on the output responses namely surface finish during AFF. Rhoades [1] studied the basic principle of AFM and delineated that the depth of cut primarily depends on extrusion pressure, abrasive grains, relative hardness, and sharpness. The axial force in AFM are highly subjected to the behaviour of the AFM medium. The flow pattern of the abrasive medium depends on the characteristics of the medium, machining parameters, as well as the shape of the work piece and the tooling. Mali and Manna [2–5] review on abrasive flow finishing processes, studied on AFM during finishing of Al/ SiCp-MMC work piece, simulation and optimized the AFM parameters for surface finish and material removal. They claimed that there was an improvement in the surface finish over conventional machining. Molla and Manna [6] studied the effects of ECG parameters on R_a (μm) during finishing of hybrid Al/(Al_2O_3 + Zr_2O_2)–MMC. Authors concluded that the R_a reduced from 1.40 μm to 96 nm by ECG process. Loveless et al. [7] discussed the effects of AFM on surfaces created by turning, milling, grinding, and wire electrical discharge machining. Authors claimed that the abrasive media viscosity considerably affected the surface finish while extrusion pressure did not have a major effect in this. Alitavoli and Mehran (2005) studied on AFM and concluded that the material removal increases with increase in media flow speed, abrasive concentration, abrasive mesh size and number of cycles. The author also concludes that material removal in aluminium is greater than brass for some input condition. Wang [8] discussed the efficiency of abrasive gel in material removal and effect of it on the surface roughness. Author proposed that the silicone rubber (or vinyl silicone polymer) as the best material media for finishing of metal surface because of its cost effectiveness. Author also concluded that 80% of the surface improvements occurs within the 5 number of passes. Smaller sized abrasives produces better surface over large sized abrasives. Song [9] examine the temperature dependence and its effect on surface roughness in AFM. Author concluded that the media will have good machining performance in the initial few cycles then media temperature rise quickly. They concluded that the best workable temperature should be below **25** °C throughout the AFM. Jain and Jain [10] use the stochastic simulation measure to analyze the active grain density in AFM and concluded that the active grain density increases with increase in abrasive concentration and active grain density are higher for a smaller grain diameter for a specific concentration. This phenomenon has great impact on surface finish by AFM. Gorana and Jain [11] studied the axial and radial forces during AFM using a two components disc dynamometer and concluded that the extrusion pressure, concentration of abrasive particles, average particle size affects the cutting forces and active grain density. Authors concluded that the ploughing and rubbing are the possible mechanism for removal of material, and reduction in the surface roughness heights is linearly proportional to the force ratio (Fr/Fa). Kar and Ravi kumar [12] developed a substitute media for AFM process from the different viscoelastic base materials i.e. butyl rubber and natural rubber, silicon carbide abrasive and naphthenic oil. Author concluded that the butyl rubber base media provides good performance as compared to natural rubber based media. The surface finish increases with increase in abrasive concentration but at high percentage (above 78%) of abrasive concentration, the flow becomes difficult as well as carrier acts as an insufficient binder to hold the abrasive. Sankar et al. [13] developed styrene butadiene rubber (SBR) media and used it in the AFF process for finishing of Al-SiC MMCs. Author concluded that the surface finish increases up to extrusion pressure 6 MPa, be hand that the surface finish decreases. Surface finishing increases gradually with increase in weight percentage of naphthenic oil up to 10% and then starts decreasing. Williams and Rajurkur [14] performed experiments on extrusion abrasive flow finishing for producing the parts of four side holes and a center hole and concluded that the metal removal rate of the middle hole was 30% more than that inside holes. Sadiq and Shunmugam [15] finished the external curved surface by rotation of pad with abrasive-mixed MR fluid is pushed up and down. Authors conducted the experiments on aluminium and austenitic stainless steel work pieces to understand the effect of magnetic field and concluded that the improvement of surface finish is better for rougher surface at high rotation speed and a reduction of surface roughness heights is consistent with process duration. Reddy et al. [16] studied on centrifugal force assisted abrasive flow machining (CFAAFM) and applied response surface methodology (RSM) to develop the mathematical relationships for material removal and surface finish during finishing of cast Al alloy workpiece. Authors identified the improvement MR and SR significantly in centrifugal force assisted abrasive flow machining as compared to abrasive flow finishing. Dabrowski et al. (1999) studied on the electrochemically assisted abrasive flow machining (ECAFM) with several solid electrolytes and various bonds Authors concluded that the abrasive properties of the electrolyte can be increased by adding the Al_2O_3 and SiC powders, again the consistence of the electrolyte can be maintained by adding the SiO_2 enhance surface finish. Omer Eyercioglu et al. [17] investigated on AFF during finishing of DIN 1.2379 tool steel through prepared and utilized different types of abrasives and claimed that the recast layer formed during WEDM was successfully removed by AFM process. Wang and Weng [18] developed low cost and effective abrasive media to increase the surface finish of the WEDM machined surface. Authors concluded that the vinyl-silicone polymer has good deformation and low flow effect; it can flow through the complicated holes easily, moreover it will not stick on the work piece surface after finishing. The abrasive particles and vinyl-silicone polymer were mixed uniformly to form the flexible media and they claimed that the surface finish improvement rate reaches to 84%. However, the polishing time might be several times longer but the polishing efficiency is very good when new abrasive media are used in AFF. Jha and Jain [19] finished the complex shape surface on developed magnetorheological abrasive flow finishing set-up in nano domain. Authors claimed that the R_a is 30 nm on stainless steel work pieces. Singh and Shan [20] developed MAFM set-up and effect of machining parameters on responses were studied. Authors concluded that the magnetic field significantly affect the material removal and change in surface roughness. Higher material removal and higher change in surface roughness heights were identified during finishing of brass material at low flow rates of medium and high magnetic flux density. They also claimed that the performance of MAFM was much better over AFM. Jain et al. [21] optimized AFM and MAF parameters using genetic algorithms. Authors developed a surface roughness model which can explain the objective function of AFM process.

11.2 Preparation of Curved Surface Workpiece Specimens

The Alloy 61S is used as workpiece material. Alloy 61S commercially available as tempered graded Al alloy. The EDX technique is employed to analyse the composition of the selected workpiece sample. Figure 11.1 shows the chemical composition of Alloy 61S used for experiments through Energy Dispersive X-Ray Analyser (EDX). Figure 11.2 shows the optical micrograph, which represents the distribution of elements in the Alloy 61S workpiece material. Utilized the SOLID WORKS 2016 and created the 3-D model of curved surface before machining on vertical milling centre. Figure 11.3 shows the generated curved surface on SOLID WORKS 2016.

The curved surface specimens were prepared as per the dimension Fig. 11.3c on CNC Vertical Milling centre ANFORD ECO VP 850. Figure 11.4 shows the CNC Vertical Milling centre ANFORD ECO VP 850 used for machining of curved surface on Alloy 61S material. The machines has vertically oriented spindles that approach

Al 5/3/2018

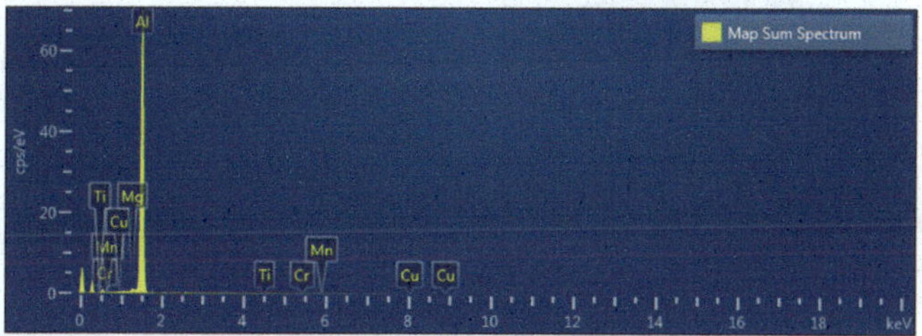

Map Sum Spectrum Element	Line Type	Weight %	Weight % Sigma	Atomic %
Al	K series	98.03	0.12	98.33
Fe	K series	0.50	0.05	0.24
Mn	K series	0.00	0.00	0.00
Mg	K series	0.66	0.03	0.74
Ti	K series	0.00	0.00	0.00
Cr	K series	0.00	0.00	0.00
Zn	K series	0.12	0.07	0.05
Cu	K series	0.04	0.06	0.02
Si	K series	0.64	0.05	0.62
Total		100.00		100.00

Fig. 11.1 EDX of workpiece material

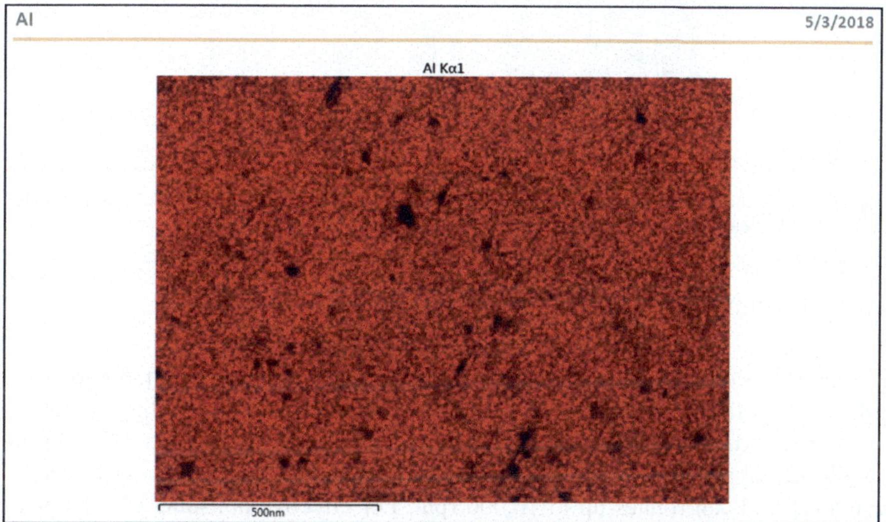

Fig. 11.2 Optical micrograph of Alloy 61S workpiece specimen

workpiece mounted on its table. The surface roughness heights, R_a (μm) and R_t (μm) of the machined curved surface were measured by utilizing the Surtronic 3 Plus curved surface measuring instrument. The average measured surface roughness heights, R_a (μm) and R_t (μm) after machined on vertical milling centre were 0.758 and 4.21 μm respectively. The SEM images were taken with different magnification to analyze the machined curved surface texture. Figure 11.5 shows the SEM images of machined surface machined by CNC Vertical Milling at different magnification.

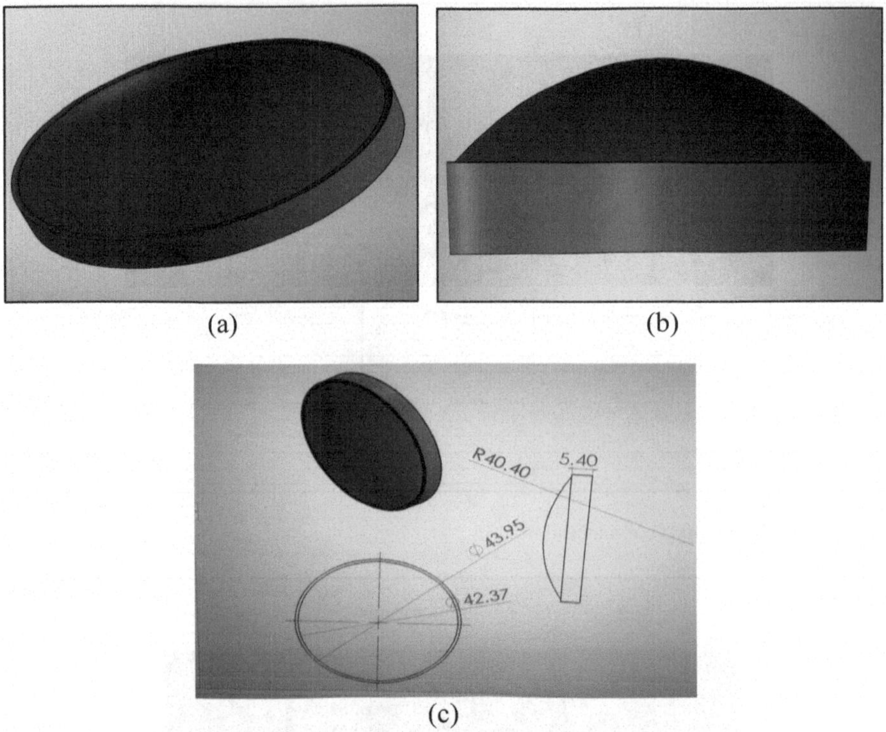

Fig. 11.3 **a**, **b** and **c** Different views of generated curved surface of the workpiece sample created on SOLID WORKS 2016

11.3 Polishing of Curved Surface

The rotary abrasive float polishing (RAFP) set-up has been designed, fabricated and utilized for polishing of curved surface prepared by CNC vertical milling operation. The fabricated RAFP setup has a motor which provides rotational motion to the spindle of capacity up to 10,000 rpm. The spindle mounted polishing brash i.e. scrubbing pad can rotates up to 10,000 rpm. For effective polishing of workpiece specimen surface, it is important to vary the spindle speed i.e. scrubbing pad rotational speed and carried out the experiments. For variation of motor spindle speed i.e. variation of scrubbing pad speed a DC potentiometer is connected with the main DC supply. The preliminary experiments were carried out with varying the scrubbing pad speed from 2000 to 10,000 rpm and polishing the curved surface. Table 11.1 represents the preliminary experiments condition used for polishing of curved surface by developed rotary abrasive float polishing set-up. Figure 11.6 shows the polishing of curved surface utilized fabricated abrasive float polishing setup.

Fig. 11.4 CNC vertical milling centre ANFORD ECO VP 850 used for machining of curved surface on Alloy 61S workpiece specimen

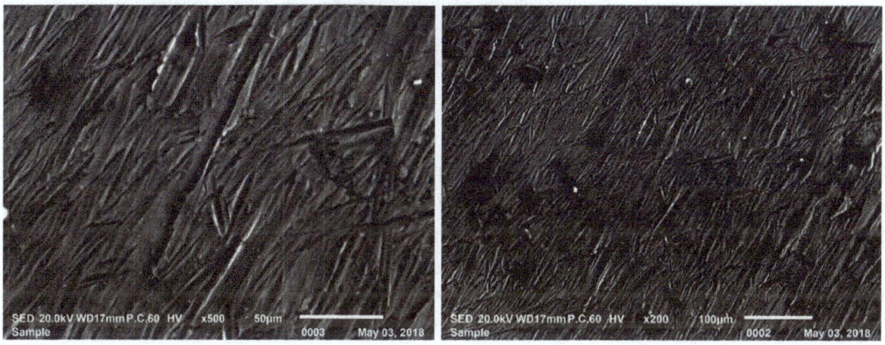

(a) (b)

Fig. 11.5 SEM images of machined curved surface machined by CNC vertical milling centre

The polishing of Alloy 61S curved surface was done by rotary abrasive polishing setup with prepared Al_2O_3 abrasive slurry. The Al_2O_3 is an electrically insulating material, however it has a moderate thermal conductivity (40 W/m K). It is normally available in crystalline shape, called corundum or aluminium oxide, high hardness makes it an abrasive. The abrasive slurry was prepared by mixing of unused transformer oil, water and pulp with Al_2O_3 abrasive powder. Three different concentration

Table 11.1 Preliminary experiments condition used for polishing

Input voltage	230 V, AC
Power	250 W
Scrubbing pad rotational speed varied	2000–10,000 rpm (2000, 4000, 6000, 8000 and 10,000)
Continuous polishing time	20–240 s (20, 40, 60, 80, 100, 120, 140, 160, 180 s)
Concentration of Al_2O_3 abrasive powder	15–50 g/l (15, 20, 25, 30, 35, 40 g/l)

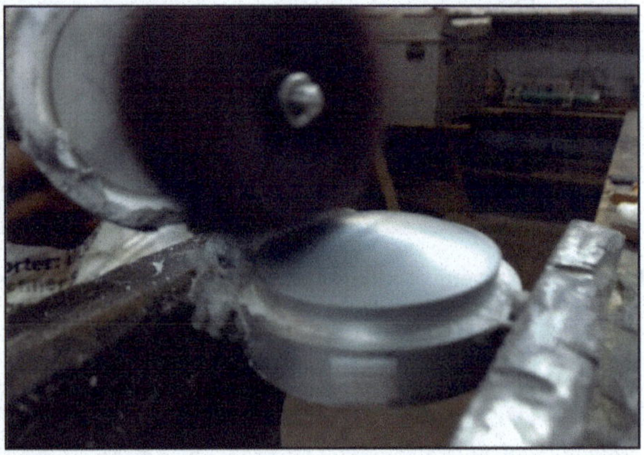

Fig. 11.6 Abrasive float polishing setup used for polishing of curved surface

abrasive slurry were prepared and used for experiments. A fixture was specially fabricated to hold the workpiece specimen. Polishing operations were carried out with different RPM on rotary abrasive float polishing set-up. A tachometer was used to measure the spindle speed during polishing. Table 11.2 represents the developed rotary abrasive float polishing set-up parameters and their levels used for Taguchi design based experiments.

Table 11.2 Parameters and their setting values

Sr. no.	Symbol	Machining parameters	Levels			Unit
1	A	Abrasive quantity	21	30	40	g/l
2	B	Rotation per minute	3500	7000	10,000	rpm
3	C	Time duration	40	80	120	s

11.4 Results and Discussions

Experiments were carried out based on Taguchi design of experiment $L_9(3^4)$ orthogonal array. Figure 11.7 shows the polished curved surface work piece specimen after polishing by fabricated developed rotary abrasive float polishing setup. The surface roughness heights, R_a (μm) and R_t (μm) of the polished curved surface were measured by utilizing the Surtronic 3 Plus curved surface measuring instrument. The lowest measured surface roughness heights, R_a and R_t after rotary abrasive float polishing were 0.079 and 0.86 μm respectively. Table 11.3 represents the devel-

Fig. 11.7 Workpiece specimen after finishing by rotary abrasive float polishing set-up

Table 11.3 Experiments based on $L_9(3^4)$ orthogonal array and experimental results

Sr. no.	Abrasive concentration	Rotation per minute	Time duration	Surface roughness height (R_a, μm)	Surface roughness height (R_t, μm)
1	21	3500	40	0.218	1.47
2	21	7000	80	0.167	1.09
3	21	10,000	120	0.101	0.94
4	30	3500	80	0.207	1.32
5	30	7000	120	0.113	1.12
6	30	10,000	40	0.091	0.98
7	40	3500	120	0.139	1.18
8	40	7000	40	0.092	0.97
9	40	10,000	80	0.079	0.86

oped rotary abrasive float polishing parameters, their setting values and experimental results for surface roughness heights, R_a (μm) and R_t (μm).

11.4.1 Effect of Developed RAFP Parameters on Surface Roughness Heights

In this section, acquired results were utilized to draw the effect graphs i.e. effects of RAFP parameters on surface roughness heights, R_a and R_t (μm). Figure 11.8 shows the effect of scrubbing pad rpm on surface roughness heights, R_a (μm) and R_t (μm). From Fig. 11.8, it is clear that with the increased of scrubbing pad rpm the surface roughness heights R_a (μm) and R_t (μm) both decreased. It is may be due to increase in kinetic energy which is essential to fracture or flatten the surface peaks and remove the debris from the curved surface.

Figure 11.9 show the effect of abrasive concentration (g/l) of slurry on surface roughness heights, R_a (μm) and R_t (μm). From Fig. 11.9, it is clear that the surface

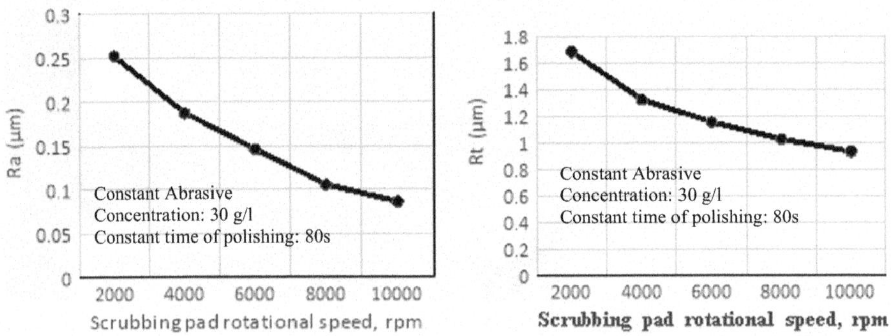

Fig. 11.8 Variation of surface roughness heights, R_a (μm) and R_t (μm) with scrubbing pad RPM

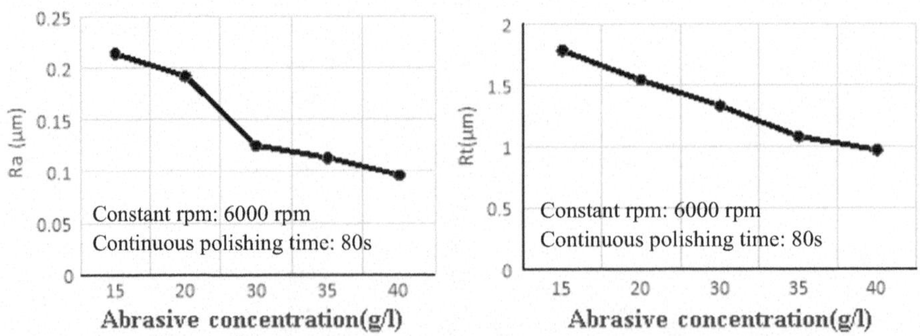

Fig. 11.9 Variation of surface roughness height, R_a (μm) and R_t (μm) with abrasive Al_2O_3 powder concentration

roughness heights, R_a (μm) and R_t (μm) both decrease with increase in abrasive concentration. It is may be due to large number of abrasive particles present in the abrasive slurry there by increases number of cutting edges and hence increase material removal as well as removal of debris from the curved machined surface, enhance surface finish. The variation of surface roughness, R_a (μm) and R_t (μm) with time of continuous polishing plots were drawn (not shown) and identified that the surface roughness heights, R_a (μm) and R_t (μm) both decrease with increase in time duration i.e. time of continuous polishing. It is may be due to by increase of time duration increases the duration of contact of cutting edges and remain in action for long duration of time, enhance the surface finish.

11.4.2 Optimal Parametric Contribution for Surface Finish

The S/N ratio was calculated based on 'Lower-the-Better (LB)' criteria for surface roughness heights, R_a & R_t (μm) and relation as follows:

$$\text{S/N Ratio for (LB)} = \eta = -10\log_{10}\left[\sum_{i=11}^{n} y^2\right] \quad (i = 1, 2, 3, \ldots, n) \qquad (11.1)$$

Utilized the above mentioned Eq. 11.1 and acquired results calculated S/N ratio (dB), and draw the S/N ratio graphs for R_a (μm) and R_t (μm). Figure 11.10 shows the S/N ratio (dB) graphs for R_a (μm). From Fig. 11.10 identified the optimal parametric combination for better surface finish, R_a (μm) is $A_3B_3C_3$ i.e. at 40 g/l abrasive slurry, 10,000 rpm scrubbing pad rotational speed and 80 s continuous polishing.

In S/N ratio graphs Fig. 11.10, Parameter A: abrasive concentration (g/l slurry), Parameter B: scrubbing pad rotational speed (rpm), Parameter C: Time in second for continuous polishing (s). Similarly, the S/N ratio was calculated based on 'Lower-the-Better (LB)' criteria and Eq. 11.1 for surface roughness height, R_t (μm) and plotted the graphs (not shown). From the S/N ratio (dB) graphs identified optimal parametric combination for better surface finish, R_t (μm) which is $A_3B_3C_3$ i.e. at 40 g/l abrasive slurry, 10,000 rpm scrubbing pad rotational speed and 80 s continuous polishing.

11.4.3 SEM Images of Polished Curved Surface

Figure 11.11a, b show the SEM images of Alloy 61S curved surface specimens after polishing. Figure 11.11a shows the polished surface on developed rotary abrasive float polishing set-up at 3500 rpm scrubbing pad rotational speed, 30 g/l abrasive concentration and for 80 s continuous polishing. Figure 11.11b shows the polished surface on developed rotary abrasive float polishing set-up at 10,000 rpm scrubbing

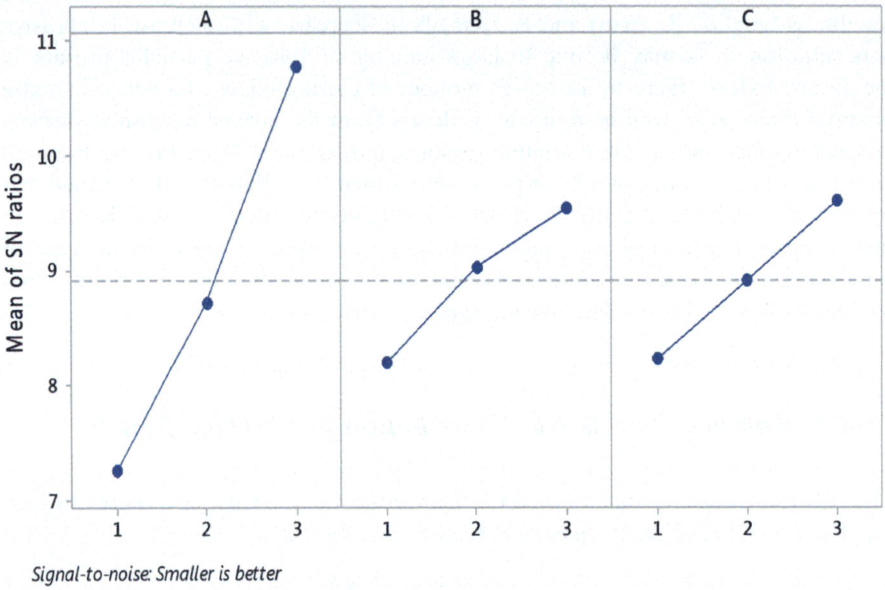

Fig. 11.10 S/N ratio graph

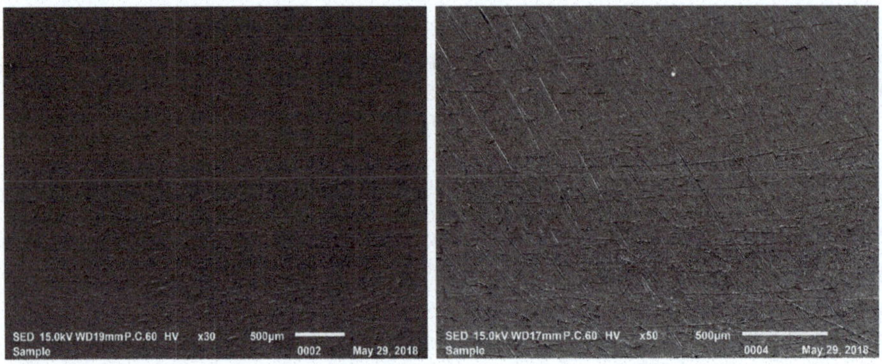

(a) (b)

Fig. 11.11 SEM micrographs images of finished curved surface after polishing

pad rotational speed, 20 g/l abrasive concentration and for 120 s continuous polishing. Figure 11.5 shows the SEM images of milled curved surface, where it is found that major part of the peaks due to ploughing during milling on CNC vertical milling centre. Comparing the SEM images before and after polishing Figs. 11.5 and 11.11, it clear that very good surface finish can be possible to achieve on curved surface by rotary abrasive float polishing technique.

11.5 Summary

Based on the experimental results during generation of curved surface by CNC milling and finishing of curved surface by developed rotary abrasive float polishing set-up, it is concluded that the CNC Vertical milling centre can be used for generation of curved surface on Alloy 61S workpiece material and rotary abrasive float polishing technique can be effectively used to polish the curved surfaces. The average measured surface roughness heights, R_a and R_t after machined by vertical milling centre were 0.758 and 4.21 μm respectively. The lowest measured surface roughness heights, R_a and R_t after rotary abrasive float polishing were 0.079 and 0.86 μm respectively. Experimental results revealed that the considerable improvement in the surface finish R_a and R_t, when finishing was done by rotary abrasive float polishing over milling of curved surface. The reduction of surface roughness height, R_a (μm) is 86.06% with respect to the milled surface roughness height 0.758 μm. The reduction of surface roughness height, R_t (μm) is 79.57% with respect to the milled surface roughness height 4.21 μm. The optimization of the developed rotary abrasive float polishing set-up can help to set the parameters in actual practice and to get the better surface finish. The Scanning Electron Microscopy (SEM) images also revealed the improvement in surface finish.

References

1. Rhoades LJ (1985) Abrasive flow machining and its use. In: Proceeding of the Non-conventional traditional machining conference, Cincinnati, OH; J Mater Process Technol 28:573–582
2. Manna A, Mali HS (2015) Optimum selection of abrasive flow machining conditions during fine finishing of Al/15 wt% SiC-MMC using taguchi method. Int J Adv Manuf Tech 50(9–12):1013–1024
3. Manna A, Mali HS (2014) An experimental investigation during finishing of particulate reinforced Al/10 wt% SiC_p-MMC on developed AFF setup. Int J Adv Manuf Tech 28:1263–1272
4. Mali HS, Manna A (2012) Simulation of surface generated during abrasive flow finishing of Al/SiCp-MMC using neural network. Int J Adv Manuf Tech 61:1263–1268
5. Mali HS, Manna A (2009) Current status and application of abrasive flow finishing processes: a review. Proc Mech E Part B J Manuf Eng 223:809–820
6. Manna A, Molla KZ (2014) An experimental investigation during nano finishing of hybrid Al/(Al_2O_3 + ZrO_2)–MMC on developed ECG setup. In: 5th international & 26th all India manufacturing technology, design and research conference (AIMTDR 2014) December 12th–14th, IIT Guwahati, Assam, India, 630:1–7
7. Loveless R, Williams RE, Rajurkur KP (1994) A study of the effect of abrasive flow finishing on various machined surfaces. Int J Adv Manuf Tech 47:133–151
8. Wang (2015) Evaluation of rheological properties of Gel abrasive in magnetic abrasive finishing. Adv Mech Eng 5:15–24
9. Songs (2008) Temperature dependence and effect on surface roughness in abrasive flow machining. Adv Mater Res 53:375–380
10. Jain RK, Jain VK (2004) Stochastic simulation of active grain density in abrasive flow machining. J Mater Process Technol 152:17–22

11. Gorana VK, Jain VK (2004) Experimental investigation into cutting forces and active grains density during abrasive flow machining. Int J Mach Tools Manuf 44:201–2011
12. Kar KK, Ravi kumar NL (2009) Performance evaluation and rheological characterization of newly developed butyl rubber based media for abrasive flow machining process. J Mater Proc Tech 209:2212–2221
13. Sankar MR, Ramkumar J, Jain VK (2009) Experimental investigation and mechanism of material removal in nano finishing of MMCs using abrasive flow finishing (AFF) process. Wear 266(7–8):688–698
14. Williams RE, Rajurkar KP (1989) Metal removal and surface finish characteristics in abrasive flow machining mechanics of deburring and surface finishing processes. In: Presented at the winter annual meet of the American SOC. Mechanical, University of Nebraska, 38:93–106
15. Sadiq, Shunmugam (2009) Investigation into magnetorheological abrasive honing (MRAH). Int J Mach Tools Manuf 49(7):554–560
16. Reddy MK, Sharma AK, Kumar P (2008) Some aspects of centrifugal force assisted abrasive flow machining of 2014 Al alloy. Proc Inst Mech Eng Part B J Eng Manuf 222(7):773–783
17. Omer Eyercioglu et al. (2016) Effects of abrasive types on the surface integrity of abrasive-flow-machined surfaces. Proc Inst Mech Eng Part B J Eng Manuf 232(6)
18. Wang AC, Weng SH (2007) Developing the polymer abrasive gels in AFM process. J Mater Process Technol 192–193:486–490
19. Jha and Jain (2004) Design and development of the magnetorheological abrasive flow finishing (MRAFF) process. Int J Mach Tools Manuf 44(10):1019–1029
20. Singh and Shan (2002) Development of magneto abrasive flow machining process. Int J Mach Tools Manuf 42(8):953–959
21. Jain NK, Jain VK, Jha S (2007) Parametric optimization of advanced fine-finishing processes. Int J Adv Manuf Tech 34(11–12):1191–1213
22. Tavoli M Ali, Zadeh N Nariman, Khakhal A, Mehran M (2007) Multi-objective optimization of abrasive flow machining processes using polynomial neural networks and genetic algorithms. Mach Sci Technol 10:491–510
23. Singh S, Ravi Sankar M, Jain VK, Ramkumar J (2014) Modeling of finishing forces and surface roughness in abrasive flow finishing (AFF) process using rheological properties. In: 5th international & 26th all India manufacturing technology, design and research conference (AIMTDR 2014) December 12th–14th, IIT Guwahati, Assam, India, 49:1–6

Index

S. Das et al. (eds.), *Advances in Abrasive Based Machining and Finishing Processes*, Materials Forming, Machining and Tribology,
https://doi.org/10.1007/978-3-030-43312-3

O

P

R

S

T

U

V

W

The manufacturer's authorised representative in the EU is Springer Nature Customer Service Centre GmbH, Europaplatz 3, 69115 Heidelberg, Germany. If you have any concerns regarding our products, please contact ProductSafety@springernature.com

Printed and bound by CPI Group (UK) Ltd, Croydon, CR0 4YY
25/09/2026
02222904-0002